HOME APPLIANCE SERVICING

by Edwin P. Anderson

THEODORE AUDEL & CO.
a division of
HOWARD W. SAMS & CO., INC.
4300 West 62nd Street
Indianapolis, Indiana 46268

683
AND

Foreword

The purpose of this book is to supply the practical information needed to repair and service modern home appliances. With the ever-increasing number of fairly complex automatic and semi-automatic appliances employed in the home, a greater amount of knowledge is required to properly and efficiently diagnose and service these appliances. This book provides the reader with the knowledge he needs to accomplish these repairs with a minimum of effort and without resorting to time-consuming guesswork.

The text is arranged in a logical sequence. The first chapters furnish the electrical fundamentals and techniques necessary to understand the construction and operation of all electrical appliances. The operation and servicing of resistance-heating appliances, such as irons, toasters, ranges, etc., are fully explained in the second group of chapters. Finally, the motor-driven appliances, both gas and electric, are treated in detail in the third breakdown of chapters. Each chapter is also divided into logical sections that deal with operation, construction, installation (of major appliances), servicing, and repairs.

The correct servicing methods are of the utmost importance to the serviceman, since time, money, and reputation can all be lost when repeat calls are required. With the aid of this book, the reader will be able to service and install any home appliance by using the detailed troubleshooting and repair methods described.

EDWIN P. ANDERSON

Contents

Chapter 9

Chapter 10

Chapter 11

Chapter 12

Chapter 13

Chapter 14

Chapter 15

Chapter 16

Chapter 17

Chapter 18

CHAPTER 1

Fundamental Electricity

In order for an appliance technician to deal intelligently with the numerous and sometimes fairly complex home appliances which today constitute an essential part of modern living, it is of primary importance that he possess a thorough knowledge of electricity and electric circuits. The purpose of the following chapters, therefore, is to provide the necessary fundamentals that will enable him to understand, correctly analyze, and remedy the defects of any home appliance in which electricity is employed.

ATOMS

It is a well-established fact that everything physical is built up of atoms. These are particles so small that they cannot be seen even through the most powerful microscope. The atom, in turn, consists of several kinds of still smaller particles, one of which is the *electron*.

There are more than 100 varieties of atoms known, each representing one of the chemical elements from which all matter is constructed. Some of the more common elements are hydrogen, oxygen, nitrogen, carbon, iron, and copper.

The most widely accepted modern physical picture of the atom corresponds roughly to a miniature of our solar system. Thus, corresponding to the sun is the nucleus of the atom, which, in general, is a small compact structure composed of a combination of extremely minute particles called *protons* and *neutrons*.

The proton has a positive charge equal in magnitude but opposite in sign to that of the electron. Its mass is extremely large compared to that of the electron. The neutron has very nearly the same mass as the proton, but is uncharged. Practically all the mass of the atom is associated with the small, dense nucleus. Revolving about the nucleus in orbits at relatively large distances from it are one or more electrons.

Atomic Weight

The relative weight of one atom of an element, referred to some other element taken as standard, is called the *atomic weight*. The hydrogen atom, being the simplest element, was formerly taken as unity, or 1, but the greater number of scientists now have assigned the atomic weight of 16 for oxygen, which gives hydrogen the atomic weight of 1.0080.

The Hydrogen Atom

The simplest of all atoms is that of the gas hydrogen, whose nucleus, Fig. 1, consists of a single proton with a single electron revolving about it. Here the two charges revolve about each other in space much like a whirling dumbbell, except that there is no rigid connection between them.

The Helium Atom

The next atom in simplicity is that of the gas helium, whose nucleus, Fig. 2, consists of two protons and two neutrons bound together in a compact central core of great electrical stability.

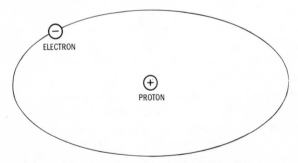

Fig. 1. Structure and electron orbit of the hydrogen atom. The proton has a positive charge, and the electron has a negative charge.

Revolving about this compact nucleus are two electrons. The neutrons seem to have the ability to hold the positively charged protons together in the nucleus, since if it were not for the neutrons, the two protons would separate from one another.

Atoms of Other Elements

Atoms of other elements become increasingly more complex by the successive addition of one electron to those revolving about the nuclei, and with the progressive addition of protons and neu-

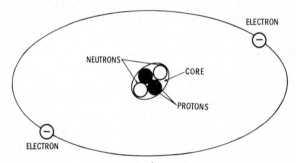

Fig. 2. Structure and electron orbit of a neutral helium atom.

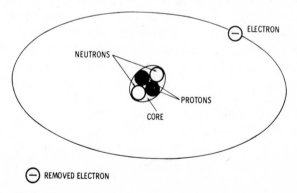

Fig. 3. Structure and electron orbit of a positive helium atom.

trons to the nuclei. In every case, however, the normal atom has an exactly equal number of positive and negative elementary charges, so that the atom as a whole is neutral; that is, it behaves toward electrified bodies as though it had no charge at all and is said to be in equilibrium.

Positively and Negatively Charged Substances

With reference to the picture of the neutral atom, it will be easy to understand what takes place when a substance is electrically charged. Assume that by some means one of the external electrons of the neutral helium atom is removed, as shown in Fig. 3. The result will be an unsatisfied atom insofar as the balance between the positive and negative charges is concerned. The excess of one proton in the nucleus gives the atom a positive charge; if the previously removed electron is permitted to return to the atom, it will again become neutral as it was in Fig. 2.

A positively charged body, therefore, is one which has been deprived of one or more of its electrons, whereas a negatively charged body is one which has a surplus (acquired more than its normal number) of electrons.

In its unbalanced state, the atom will tend to attract any free electrons that may be in the vicinity. This is exactly what takes place when a stick of sealing wax or amber is rubbed with a piece of flannel. The wax becomes negatively charged, and the flannel becomes positively charged. During the rubbing process, the force of friction rubs off some of the electrons from the atoms composing the flannel and leaves these electrons on the surface of the wax. The surface atoms of the flannel are left deficient in electrons, and the surface atoms of the wax have a surplus. If the wax and the flannel are touched together after being rubbed, there will be a readjustment of electrons, the excess on the wax returning to the deficient atoms of the flannel, as shown in Fig. 4.

Most of the electrons in the universe exist as component parts of atoms, but it is possible for an electron to exist in the free state apart from the atom, temporarily at least. Free electrons exist to some extent in gases, in liquids, and in solids, but are much more plentiful in some substances than in others.

ELECTRICITY IN MOTION

Flow of Electric Current

The presence of free electrons in substances enables us to account for the flow of electricity. The more free electrons a substance contains, the better conductor of electricity it is; it is because of the great numbers of free electrons in metals that they are such good conductors of electricity. Such substances as glass, porcelain, rubber, mica, etc., with their comparatively few free electrons, are used as insulators. These free electrons are always in a state of continual rapid motion, or thermal agitation. The situation is analogous to that in a gas where it is known that the molecules, according to the kinetic theory, are in a state of rapid motion with a random distribution of velocity. If it were possible at a given instant to examine the individual molecules or electrons,

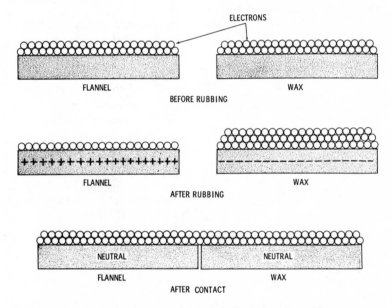

Fig. 4. Positive and negative charged bodies. In the rubbing process, electrons are removed from the flannel and deposited on the wax. Later contact between the two bodies results in a redistribution of the electrons, and a neutral condition is produced.

it would be found that their velocities vary enormously and as a function of temperature. The higher the temperature of a substance, the higher the velocity of the atoms and electrons contained in that substance.

If by some means the random movement of the electrons in a conductor could be controlled and made to flow in one direction, there would result what is called a *flow of electric current*. Means of controlling or directing the electron motion are provided by chemical energy, as in a battery, or by supplying mechanical energy to a generator, as shown in Fig. 5.

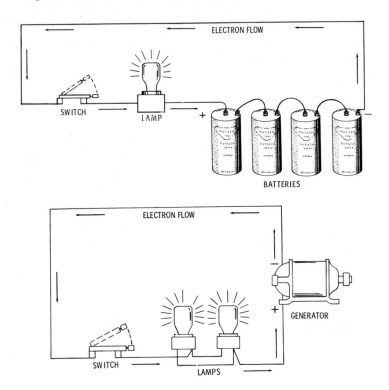

Fig. 5. Methods used in generation of electron flow.

Direct and Alternating Currents

There are two types of electric currents which must be considered when dealing with electrical appliances. When electrons move in the same direction along a conductor or circuit, it is said that a direct current flows through the conductor or circuit. This type of current is furnished by batteries and certain types of generators, as shown in Fig. 5.

When, on the other hand, the electrons are made to move first in one direction and then in the opposite direction in a circuit, the current is said to be alternating. The reversals may occur at any rate from a few per second up to a large number, depending on the medium of utilization and the method of generation.

In the practical utilization of electricity, such as that employed in homes and factories for lighting and power purposes, the number of reversals, or alternations, per second is usually 120. Two alternations of current constitute one cycle of alternating current, and the number of cycles per second is termed the *frequency* of the current. The usual frequency employed is 60 cycles.

The reason for the universal use of alternating current in homes and factories is that it is easier to use and transmit. Alternating-current generators can be built for greater capacities and higher voltages than direct-current generators. In addition, with alternating current, the voltage may be raised or lowered more economically by means of transformers.

In most cases, electrical energy must be used at low voltages for reasons of safety. When energy is to be used at a distance from the generator, it must be transmitted at high voltage, which usually amounts to about 1000 volts per mile of distance involved. In this manner, the electrical energy is transmitted at a low current value, which means that smaller conductors are required; because of this higher transmission voltage, the line losses are also smaller. Thus, for the sake of efficiency, the voltage is stepped up by means of one transformer at the power station, or generator end of the line, and it is then reduced to a safe value by means of another transformer at the consumer end.

CHAPTER 2

Measurement of Electricity

CURRENT

Since the electron is a minute particle of electricity, a great number of particles is required to light a lamp. The practical measurement of electron flow, or electric current, is the ampere. One ampere is equivalent to the movement of many billions of electrons past a given point in a circuit in one second.

An approximate appreciation of current requirements will be had if it is mentioned that the familiar 50-watt electric lamp takes about one-half ampere, whereas the average electric iron requires about ten amperes, as shown in Fig. 1. One milliampere is equal to one-thousandth of an ampere, or 1000 milliamperes equals one ampere.

VOLTAGE

It has been previously mentioned that the directed motion of free electrons in a conductor constitutes an electric current. Also, the larger the number of free electrons through the conductor, the larger will be the current flowing through it. It can similarly be shown that the greater the electron flow, the higher will be the repulsive force, or pressure, between the electrons; also, the higher the pressure, the more electrons will flow through the conductor. Electric pressure is variously called difference in potential, voltage, or electromotive force. The unit of measurement is the volt. Voltage is simply a term used to indicate the electrical pressure in a circuit through which electrons are flowing.

The electric force (or electromotive force, abbreviated emf) that causes current to flow may be developed in several ways. The action of certain chemical solutions on dissimilar metals will bring

15

about an emf between the terminals of the metals. Such a combination is called a cell, and a group of cells joined together in an electric circuit forms an electric battery. The amount of current

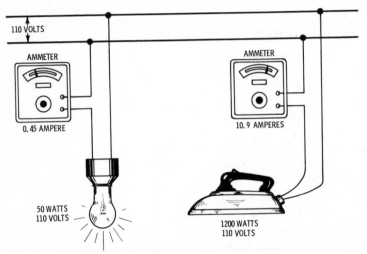

Fig. 1. The method of current measurement and ammeter reading when connected in the circuit of a 50-watt lamp and a 1200-watt iron, respectively.

such cells can generate is limited, and in the course of current flow, one of the metals is eaten away. As a consequence, the electrical energy that can be taken from a battery is rather small.

Where a large amount of energy is required, it is usually furnished by an electric generator, as shown in Fig. 2, which develops an emf by a combination of magnetic and mechanical means. Such generators are used to supply the electrical energy that is distributed to homes and factories for lighting and power.

RESISTANCE

The resistance offered to an electric current varies with the material, shape, and dimensions of the conductor. Thus, two conductors having the same size, shape, and dimensions but of different material will vary in the amount of current flow even though

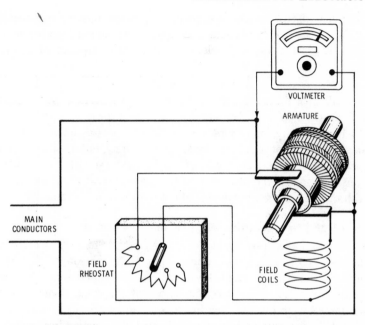

Fig. 2. Measuring voltage across the terminals of a DC generator.

a similar emf is applied to both. This is due to the difference in *resistance* of the conductors. The ability of a conductor to carry an electric current depends on the resistance of the material in it;

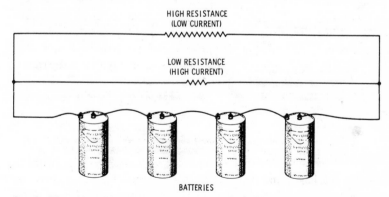

Fig. 3. The effect of resistance on current flow. The amount of resistance in a single branch is inversely proportional to the amount of current flowing through that branch.

17

the lower the resistance, the greater the current for a given emf, as shown in Fig. 3. One of the best conductors of electricity is copper, which is the reason for this metal being so widely used in electrical circuits.

The resistance of a conductor or wire is measured in ohms. A circuit has a resistance of one ohm when an applied emf of one volt causes a current of one ampere to flow through it.

The resistivity, or specific resistance, of a material is the resistance, in ohms, of a cube of material measuring one centimeter on each edge. It is frequently convenient in making resistance calculations to compare the resistance of the material under consideration with that of a copper conductor of the same size and shape.

Effect of Wire Length and Size on Resistance

The resistance to current flow in a given conductor is directly proportional to its cross-sectional area. Thus, the longer the path through which the current flows, the higher the resistance of that conductor. Also, given two conductors of the same material and the same length but differing in cross-sectional areas, the one with the larger area will have the lower resistance, as shown in Fig. 4.

It is readily possible to combine the foregoing statements concerning resistance into a single formula that will enable us to cal-

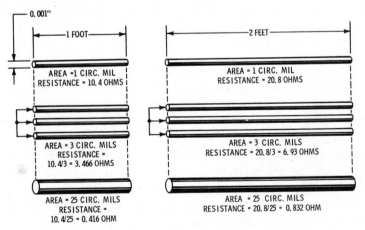

Fig. 4. The resistance of a conductor decreases with an increase in the area through which the current flows.

culate the resistance of conductors of any size, shape, and material. For example, if the resistance of a mil-foot of wire given in ohms is multiplied by the total length in feet and divided by its cross section in circular mils, the result will be the total resistance of the wire in ohms. This is expressed as:

$$R = \rho \frac{L}{A} \qquad (1)$$

where,

R = resistance in ohms,
ρ = resistance of one mil-foot (10.4 for copper),
L = length in feet,
A = area in circular mils.

Example—Determine the resistance of a circular copper wire having a diameter of one-eighth inch and a length of 1000 feet.
Solution—A substitution of values in Formula (1) produces:

$$R = 10.4 \times \frac{1000}{125 \times 125} = 0.665 \ ohm$$

In most practical cases, however, the problem will be to determine the resistance of a circular wire of a given gauge number and length, and such problems are most easily solved with the help of Table 1. From Table 1, it will be noted that a 1000-foot length of No. 8 wire has a diameter of 0.1285 inch and a resistance of 0.640 ohm. A 500-foot length of the same size wire would consequently have a resistance of 0.640/2 or 0.320 ohm.

Effect of Temperature on Resistance

It is known that the resistance and temperature of most conductors will increase proportionally with the amount of current flowing through them. Carbon and liquid electrolytes form an important exception to this rule.

The amount by which the resistance of a substance will change with a change in temperature of one degree is usually expressed as a percentage of the known resistance at 75° F. This percentage is known as the *temperature coefficient of resistance* and is approximately 0.0022 for all pure metals.

The relation of resistance changes due to changes in temperature is written:

Table 1. Bare Copper Wire Data

No. AWG	Diameter, Mils	Area, Circular Mils	Pounds per 1000 Ft.	Pounds per Mile	Ohms per 1000 Ft.	Ohms per Mile	Feet per Ohm
0000	460	211600	640	3382	0.0500	0.2639	20010
000	410	167800	508	2682	0.0630	0.3327	15870
00	364.8	133100	403	2127	0.0792	0.4196	12580
0	324.9	105500	319.5	1687	0.1002	0.529	9980
1	289.3	83690	253.3	1337	0.1264	0.667	7914
2	257.6	66370	200.9	1061	0.1593	0.841	6276
3	229.4	52640	159.3	841	0.2009	1.061	4977
4	204.3	41740	126.4	667	0.2533	1.337	3947
5	181.9	33100	100.2	529	0.3195	1.687	3130
6	162.0	26250	79.5	419	0.403	2.127	2482
7	144.3	20820	63.0	332.7	0.508	2.682	1969
8	128.5	16510	50.0	263.9	0.640	3.382	1561
9	114.4	13090	39.63	209.2	0.808	4.26	1238
10	101.9	10380	31.43	166.0	1.018	5.38	982
11	90.7	8234	24.92	131.6	1.284	6.78	779
12	80.8	6530	19.77	104.4	1.619	8.55	618
13	72.0	5178	15.68	82.8	2.042	10.78	490
14	64.1	4107	12.43	65.6	2.575	13.60	388.3
15	57.1	3257	9.86	52.1	3.247	17.14	308.0
16	50.8	2583	7.82	41.3	4.09	21.62	244.2
17	45.3	2048	6.20	32.74	5.16	27.26	193.7
18	40.3	1624	4.92	25.96	6.51	34.37	153.6
19	35.89	1288	3.899	20.59	8.21	43.3	121.8
20	31.96	1022	3.092	16.33	10.35	54.6	96.6
21	28.46	810	2.452	12.95	13.05	68.9	76.6
22	25.35	642	1.945	10.27	16.46	86.9	60.8
23	22.57	510	1.542	8.14	20.76	109.6	48.2
24	20.10	404	1.223	6.46	26.17	138.2	38.21
25	17.90	320.4	0.970	5.12	33.00	174.2	30.30
26	15.94	254.1	0.769	4.06	41.6	219.8	24.03
27	14.20	201.5	0.610	3.220	52.5	277.1	19.06
28	12.64	159.8	0.484	2.554	66.2	349.4	15.11
29	11.26	126.7	0.3836	2.025	83.4	441	11.98
30	10.03	100.5	0.3042	1.606	105.2	555	9.50
31	8.93	79.7	0.2413	1.274	132.7	701	7.54
32	7.95	63.21	0.1913	1.010	167.3	883	5.98
33	7.08	50.13	0.1517	0.801	211.0	1114	4.74
34	6.30	39.75	0.1203	0.635	266.0	1404	3.759
35	5.62	31.52	0.0954	0.504	335.5	1771	2.981
36	5.00	25.00	0.0757	0.400	423	2233	2.364
37	4.45	19.83	0.0600	0.3169	533	2816	1.875
38	3.965	15.72	0.0476	0.2513	673	3551	1.487
39	3.531	12.47	0.03774	0.1993	848	4478	1.179
40	3.145	9.89	0.02993	0.1580	1069	5644	0.935

$$R_h = R_l \left[1 + \alpha \left(T_h - T_l \right) \right] \qquad (2)$$

where,

R_h = resistance in ohms at the higher temperature,

R_l = resistance in ohms at the lower temperature,

T_l = lower temperature of conductor in degrees Fahrenheit,

T_h = higher temperature of conductor in degrees Fahrenheit,

α = temperature coefficient of resistance per degree Fahrenheit (0.0022 for copper).

Thus, for example, if a copper conductor has a resistance of 0.5 ohm at 40°F., its resistance at 160°F. becomes :

$$R_h = 0.5 [1 + 0.0022 (160 - 40)] = 0.5 + 0.132 = 0.632 \ ohm$$

Effect of Size on Current-Carrying Capacity

Conductors must have adequate mechanical strength, insulation, and current-carrying capacity for the particular conditions under which they are to be used. Table 2 shows the various sizes of copper conductors and the amount of current they are permitted to carry for the different type of commercial insulation.

Effect of Temperature on Expansion

In the foregoing, it was observed that the electrical resistance of metals increases with a rise in temperature. In addition to a change in their electrical properties, heat also causes metals to expand.

The expansion of a unit length for one degree Fahrenheit is called the *linear coefficient of expansion*. The increase per degree for a unit *surface* is termed surface, or area, expansion, while the increase per degree for a unit of *volume* is termed cubic expansion. Although the expansion per inch of a metal is quite small for each degree increase in temperature, a steel joist 100 feet in length will increase approximately one inch in length for an increase in temperature of 100° F.

It is because of expansion and contraction, due to temperature changes, that long structures must not be rigidly fixed at both ends. Steel train rails, for example, are laid about one-half inch apart to allow for expansion. Not only metals but other solids, such as concrete and glass, and other nonmetallic substances such as liquids and gases, also expand when heated.

Application of Expansion—The expansion and contraction of metals and fluids are utilized in several ways. The ordinary glass-stemmed thermometer, Fig. 5, is a familiar application of expansion of a fluid when heated. A thermometer consists of a cylindrical glass tube of uniform bore and diameter that is sealed

Table 2. Allowable Current-Carrying Capacities of Insulated Copper Conductors in Amperes

Size AWG MCM	Rubber Type R Type RW / Type RU Type RUW / Type RH-RW / Thermoplastic Type T Type TW	Rubber Type RH / RUH / Type RH-RW / Type RHW Thermoplastic Type THW / THWN	Paper / Thermoplastic Asbestos Type TA / Thermoplastic Type TBS / Silicone Type SA / Var-Cam Type V / Asbestos Var-Cam Type AVB / MI Cable / RHH	Asbestos Var-Cam Type AVA Type AVL	Impregnated Asbestos Type AI Type AIA	Type AA Asbestos Type A
14	15	15	25	30	30	30
12	20	20	30	35	40	40
10	30	30	40	45	50	55
8	40	45	50	60	65	70
6	55	65	70	80	85	95
4	70	85	90	105	115	120
3	80	100	105	120	130	145
2	95	115	120	135	145	165
1	110	130	140	160	170	190
0	125	150	155	190	200	225
00	145	175	185	215	230	250
000	165	200	210	245	265	285
0000	195	230	235	275	310	340
250	215	255	270	315	335	...
300	240	285	300	345	380	...
350	260	310	325	390	420	...
400	280	335	360	420	450	...
500	320	380	405	470	500	...
600	355	420	455	525	545	...
700	385	460	490	560	600	...
750	400	475	500	580	620	...
800	410	490	515	600	640	...
900	435	520	555	...	...	...
1000	455	545	585	680	730	...
1250	495	590	645	...	...	...
1500	520	625	700	785	...	...
1750	545	650	735	...	...	...
2000	560	665	775	840	...	...

at one end. A fluid (usually mercury or one of the alcohols) is first placed in the tube, which is then heated until the fluid expands and fills the tube, thereby driving out the air. It is necessary to create a vacuum, otherwise the air would prevent the fluid from expanding in the closed tube. After the air has been driven out, the tube is sealed. It is then placed in an atmosphere of free steam representing the boiling point of water, and next in an ice bath consisting of broken pieces of ice floating in water. The positions of the liquid at both of these points are marked on the tube; the boiling point represents 212° F., while the freezing point represents 32° F. The intervening distance between these two points is divided into 180 divisions, and each division is called a degree. The centigrade thermometer has 100 divisions

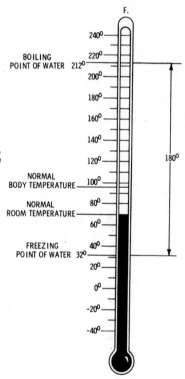

F.

240°

BOILING
POINT OF WATER 212° 220°

200°

180°

160°

140°

120°   180°

NORMAL
BODY TEMPERATURE 100°

NORMAL
ROOM TEMPERATURE 80°

60°

FREEZING
POINT OF WATER 32° 40°

20°

0°

-20°

-40°

Fig. 5. A thermometer, graduated in degrees Fahrenheit, showing several useful temperatures.

between these two points, which are marked 0° for freezing and 100° for boiling.

Thermostats used in the regulation of temperature represent another common application of the principle of expansion with an increase in temperature. One type of thermostat usually employed in the regulation of temperature in electrical home appliances depends for its operation on the expansion of metals. The working principles of such a thermostat (often called a bimetallic thermostat) are shown in Fig. 6.

Two metals of different coefficients of expansion are welded together to form a bimetallic unit, or blade. With the blade securely anchored at one end, a circuit is formed, and the two contact points are closed to the passage of an electric current. Because an electric current produces heat in its passage through the bimetallic blade, the metals in the blade begin to expand, but at different rates. The metals are so arranged that the one which has a greater coefficient of expansion is placed at the bottom of the unit. After a certain time interval, the operating temperature will be reached, and the contact points will become separated due to the bending of the blades, thus disconnecting the heating element from its source of power. After a short period, the contact blades will become sufficiently cooled and will cause the contact points again to be joined, thus re-establishing the circuit. The current will then be permitted

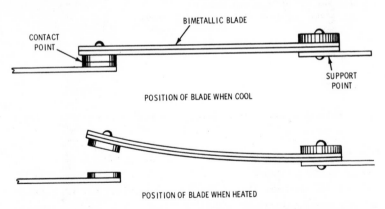

Fig. 6. Working principles of a simple bimetallic thermostat.

to heat the appliance. The foregoing cycle is repeated over and over again and in this manner prevents the temperature from rising too high or from dropping too low.

Temperature Scale Relations—A change from one temperature scale to another can conveniently be made by using the following formulas:

Fahrenheit = (9/5 × centigrade) + 32
centigrade = 5/9 (Fahrenheit − 32)

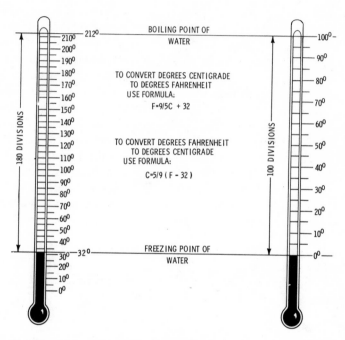

Fig. 7. Fahrenheit and centigrade thermometers, showing the conversion from one scale to the other.

Thus, if the temperature of a certain heating element is 400°C., its corresponding temperature on the Fahrenheit scale is (9/5 × 400) + 32 or 752 degrees. Fig. 7 illustrates the comparison of Fahrenheit and centigrade temperature scales.

CHAPTER 3

Ohm's Law—Circuit Fundamentals

In any electric circuit through which a direct current flows, there exists a simple mathematical relationship between current, voltage, and resistance. This relationship is called Ohm's law, and simply expresses that:

$$Voltage = Current \times Resistance$$

Since it has previously been shown that the unit for current is amperes, the unit for resistance is ohms, and the unit for voltage is volts, we may write:

$$Amperes = \frac{Volts}{Ohms}$$

which is usually written,

$$I = \frac{E}{R}$$

$$R = \frac{E}{I}$$

$$E = I \times R$$

The foregoing formulas are of the utmost importance to any electrical appliance serviceman, since they contain the means whereby current, voltage, and resistance in a circuit may be determined. Ohm's law also shows that, for a given voltage, the lower the resistance, the larger the current.

SIMPLE ELECTRIC CIRCUITS

Circuits in most household appliances may be divided into three classes:

OHM'S LAW — CIRCUIT FUNDAMENTALS

1. Series circuits.
2. Parallel circuits.
3. Series-parallel or parallel-series circuits.

The following will show how resistances may be joined together to form an electric circuit of the foregoing classifications and the use of the methods most commonly employed in simple resistance, current, and voltage calculations.

Series Circuits

A series circuit may be defined as one in which the current is of a constant value throughout. Thus, an ammeter placed at any point in the circuit would give the same reading. Fig. 1 represents such a circuit.

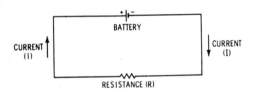

Fig. 1. A simple series circuit in which a battery supplies the current flow through a single resistance.

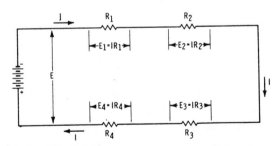

Fig. 2. A typical series circuit in which a battery supplies the current flow through four series-connected resistances.

When a circuit has a number of resistances connected in series, as in Fig. 2, the total resistance of the circuit is equal to the sum of the individual resistances. If these resistances are numbered R_1, R_2, R_3, etc., then:

$$R_{total} = R_1 + R_2 + R_3 + \ldots + R_N$$

Also, the drops in voltage across the individual resistances in a series circuit are equal to the total impressed voltage. With reference to Fig. 2, this may be written as follows:

$$E = E_1 + E_2 + E_3 + E_4$$
$$E = IR_1 + IR_2 + IR_3 + IR_4$$
$$E = I (R_1 + R_2 + R_3 + R_4)$$

Example—What voltage must be furnished by the battery in Fig. 2 in order to force a current of 2 amperes through the circuit, if R_1, R_2, R_3, and R_4 are 5, 10, 15, and 20 ohms, respectively?
Solution—The total resistance is:

$$R = 5 + 10 + 15 + 20 = 50 \ ohms$$

The total voltage is:

$$IR = 2 \times 50 = 100 \ volts$$

The total voltage must equal the sum of the individual voltage drops. This fact may conveniently be used as a check.
Thus,

$$E_1 = 2 \times 5 = 10 \ volts$$
$$E_2 = 2 \times 10 = 20 \ volts$$
$$E_3 = 2 \times 15 = 30 \ volts$$
$$E_4 = 2 \times 20 = 40 \ volts$$

Hence, $10 + 20 + 30 + 40 = 100$ volts, as before.
Example—A certain electric iron draws a current of 5 amperes when connected across a 120-volt circuit. What is its total resistance?
Solution—According to Ohm's law, the resistance of the iron is:

$$R = \frac{E}{I} = \frac{120}{5} = 24 \ ohms$$

Parallel Circuits

A parallel circuit may be defined as one in which there are two or more parts connected between two points in a circuit. Fig. 3

shows a simple parallel circuit consisting of three resistances—R_1, R_2, and R_3—connected between points A and B. The voltage across the various resistances is the same, and the current flowing through each resistance varies inversely with the value of the resistance. The sum of all the currents, however, is equal to the amount of current leaving the battery. Thus,

$$E = I_1 R_1 = I_2 R_2 = I_3 R_3$$

and,

$$I = I_1 + I_2 + I_3$$

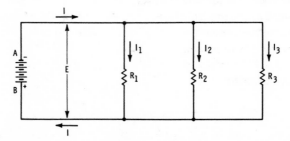

Fig. 3. A parallel circuit with three resistances connected across the battery terminals.

When Ohm's law is applied to the individual resistances, the following equations are obtained:

$$I_1 = \frac{E}{R_1} \qquad I_2 = \frac{E}{R_2} \qquad I_3 = \frac{E}{R_3}$$

From the foregoing, it follows that:

$$I = \frac{E}{R_1} + \frac{E}{R_2} + \frac{E}{R_3}$$

$$I = E\left(\frac{1}{R_1} + \frac{1}{R_2} + \frac{1}{R_3}\right)$$

and since $I = E/R$, the effective resistance R of several resistances connected in parallel is:

$$\frac{1}{R} = \frac{1}{R_1} + \frac{1}{R_2} + \frac{1}{R_3}$$

This formula may also be written:

$$R = \frac{1}{\dfrac{1}{R_1} + \dfrac{1}{R_2} + \dfrac{1}{R_3}}$$

When this formula is applied to two resistances in parallel, it becomes:

$$R = \frac{R_1 \times R_2}{R_1 + R_2}$$

A proper application of these formulas will be obtained if the following problems are considered.

Example—A parallel circuit consists of three branches in which the individual resistances are 4, 5, and 10 ohms, respectively. What is the effective resistance of the circuit?

Solution—If the resistance values are substituted in the proper formula, we obtain:

$$R = \frac{1}{\dfrac{1}{4} + \dfrac{1}{5} + \dfrac{1}{10}} = \frac{1}{0.55} = 1.82 \ ohms$$

Example—What is the effective resistance of a parallel circuit whose individual resistances are 2 and 4 ohms, respectively?

Solution—By using the proper formula, we obtain:

$$R = \frac{2 \times 4}{2 + 4} = \frac{8}{6} = 1.33 \ ohms$$

Example—In a parallel-resistance combination, such as that shown in Fig. 3, R_1, R_2, and R_3 are 5, 15, and 20 ohms, respectively. What will be the total current and the current flowing through each resistance if the applied potential is 10 volts across the battery terminals?

Solution—The total resistance R for the combination is found as follows:

$$\frac{1}{R} = \frac{1}{5} + \frac{1}{15} + \frac{1}{20} = \frac{19}{60}$$

$$R = \frac{60}{19} = 3.16 \ ohms$$

31

$$\text{The total current} = \frac{10}{3.16} = 3.16 \text{ amperes}$$

$$\text{The current in the 5-ohm resistance} = \frac{10}{5} = 2 \text{ amperes}$$

$$\text{The current in the 15-ohm resistance} = \frac{10}{15} = 0.66 \text{ ampere}$$

$$\text{The current in the 20-ohm resistance} = \frac{10}{20} = 0.5 \text{ ampere}$$

The currents through the resistances may conveniently be added as a check. Thus, $2 + 0.66 + 0.5 = 3.16$ amperes.

Series-Parallel Circuits

A series-parallel, or parallel-series, circuit comprises a combination of series and parallel branches, or parts. Fig. 4 serves to illustrate how resistances may be connected in such combinations. Circuits of this type, although somewhat more complex than the straight series and parallel circuits previously discussed, should not cause too much trouble in analyzing them if each parallel-resistance combination is reduced to its equivalent series resistance before combining it with the remainder of the circuit. The method used in calculating series-parallel resistance combinations is shown in the following examples.

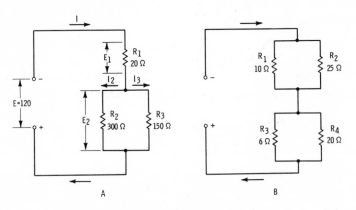

Fig. 4. Typical series-parallel circuits.

Example—What is the equivalent resistance and current flow in the circuit of Fig. 4A, if the applied potential is 120 volts?

Solution—The total equivalent resistance is obtained by a simple substitution in the parallel-resistance formula. That is,

$$R = R_1 + \left(\frac{R_2 \times R_3}{R_2 + R_3}\right) = 20 + \left(\frac{300 \times 150}{300 + 150}\right)$$
$$= 20 + 100 = 120 \ ohms$$
$$I = \frac{E}{R} = \frac{120}{120} = 1 \ ampere$$

By inspection, it will be observed that twice as much current will flow in the 150-ohm branch as in the branch having a resistance of 300 ohms. Thus, the current flow through the parallel branch will be two-thirds and one-third ampere, respectively.

The voltage drop across each resistance combination is obtained as follows:

$$E_1 = IR_1 = 1 \times 20 = 20 \ volts$$
$$E_2 = I \times R_{eqv} = 1 = 100 = 100 \ volts$$

Example—What is the total resistance value of the circuit illustrated in Fig. 4B?

Solution—Since this circuit consists of two parallel branches, it will again be necessary to reduce each of the parallel circuits to its individual equivalent. Thus, the upper parallel branch has an equivalent resistance of:

$$R_{eqv} = \frac{R_1 \times R_2}{R_1 + R_2} = \frac{10 \times 25}{10 + 25} = \frac{250}{35} = 7.15 \ ohms$$

The equivalent resistance of the lower combination is:

$$R_{eqv} = \frac{R_3 \times R_4}{R_3 + R_4} = \frac{6 \times 20}{6 + 20} = \frac{120}{26} = 4.62 \ ohms$$

The total resistance of the circuit then is:

$$R_{total} = 7.15 + 4.62 = 11.77 \ ohms$$

POWER AND ENERGY

Power is the rate of doing work and is measured in watts, kilowatts, and horsepower. In the transmission of electricity through a circuit, the electrons in their movement through the conductor do not have a clear path but are in constant collision with atoms of the conductor; in this manner, they cause the material to heat up. The heat thus developed varies with the number of collisions and increases with the increase in current flow. It has been found that this developed heat, or power loss, varies directly as the resistance of the conductor in question and as the square of the current. This important relationship is written as

$$P = I^2R \; watts$$

Since, according to Ohm's law, $I = \dfrac{E}{R}$ we may multiply the left side of this formula by I and the right side by E/R to obtain:

$$I^2 = \frac{E^2}{R^2}$$

A substitution in the power formula produces:

$$P = \frac{E^2 \times R}{R^2} = \frac{E^2}{R} \; watts$$

Similarly,

$$P = E \times I \; watts$$

In the foregoing formulas,

P = the power in watts,
E = the voltage,
I = the current,
R = the resistance of the circuit.

It should be noted that in the transmission of electrical power over long distances, the conductor heat loss, although quite small per foot of conductor, is a serious economic detriment when large amounts of power are to be transmitted from one area to another. On the other hand, the heating effect of the current is desirable in resistance-heating appliances, which are equipped with special heating elements (resistance elements). Incandescent lamps are devices in which the heating of the filament to incandescence provides light energy.

Electrical power is not always turned into heat, although in all processes in which electrical power is converted into mechanical power, and vice versa, a certain amount of power is dissipated in the form of heat. For example, the power used to operate a motor is converted into mechanical motion. The power applied to a speaker in a radio or television system is converted into sound waves.

Since every electrical device has some resistance, a part of the electrical power supplied is dissipated in that resistance and hence appears as heat, even though the major part of the power may be converted into another form.

Example—A 150-watt incandescent lamp is connected to a 110-volt circuit. What is the amount of current flow through the lamp?

Solution—According to the formula $P = E \times I$, the current is,

$$I = \frac{P}{E} = \frac{150}{110} = 1.36 \text{ amperes}$$

Example—An electric soldering iron has a resistance of 425 ohms. If the potential of the source is 220 volts, what are the power consumption and energy requirements of the iron when connected to the circuit for one-half hour?

Solution—According to the formula $P = E^2/R$,

$$P = \frac{220^2}{425} = 114 \text{ watts} = 0.114 \text{ kilowatt}$$

The hourly energy consumed is 0.114 kilowatt-hour. During one-half hour, the energy consumption will be $0.5 \times 0.114 = 0.057$ kilowatt-hour or 57 watt-hours.

Efficiency

The term *efficiency*, as used in electrical devices, means the useful power output (in its converted form) divided by the power input to the device. In devices such as motors and generators, the object is to obtain power in some form other than heat, since power dissipated in the form of heat is considered as a loss, because it is not useful power. The formula for efficiency is,

$$efficiency = \frac{output}{input}$$

An application of this formula may be seen by studying the following examples.

Example—A 7-horsepower, direct-current motor requires 6.3 kilowatts at full load. What is its efficiency?

Solution—As the first step in solving the problem, convert horsepower and kilowatts into watts. That is,

$$output = 7 \times 746 = 5222 \; watts$$
$$input = 1000 \times 6.3 = 6300 \; watts$$
$$efficiency = \frac{output}{input} = \frac{5222}{6300} = 0.829, \; or \; 83\%$$

Example—A one-half horsepower, direct-current motor requires 2.5 amperes when operating at full load, and is connected to a 115-volt source. What is the efficiency of the motor?

Solution—The output power of the motor is obtained by multiplying the current drawn from the line by the voltage across the motor terminals. The product in watts is, therefore, 2.5 × 115, or 287.5. The power input equals 746/2, or 373 watts. Therefore,

$$motor \; efficiency = \frac{output}{input} = \frac{287.5}{373} = 0.77, \; or \; 77\%$$

Energy

Power and energy are two terms which are often not clearly understood. Power is the ability to do work, whereas energy is work per unit time; thus, while power is measured in watts, kilowatts, and horsepower, energy is measured in watt-hours, kilowatt-hours, and horsepower-hours.

In residences, the power company's bill is for electrical energy, not for power. Thus, if in a typical residence, six 50-watt lamps are lit during five hours in a day, their energy consumption will be $6 \times 50 \times 5 = 1500$ watt-hours, or 1.5 kilowatt-hours. Electrical energy, therefore, is equal to power multiplied by time. The common unit is the watt-hour, which means that one watt has been used for one hour. It may be written,

$$W = PT$$

where,
    W = energy in watt-hours,
    P = power in watts,
    T = time in hours.

Example—If the rate of electrical energy in a certain location is 5.5 cents per kilowatt-hour, and the monthly bill is $3.75, how many kilowatt-hours have been consumed?

Solution—The number of kilowatt-hours consumed is 375/5.5, or 68.18 = 68,180 watt-hours.

Example—If in the foregoing problem, electricity is supplied from a 110-volt line for 60 hours, what is the average current flow through the watt-hour meter, assuming that only noninductive devices are used?

Solution—The average power and current are obtained as follows:

$$P = \frac{W}{T} = \frac{68,180}{60} = 1136.4 \; watts$$

$$I = \frac{P}{E} = \frac{1136.4}{110} = 10.3 \; amperes$$

Example—A direct-current motor draws a current of 16 amperes from a 220-volt line. The full-load nameplate rating of the motor is 4 horsepower. What is the efficiency of the motor and cost of operation per eight-hour day when the cost of electrical energy is 5 cents per kilowatt-hour?

Solution—The full-load output of the motor is 4×746, or 2984 watts. The power input is 16×220, or 3520 watts. Therefore,

$$motor \; efficiency = \frac{output}{input} = \frac{2984}{3520} = 0.85, \; or \; 85\%$$

The energy consumption during 8 hours is,

$$W = PT = 3520 \times 8 = 28,160 \text{ watt-hours, or } 28.16 \text{ kilowatt-hours}$$

Since the cost of electrical energy is 5 cents per kilowatt-hour, the total cost of energy will consequently be,

$$0.05 \times 28.16, \text{ or } \$1.41$$

Measurement of Heat

Heat energy is measured in British thermal units (abbreviated Btu) and is defined as the amount of heat necessary to raise one pound of water one degree Fahrenheit. When it is desired to find the amount of heat necessary to raise a quantity of water to a certain temperature, it is only necessary to multiply the weight of the water by the temperature rise in degrees Fahrenheit. This statement may be expressed by the following formula:

$$H = Qt$$

where,
H = the amount of heat in Btu,
Q = the quantity of water in pounds,
t = the change in temperature in degrees Fahrenheit.

Example—How many Btu of heat are required to raise one quart of water from 75°F. to the boiling point (212°F.)? Note: One gallon of water weighs 8.33 pounds.

Solution—By substituting values in the formula $H = Qt$,

$$H = (8.33/4)(212 - 75) = 285.4 \text{ Btu}$$

Heating Effect of an Electric Current

A convenient method of measuring the amount of heat produced by an electric current passing through a resistance is by means of a calorimeter, the construction principles of which are shown in Fig. 5. This is simply an insulated vessel containing water in which a resistance of known value and a thermometer are placed. It has been found that when one ampere flows through a one-ohm resistance for one minute, 0.057 Btu of heat will be generated. Hence, one watt produces 0.057 Btu per minute. It is evident that the amount of heat produced per minute by a current passing through a resistance R is $0.057I^2R$. An expression giving the relations between the electrical energy in a circuit and the heat in Btu is therefore expressed as,

$$H = 0.057I^2Rt$$

where,

H = the amount of heat in Btu,
I = the current in amperes,
R = the resistance in ohms,
t = the time in minutes.

Example—How many heat units are generated in 1 hour by a heating element which carries 20 amperes and has a 2-ohm resistance?

Solution

$$H = 0.057 \times 20 \times 20 \times 2 \times 60 = 2736 \; Btu$$

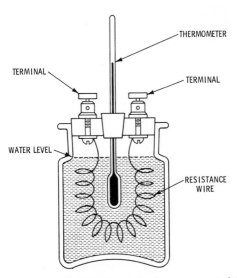

Fig. 5. A calorimeter, used to measure the heat generated by an electric current. The apparatus is connected to a suitable electric source; by reading the temperature rise and current consumption for a certain period of time, the amount of heat, in Btu's, generated by the current can be obtained.

Example—A resistance heating element of 145 ohms is placed in a one-quart container of water. How long must a current of 4 amperes flow through the element in order to raise the temperature of the water from 50°F. to boiling (212°F.)? Assume the efficiency of the process is 75%.

Solution—One gallon of water weighs 8.33 pounds; one-fourth of a gallon, consequently, weighs 2.08 pounds. From the heat formula we obtain,

$$t = \frac{H}{0.057I^2R} = \frac{2.08(212 - 50)}{0.057 \times 16 \times 145 \times 0.75} = 3.4 \text{ minutes}$$

CHAPTER 4

Fractional-Horsepower Motors

Fractional-horsepower motors are manufactured in a large number of types to suit various applications. Because of its use in a great variety of household appliances, the fractional-horsepower motor is perhaps better known than any other type. It is nearly always designed to operate on single-phase AC at standard frequencies, and is reliable, easy to repair, and comparatively low in cost.

Single-phase motors were one of the first types developed for use on alternating current. They have been perfected through the years from the original repulsion type into many improved types, such as:

1. Split-phase.
2. Capacitor-start.
3. Permanent-capacitor.
4. Repulsion.
5. Shaded-pole.
6. Universal.

SPLIT-PHASE INDUCTION MOTORS

The split-phase induction motor is one of the most popular of the fractional-horsepower types. As shown in Fig. 1, the motor consists essentially of a squirrel-cage rotor and two stator windings (a main winding and a starting winding). The main winding is connected across the supply line in the usual manner, and has a low resistance and a high inductance. The starting or auxiliary winding, which is

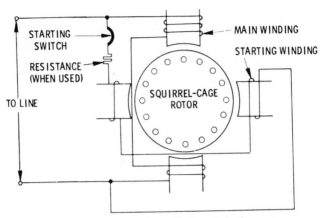

Fig. 1. Schematic diagram of a split-phase induction motor.

physically displaced in the stator from the main winding, has a high resistance and a low inductance. This physical displacement, in addition to the electrical phase displacement produced by the relative electrical resistance values in the two windings, produces a weak rotating field which is sufficient to provide a low starting torque.

After the motor has accelerated to 75 or 80% of its synchronous speed, a starting switch (usually centrifugally operated) opens its contacts to disconnect the starting winding. The function of the starting switch (after the motor has started) is to prevent the motor from drawing excessive current from the line and to protect the starting winding from damage due to heating. The motor may be started in either direction by reversing the connections to either the main or auxiliary winding, *but not to both*.

The characteristics of a resistance-type split-phase motor are shown in Fig. 2. The split-phase motor is most commonly used in sizes ranging from 1/30 to 1/2 HP for applications such as fans, business machines, automatic musical instruments, buffing machines, etc.

Resistance-Start Motors

A resistance-start motor is a form of split-phase motor having a resistance connected in series with the auxiliary winding. The auxiliary circuit is opened by a starting switch when the motor has attained a predetermined speed.

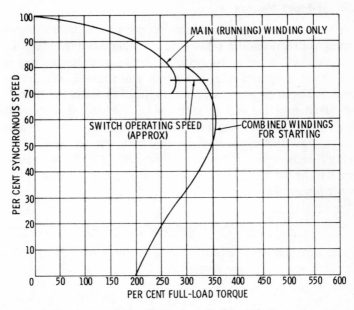

Fig. 2. Speed-torque characteristics of a split-phase induction motor.

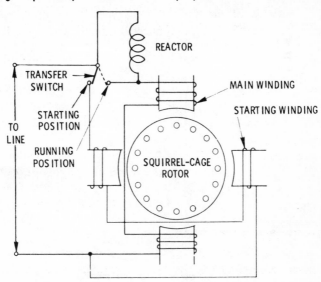

Fig. 3. Schematic diagram of a reactor-start motor.

Reactor-Start Motors

A reactor-start motor is a form of split-phase motor designed for starting with a reactor in series with the main winding. The reactor is short-circuited, or otherwise made ineffective, and the auxiliary (starting) circuit is opened when the motor has attained a predetermined speed. A circuit arrangement for this type of motor is shown in Fig. 3. The function of the reactor is to reduce the starting current and to increase the angle of lag of the main-winding current behind the voltage. This motor will develop approximately the same torque as the split-phase motors discussed previously. The centrifugally-operated starting switch must be of the single-pole double-throw type for proper functioning.

CAPACITOR-START MOTORS

The capacitor-start motor is another form of split-phase motor having a capacitor or condenser connected in series with the auxiliary winding. The auxiliary circuit is opened when the motor has attained a predetermined speed. The circuit in Fig. 4 shows the winding arrangement.

The rotor is of the squirrel-cage type, as in other split-phase motors. The main winding is connected directly across the line, while

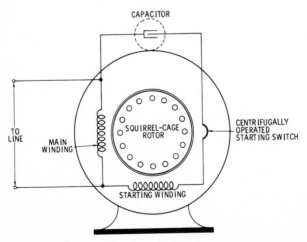

Fig. 4. Schematic diagram of a capacitor-start motor.

the auxiliary or starting winding is connected through a capacitor which is connected into the circuit through a centrifugally-operated starting switch. The two windings are approximately 90° apart electrically.

This type of motor has certain advantages over the previously described types in that it has a considerably higher starting torque accompanied by a high power factor.

PERMANENT-CAPACITOR MOTORS

A permanent-capacitor type of motor has its main winding connected directly to the power supply, and the auxiliary winding connected in series with a capacitor. Both the capacitor and auxiliary winding remain in the circuit while the motor is in operation. There are several types of permanent-capacitor motors, differing from one another mainly in the number and arrangement of capacitors employed. The running characteristics of this type of motor are extremely favorable, and the torque is fixed by the amount of additional capacitance, if any, added to the auxiliary winding during starting.

The simplest of this type of motor is the low-torque, permanent-capacitor motor shown in Fig. 5. Here, a capacitor is permanently

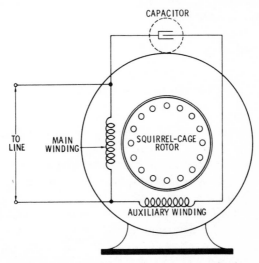

Fig. 5. Schematic diagram of a permanent-capacitor split-phase motor.

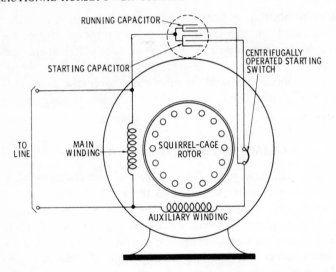

Fig. 6. Schematic diagram of a high-torque capacitor motor.

connected in series with the auxiliary winding. This type of motor can be arranged for an adjustable speed by the use of a tapped winding or an autotransformer regulator.

High-torque motors are usually provided with one running and one starting capacitor connected as shown in Fig. 6, or with an autotransformer connected to increase the voltage across the capacitor during the starting period, as indicated in Fig. 7.

REPULSION MOTORS

Repulsion motors have a stator winding arranged for connection to the source of power, and a rotor winding connected to a commutator. As shown in Fig. 8, brushes on the commutator are short-circuited and are so placed that the magnetic axis of the rotor winding is inclined to the magnetic axis of the stator winding. This type of motor has a varying speed characteristic.

Principally, it has a stator like that of a single-phase motor, but has a rotor like the armature of a DC motor, with the opposite brushes on the armature short-circuited—that is, connected together by a connector with a negligible resistance.

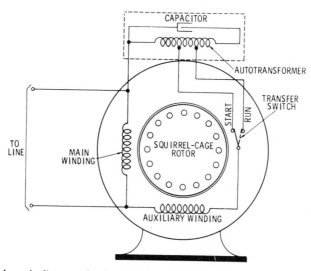

Fig. 7. Schematic diagram of a high-torque capacitor motor using an autotransformer.

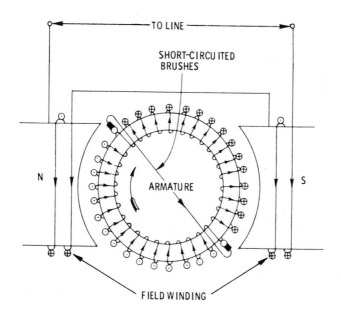

Fig. 8. Simplified schematic of a series repulsion motor.

The brushes are placed so that a line connecting them makes a small angle with the neutral axis of the magnetic field of the stator. The stator induces a current in the armature; the current produces an armature field with poles in the neighborhood of the brushes. These armature fields have the same polarity as the adjacent field poles, and are repelled by them so that this repulsion causes the armature to revolve. The motor derives its name from this action.

A repulsion motor of this original type has characteristics (Fig. 9) similar to those of the series motor. It has a high starting torque and

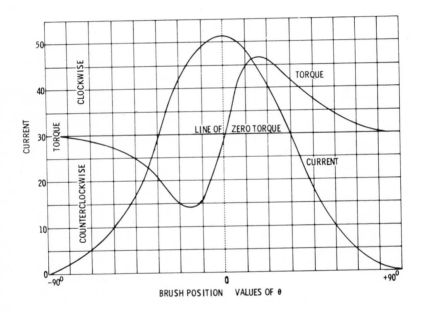

Fig. 9. Characteristic curves of a repulsion motor showing the effect of shifting brushes on current and torque.

moderate starting current. It has a low power factor, except at high speeds. For this reason, it is often modified into the *compensated repulsion motor,* which has another set of brushes placed midway between the short-circuited set and connected in series with the stator winding.

48

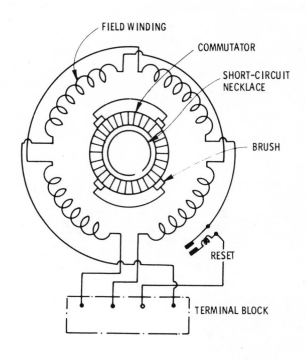

FIELD WINDING

COMMUTATOR

SHORT-CIRCUIT NECKLACE

BRUSH

RESET

TERMINAL BLOCK

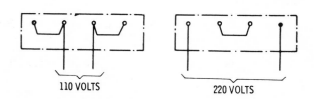

110 VOLTS

220 VOLTS

NOTE: TOP TERMINAL SHOWS CONNECTION TO MOTOR WINDINGS ONLY. FOR PROPER CONNECTION TO AC CIRCUIT CHECK VOLTAGE SOURCE AND CONNECT AS INDICATED ON LOWER TERMINALS

Fig. 10. Coil relationship in a single-phase repulsion-start induction motor.

49

Repulsion-Start Induction Motors

A repulsion-start induction motor is a single-phase motor having the same windings (Fig. 10) as a repulsion motor. At a predetermined speed, however, the rotor winding is short-circuited, or otherwise connected, to give the equivalent of a squirrel-cage winding. This type of motor starts as a repulsion motor, but runs as an induction motor with constant-speed characteristics.

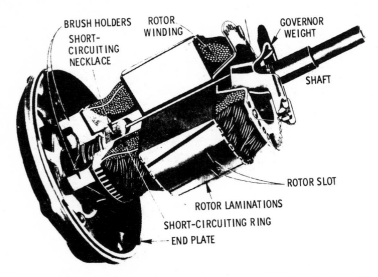

Fig. 11. The position of the short-circuiting necklace and brush mechanism while the motor is starting.

The repulsion-start induction motor has a single-phase distributed-field winding, with the axis of the brushes displaced from the axis of the field winding. The armature has an insulated winding. The current induced in the armature, or rotor, is carried by the brushes and commutator, resulting in a high starting torque. When nearly synchronous speed is attained, the commutator is short-circuited, so that the armature is then similar in function to a squirrel-cage armature.

The photographs in Figs. 11 and 13 show the working principles of the mechanism for simultaneously lifting the brushes and short-circuiting the commutator to change the operation from repulsion to induction. The object of lifting the brushes is to eliminate wear on the

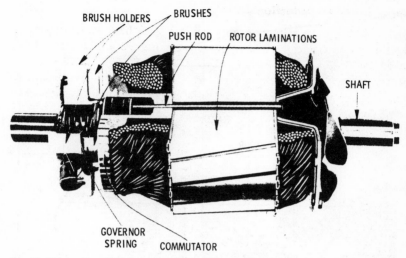

Fig. 12. The position of the short-circuiting necklace and brush mechanism after the governor weights have operated.

commutator during the running periods, since it makes no difference electrically whether the brushes are in contact or not after the motor comes up to speed.

This type of motor has gone through many stages of improvement since its first appearance on the market, although its general principle has remained the same. The general reliability of this type of motor is largely governed by the reliability of the short-circuiting mechanism. For this reason, it has been the constant aim of engineers to improve on the principle and construction of the short-circuiting switch.

Since the motor starts on the repulsion principle, it has the same starting characteristics as the repulsion motor described previously —namely, high starting torque and low starting current. As the motor speeds up, the torque falls off rapidly. At some point on the speed-torque curve, usually at about 80% of synchronism, the commutator is automatically short-circuited, producing the effect of a cage winding in the armature, and the motor comes up to speed as an induction motor. After the commutator has been short-circuited, the brushes do not carry current and therefore may be lifted from the commutator, but lifting is not absolutely necessary.

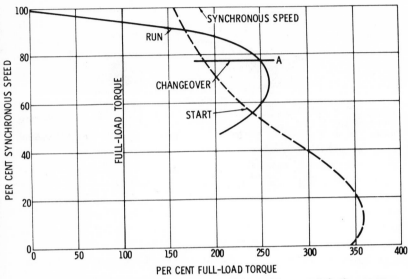

Fig. 13. Speed-torque characteristics of a typical repulsion-start induction motor.

The curve in Fig. 13 shows the speed-torque characteristics of a typical repulsion-start induction motor. The short-circuiting mechanism operates at point A, at which time the induction-motor torque is greater than the repulsion-motor torque. This means that if the repulsion winding has sufficient torque to bring the load up to this speed, there will be sufficient torque as an induction motor to bring the load up to full speed. The higher the speed at which the short-circuiting mechanism operates, the lower will be the induction-motor current at that point and, consequently, the less disturbance to the line. After the commutator has been short-circuited, the motor has the same characteristics as the single-phase induction motor previously described.

If the short-circuiting mechanism operates before the repulsion curve crosses the induction-motor curve, and if the torque of the induction motor is less than that required to accelerate the load, the motor may slow down until the short circuit is removed from the commutator, in which case the motor will again operate repulsion. The armature will then speed up until the commutator is again short-circuited, after which the armature will slow down until it again becomes repulsion. This cycle will be repeated over and over again until some change takes place.

The efficiency and the maximum running torque of the repulsion-start induction motor are usually less than those of a cage-wound induction motor of comparative size. In other words, a repulsion-start induction motor must be larger than a cage-wound motor of the same rating to give the same performance.

Repulsion-Induction Motors

A repulsion-induction motor is another form of repulsion motor which has a squirrel-cage winding in the rotor in addition to the repulsion winding. A motor of this type may have either a constant-speed or a varying-speed characteristic. Specifically, this motor is a combination of the repulsion and induction types and operates on the combined principle of both repulsion and induction. It is sometimes termed a squirrel-cage repulsion motor. In this motor, the desirable starting characteristics of the repulsion motor and the con-

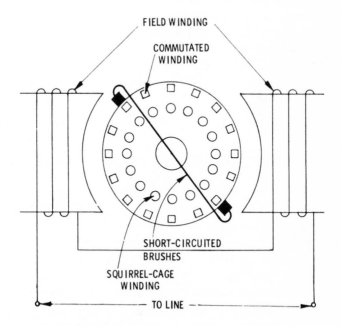

Fig. 14. Scehmatic diagram of a repulsion-induction motor.

stant-speed characteristic of the induction motor are obtained. It is, of course, impossible to combine the two types of motors and obtain only the desirable characteristics of each.

As shown in Fig. 14, the field has the same type of windings used in the repulsion-start induction motor, and the armature has two separate and independent windings. They are:

1. Squirrel-cage winding.
2. Commutated winding.

Both of these armature windings function during the entire period of operation of the motor. There are no automatic devices, such as the starting switch of the split-phase motor or the short-circuiting device of the repulsion-start induction motor.

The cage winding is located in slots below those which contain the commutated winding. The slots which contain the two windings may or may not be connected by a narrow slot. Usually, there are the same number of slots for the two windings. It is not, however, absolutely essential that they be the same. Because of its construction, the squirrel-cage winding inherently has a high inductance. Its reactance, with the armature at rest, is therefore high.

The commutated winding has a low reactance, with the result that most of the current will flow in this winding. The ideal condition at starting would be for all of the flux to pass beneath the commutated winding and none of it beneath the cage winding. If this condition could be obtained, this motor would have the same starting characteristics as the repulsion-start induction motor. However, at full-load speed, which is slightly below synchronism, the reactance of the cage winding is low, and most of the mutual flux passes beneath the cage winding. Both windings produce torque, and the output of the motor is the combined output of the cage winding and the commutated winding.

The commutation of the motor is good at all speeds. The no-load speed is above synchronism and is limited by the combined effect of the field winding on the commutated winding and the cage winding, and by the action of the two armature windings on each other. At synchronous speed, a squirrel-cage motor has no torque. At synchronous speed, and for a short distance above synchronous speed, the torque of a repulsion-induction motor is greater than that of the commutated winding alone. This shows that because of the interaction between the two armature windings, the squirrel cage supplies torque instead of acting as a brake.

At full-load speed, and up to about the maximum running torque point, the torque of this motor is greater than the sum of the torques of the cage winding and the commutated winding. The inherent locked-torque curve of a repulsion-induction motor, shown in Fig. 15,

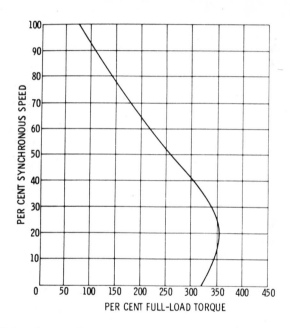

Fig. 15. Speed-torque characteristic of a typical repulsion-induction motor.

is similar to that of the repulsion motor. At soft neutral, the primary winding carries the squirrel-cage current in addition to the exciting current of the motor.

Since its starting current is low, the repulsion-induction motor may be operated from lighting circuits when driving frequently started devices. The repulsion-induction motor is especially suitable for such applications as household refrigerators, water systems, garage air pumps, gasoline pumps, compressors, etc.

SHADED-POLE MOTORS

A shaded-pole motor is a single-phase induction motor equipped with an auxiliary winding displaced magnetically from, and connected in parallel with, the main winding. This type of motor is manufactured in fractional-horsepower sizes, and is used in a variety of household appliances, such as fans, blowers, hair driers, and other applications requiring a low starting torque. It is operated only on alternating current, is usually nonreversible, is low in cost, and is extremely rugged and reliable. The diagram of a typical shaded-pole motor is shown in Fig. 16.

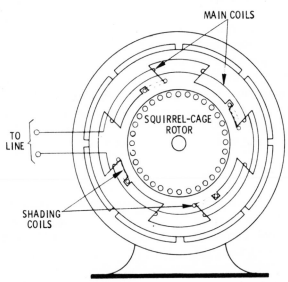

Fig. 16. A single-phase motor with shading coils for starting.

The shading coil (from which the motor has derived its name) consists of low-resistance copper links embedded in one side of each stator pole, which are used to provide the necessary starting torque. When the current increases in the main coils, a current is induced in the shading coil. This current opposes the magnetic field building up in the part of the pole pieces they surround. This produces the condition shown in Fig. 17, where the flux is crowded away from that portion of the pole piece surrounded by the shading coil.

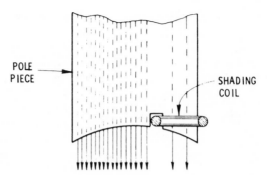

Fig. 17. Action of a shading coil in an AC motor.

When the main-coil current decreases, the current in the shading coils also decreases until the pole pieces are uniformly magnetized. As the main-coil current and the magnetic flux of the pole piece continue to decrease, the current in the shading coils reverses and tends to maintain the flux in part of the pole pieces. When the main-coil current drops to zero, current still flows in the shading coils to give the magnetic effect which causes the coils to produce a rotating or magnetic field that makes the motor self-starting.

UNIVERSAL MOTORS

A universal motor is a series-wound or compensated series-wound motor which may be operated either on direct current or on single-phase alternating current at approximately the same speed and output. These conditions must be met when the DC and AC voltages are approximately the same, and the AC frequency is not greater than 60 cycles per second.

Universal motors are commonly manufactured in fractional-horsepower sizes, and are preferred because of their use on either AC or DC, particularly in areas where power companies supply both types of current.

As previously noted, all universal motors are series wound—therefore, their performance characteristics are very much like those of the usual DC series motor. The no-load speed is quite high, but seldom high enough to damage the motor, as in the case with larger DC series motors. When a load is placed on the motor, the speed decreases and continues to decrease as the load increases. Although universal motors of several types of constructon are manufactured, they all have the varying-speed characteristics just mentioned.

Because of the difficulty in obtaining like performance on AC and DC at low speeds, most universal motors are designed for operation at speeds of 3500 rpm and higher. Motors operating at a load speed of 8000 to 10,000 rpm are common. Small stationary vacuum cleaners and the larger sizes of portable tools have motors operating at 3500 to 8000 rpm.

The speed of a universal motor can be adjusted by connecting a resistance of proper value in series with the motor. The advantage of this characteristic is obvious in an application such as a motor-driven sewing machine, where it is necessary to operate the motor over a wide range of speeds. In such applications, adjustable resistances are used by which the speed is varied at will.

When universal motors are to be used for driving any apparatus, the following characteristics of the motor must be considered:

1. Change in speed with change in load.
2. Change in speed with change in frequency of power supply.
3. Change in speed due to change in applied voltage.

Since most small motors are connected to lighting circuits, where the voltage conditions are not always the best, this last item is of the utmost importance. This condition should also be kept in mind when determining the proper motor to use for any application, regardless of type. In general, the speed of the universal motor varies with the voltage. The starting torque of universal motors is usually much more than required in most applications, and not to be considered.

Universal motors are manufactured in two general types. They are:

1. Concentrated-pole, noncompensated.
2. Distributed-field, compensated.

Most motors of low-horsepower rating are of the concentrated-pole, noncompensated type, while those of higher ratings are of the distributed-field, compensated type. The dividing line is approximately 1/4 HP, but the type of motor to be used is determined by the severity of the service and the performance required. All of the motors have wound armatures similar in construction to an ordinary DC motor.

The concentrated-pole, noncompensated motor is exactly the same in construction as a DC motor except that the magnetic path is made up of laminations. The laminated stator is made necessary because the magnetic field is alternating when the motor is operated on alternating current. The stator laminations are punched, with the poles and the yoke in one piece.

The compensated type of motor has stator laminations of the same shape as those in an induction motor. These motors have stator windings in one of two different types.

The noncompensated motor is simpler and less expensive than the compensated motor and would be used over the entire range of ratings if its performance were as good as that of the compensated motor. The noncompensated type is used for the higher speeds and lower horsepower ratings only. Figs. 18 and 19 show the speed-torque curves for a compensated and noncompensated motor, respectively.

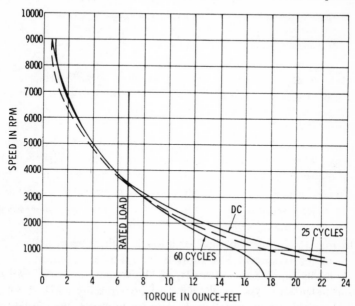

Fig. 18. Speed-torque characteristics of a typical ¼ HP, 3400 rpm, compensated universal motor.

It will be noted in Fig. 18 that, although the rated speed is relatively low for a universal motor, the speed-torque curves for various frequencies lie very close together up to 50% above the rated-torque load.

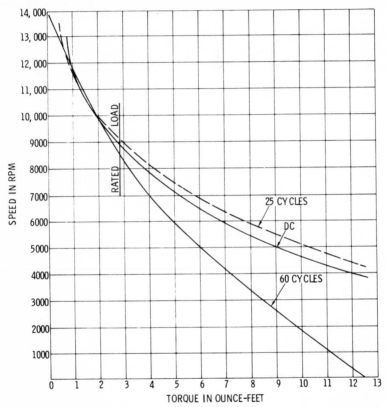

Fig. 19. Speed-torque characteristic of a typical ¼ HP, 8700 rpm, noncompensated universal motor.

In Fig. 19, the performance of a much higher-speed, noncompensated motor is shown. For most universal-motor applications, the variation in speed at rated loads, as shown on this curve, is satisfactory. However, the speed curves separate rapidly above full load. If this motor had been designed for a lower speed, the tendency of the speed-torque curves to separate would have been more pronounced. The chief cause of the difficulty in keeping the speeds the same is the reactance voltage which exists when the motors are operated on AC current. Most of this reactance voltage is produced in the field windings by the main working field. However, in the noncompensated motor, some of it is produced in the armature winding by the field produced by the armature ampere-turns. The true working voltage is obtained by subtracting the reactance voltage vectorially from the line voltage. If the reactance is high, the performance at a given load will be the same as if there were no reactance voltage and as if the applied voltage had been reduced, with consequent reduction in speed.

SPEED CONTROL

The control method used in single-phase motors depends on the type of motor selected for a certain application. Speed control is

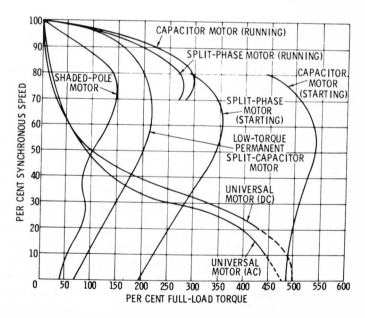

Fig. 20. Speed-torque characteristics of common types of fractional-horsepower motors.

usually accomplished by means of a centrifugal switch on motors having speeds above 900 rpm. A comparison of typical speed-torque characteristics of fractional-horsepower, single-phase motors is given in Fig. 20.

Where speeds of 900 rpm and below are involved, capacitor motors utilizing voltage relays for changeover are preferable to motors using centrifugal switches, because of the sluggishness of a centrifugal switch of normal design at the lower speeds. A voltage relay for changeover is not subject to the same limitations.

Speed control, because of its more general use, has brought about the standardization of fan speeds, permitting the use of high-slip motors. A squirrel-cage motor with 8 to 10% slip and with low maximum torque operates satisfactorily, not only for constant-speed drives, but also for adjustable-speed drives through the use of voltage control on the primary winding. Within the working range of a high-slip squirrel-cage motor having adjustable voltage applied to the primary winding, the speed-torque characteristics resemble the characteristics of a wound-rotor motor having different amounts of external resistance in the secondary circuit.

A tapped autotransformer connected to the motor through a multi-position snap switch offers the simplest form of control. The transformer may have a number of taps, only two or three of which are brought out to the switch. This provides two or three speeds, any or all of which may be changed to suit the needs of the application by the proper selection of taps to connect to the switch. Additional contacts on the switch make it a complete starter and speed regulator. Two such transformers connected in open-delta for three-phase, or in each phase of a two-phase circuit, may be used to control the speed of polyphase motors. For polyphase use, the snap switch must have duplicate contacts for the transformers.

In applying speed control by means of line-voltage adjustment, it must be remembered that the speed of the drive will vary with the load. Therefore, this method of control is not suitable for centrifugal fans, especially those with damper control which affect the fan load. Two-speed pole-changing motors offer a solution for centrifugal-fan drives, inasmuch as the speed on either pole combination is affected only slightly by the change in load. Motor speeds in the ratio of 2 to 1 are obtainable as 3600/1800 and 1800/900 rpm; 3 to 2 as

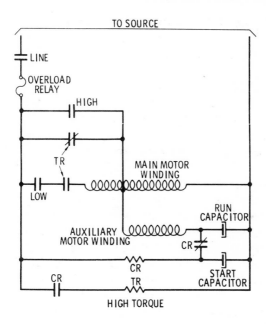

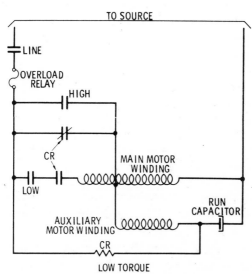

Fig. 21. Starting and running circuits of low-torque and high-torque adjustable-speed capacitor motors with tapped main windings.

1800/1200 rpm; and 4 to 3 as 1200/900 rpm. Other pole combinations are possible.

There is still another means of obtaining speed reduction on induction motors. The equivalent of building the transformer into the motor is obtained by a suitable tap on the primary winding, as shown in Fig. 21. By this means, normal speed and a single reduced speed are provided. A simple pole-changing controller is all that is required to complete the installation. Such motors are not generally available, but are provided through propeller-fan manufacturers. Each motor is designed for the characteristics of the fan it is to drive, and the tap on the winding is located to give the desired low speed.

When speed controllers are used with low-torque capacitor motors, it must be remembered that the available 50% starting torque, with full voltage applied, is reduced in proportion to the square of the voltage involved. The result is low breakaway and low accelerating torque on the low-speed setting. In manually operated control, one remedy for this limitation is a progressive starting switch which provides full voltage on the first (high-speed) position and reduced voltage on the intermediate and low-speed positions (Fig. 22).

Where temperature control is used to start and stop the motor, a preset full-voltage starting controller is usually provided. In this case, a relay is energized by the rising voltage of the motor auxiliary winding as it accelerates; the relay disconnects the motor from the direct connection to the line and connects it to the transformer tap for which the controller is set. In high-torque motors, this relay serves to disconnect the starting capacitor and to change over from full to reduced voltage on the motor.

Under the general subject of adjustable-speed fan motors, the question of the amount of speed reduction is often debated, and although it is generally agreed that 30 to 35% speed reduction will meet almost any air-conditioning requirement, as much as 50% speed reduction is sometimes specified. In most cases, a lower percentage of speed reduction is found to be acceptable. It is fairly easy to obtain 50% reduction on a fan if the characteristic curve follows the law of horsepower load, increasing as the cube of the speed, and if the fan fully loads the motor at full speed. It is quite difficult to obtain 50% speed reduction with stability in cases where the drive is 20 to 25% overmotored.

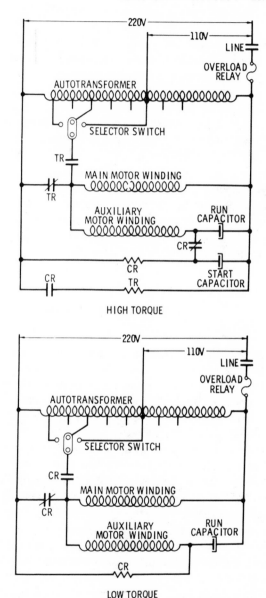

HIGH TORQUE

LOW TORQUE

Fig. 22. Starting and running circuits of low-torque and high-torque adjustable-speed capacitor motors with manually operated transformer speed regulator.

In considering the wide range of air-conditioning equipment, from unit heaters and coolers to barn ventilators and incubator ventilating fans, with the many types of equipment between these extremes, the foremost impression is one of economy coupled with simplicity and safety. Simplicity promotes economy. Any additional expense imposed in the interest of safety is abundantly justified from the standpoint of good operating practice. It is not advisable to use a motor of the "long annual service" classification where the operation is infrequent, when a motor for short annual service will meet the requirements at a lower first cost with no appreciable increase in operating cost. Similarly, a fractional-horsepower split-phase motor is less expensive than a capacitor motor.

It is essential to avoid misapplications such as low-torque motors on belted drives, or tapped-winding and transformer speed control in place of multispeed motors for centrifugal drives. The drive should always be closely motored with the correct size and type of motor to give the best results.

MOTOR APPLICATION

Although single-phase motors are most commonly used in fractional-horsepower sizes, certain applications and power-supply conditions make use of single-phase motors in integral horsepower sizes. For extremely small capacities up to 1/40 HP, the shaded-pole type of motor is most frequently used. It provides sufficient torque for fans, blowers, and other similar equipment, and the starting current is not objectionable on lighting lines.

The split-phase motor is the most commonly used in sizes ranging from 1/30 to 1/2 HP, particularly for fans and similar drives where the starting torque is low. In this type of motor, a built-in centrifugal switch is usually provided to disconnect the starting winding as the motor comes up to speed.

In sizes above 1/4 HP, a capacitor motor with 300% starting torque may be used to advantage, especially for pump and compressor drives. It may also be used in the overlapping ratings for fan drives in the lower capacities at higher speeds. However, it does not offer sufficient advantages over the split-phase motor to warrant its higher cost.

At low speeds (below 900 rpm, where centrifugal switches are less successful), and in ratings from 3/4 to 3 HP at all speeds, the capacitor motor finds its widest field of application in fan drives. The running characteristics of this motor are extremely good, and the starting torque is fixed by the amount of additional capacitance added to the auxiliary motor winding during the starting period.

A low-torque capacitor motor, in which no capacitance is added to the auxiliary winding during the starting period, provides approximately 50% starting torque. This is considered sufficient for directly connected fans, if the unit is one of constant speed or is always started on the high-speed position of the starting switch. Where the fan is coupled or belted to the motor, the high-torque type of motor with the additional starting capacitance is preferable. The changeover from start to run of the high-torque type may be accomplished either through the use of a centrifugal switch, or by electrical means responsive to current decay or voltage rise as the motor approaches normal running speed.

The repulsion-induction and the repulsion-start induction motors have ratings paralleling those of the capacitor types. These motors have commutators to provide the starting torque. Under normal running conditions, the commutator of the repulsion-start, induction-run type is short-circuited, and the motor operates as an induction motor. This differs from the repulsion-induction motor in which the commutator is not short-circuited. A squirrel-cage winding deep in the rotor is inactive at starting, but takes up the load as the rotor accelerates to full speed, where the normal load is about equally divided between the repulsion and the squirrel-cage winding. The starting efficiency of both types is high, and the 300% or more starting torque that is available makes them suitable for compressor drives.

The universal-type motor, on account of its ability to run on DC as well as on AC, is preferred on appliances which are required to operate on either an AC or a DC circuit. Popular applications for the universal-type motor are in vacuum cleaners, sewing machines, portable drills, saws, routers, home motion-picture projectors, and business machines of all kinds.

Whenever it becomes necessary to substitute a motor on any appliance, a good deal of grief and disappointment may be avoided if a motor of the same identical size and type is reinstalled. Particular attention should be observed with respect to the nameplate data giving all the necessary information. A sound rule to follow is to copy the entire nameplate reading. This observation holds true whether it is a complete motor, or whether only a spare part is required.

FRACTIONAL-HORSEPOWER MOTOR TROUBLE CHART

(Split-Phase Induction Motor)

Symptom and Possible Cause   *Possible Remedy*

Failure to Start

(a) No voltage	(a) Check for voltage at motor terminals with test lamp or voltmeter. Check for blown fuses or meter. Check for blown fuses or open overload device in starter. If motor is equipped with a slow-blow fuse, see that the fuse plug is not open and that it is screwed down tight.
(b) Low voltage	(b) Measure the voltage at the motor terminals with the switch closed. Voltage should read within 10% of the voltage stamped on the motor nameplate. Overload transformers or circuits may cause low voltage. If the former, check with the power company. Overloaded circuits in the building can be found by comparing the voltage at the meter

Symptom and Possible Cause   *Possible Remedy*

with the voltage at the motor terminals with the switch closed.

(c) Faulty cutout switch operation

(c) Cutout switch operation may be observed by removing the inspection plate in the front end bracket. The mechanism consists of a cutout switch mounted on the front end bracket, and a rotating part called the governor weight assembly which consists of a bakelite disc so supported that it is moved back and forth along the shaft by the operation of the governor weights. At standstill, the disc holds the cutout switch closed. If the disc does not hold the switch closed, the motor cannot start. This may call for adjustment of the end-play washers. Dirty contact points may also keep the motor from starting. See that the contacts are clean. After the motor has accelerated to a predetermined speed, the disc is w i t h d r a w n from the switch, allowing it to open. With the load disconnected from the mo-

69

FRACTIONAL-HORSEPOWER MOTORS

Symptom and Possible Cause   *Possible Remedy*

tor, close the starting switch. If the motor does not start, start it by hand and observe the operation of the governor as the motor speeds up, and also when the switch has been opened and the motor slows down. If the governor fails to operate, the governor weights m a y have become clogged. If it operates too soon or too late, the spring is too weak or too strong. Remove motor to service shop for adjustment. Governor weights are set to operate at about 75% of synchronous speed. Place rotor in balancing machine and, with a tachometer, determine if the governor operates at the correct speed.

(d) Open overload device

(d) If the motor is equipped with a built-in micro switch, or similar overload device, remove the cover plate in the end bracket on which the switch is mounted and see if the switch contacts are closed. Do not attempt to adjust this switch or to test its operation with a

Symptom and Possible Cause   *Possible Remedy*

match. Doing so may destroy it. If the switch is permanently open, remove the motor to the service shop for repairs.

(e) Grounded field

(e) If the motor overheats, produces s h o c k when touched, or if idle watts are excessive, test for a field ground with a test lamp across the field leads and frame. If grounded, remove the motor to the service shop for repairs.

(f) Open-circuited field

(f) These motors have a main and a phase (starting) winding. Apply current to each winding separately with a test lamp. Do not leave the windings connected too long while rotor is stationary. If either winding is open, remove the motor to the service shop for repairs.

(g) Short-circuited field

(g) If the motor draws excessive watts and, at the same time lacks torque, overheats, or hums, a shorted field is indicated. Remove to the service shop for repairs.

(h) Incorrect end play

(h) Certain types of motors have steel-enclosed cork washers at each end to cushion the end thrust.

Symptom and Possible Cause   *Possible Remedy*

Too great an end thrust, hammering on the shaft, or excessive heat may destroy the cork washers and interfere with the operation of the cutout switch mechanism. If necessary, install new end-thrust cushion bumper assemblies. End play should not exceed 0.01 inches; if it does, install additional steel end-play washers. End play should be adjusted so that the cutout switch is closed at standstill and open when the motor is operating.

(i) Excessive load

(i) This may be approximately determined by checking the ampere input with the nameplate marking. Excessive load may prevent the motor from accelerating to the speed at which the governor acts, and cause the phase winding to burn up.

(j) Tight bearings

(j) Test by turning armature by hand. If adding oil does not help, bearings must be replaced.

Motor Overheats

(a) Grounded field

(a) Test for a field ground with a test lamp between

Symptom and Possible Cause	*Possible Remedy*
	the field and motor frame. If grounded, remove the motor to the service shop for repair.
(b) Short-circuited field	(b) Test for excessive current draw, lack of torque, and presence of hum. Any of these symptoms indicates a shorted field. Remove the motor to the service shop for repair.
(c) Tight bearings	(c) Test by turning armature by hand. If oiling does not help, new bearings must be installed.
(d) Low voltage	(d) Measure voltage at motor terminals with s w i t c h closed. Voltage should be within 10% of nameplate voltage. Overloaded transformers or power circuits may cause low voltage. Check with power company. Overloaded building circuits can be found by comparing the voltage at the meter with the voltage at the motor terminals with the switch closed.
(e) Faulty cutout switch	(e) See Paragraph (c) under **Failure to Start.**
(f) Excessive load	(f) See Paragraph (i) under **Failure to Start.**

Motor Does Not Come Up to Speed

Same possible causes and possible remedies as under **Motor Overheats.**

Symptom and Possible Cause   *Possible Remedy*

Excessive Bearing Wear

(a) Belt too tight

(b) Pulleys out of alignment

(c) Dirty, incorrect, or insufficient oil

(d) Dirty bearings

(a) Adjust belt to tension recommended by manufacturer.

(b) Align pulleys correctly.

(c) Use type of oil recommended by manufacturer.

(d) Clean thoroughly. Replace worn bearings.

Excessive Noise

(a) Worn bearings

(b) Excessive end play

(c) Loose parts

(d) Misalignment

(e) Worn belts

(f) Bent shaft

(g) Unbalanced rotor

(h) Burrs on shaft

(a) See Paragraphs (a), (b), (c), and (d) under **Excessive Bearing Wear.**

(b) If necessary, add additonal end-play washers.

(c) Check for loose holddown bolts, loose pulleys, etc.

(d) Align pulleys correctly.

(e) Replace belts.

(f) Straighten shaft, or replace armature or motor.

(g) Balance rotor.

(h) Remove burrs.

Motor Produces Shock

(a) Grounded field

(b) Broken ground strap

(c) Poor ground connection

(a) See Paragraph (e) under **Failure to Start.**

(b) Replace ground strap.

(c) Inspect and repair ground connection.

Rotor Rubs Stator

(a) Dirt in motor

(b) Burrs on rotor or stator

(a) Thoroughly clean motor.

(b) Remove burrs.

Symptom and Possible Cause	*Possible Remedy*
(c) Worn bearings	(c) Replace bearings and inspect shaft for scoring.
(d) Bent shaft	(d) Repair and replace shaft or rotor.

Radio Interference

(a) Poor ground connection	(a) Check and repair any defective grounds.
(b) Loose contacts or connections	(b) Check and repair any loose contacts on switches or fuses, and loose connections on terminals.

FRACTIONAL-HORSEPOWER MOTOR TROUBLE CHART

(Repulsion-Start Induction Brush-Lifting Motors)

Failure to Start

(a) Fuses blown	(a) Check capacity of fuses. They should not be greater in ampere capacity than recommended by the manufacturer, and in no case smaller than the full-load ampere rating of the motor, and with a voltage capacity equal to or greater than the voltage of the supply circuit.
(b) No voltage or low voltage	(b) Measure voltage at motor terminals with switch closed. See that it is within 10% of the voltage stamped on the nameplate of the motor.

Symptom and Possible Cause	*Possible Remedy*
(c) Open-circuited field or armature	(c) Indicated b y excessive sparking in starting, refusal to start at certain positions of the rotor, or by a humming s o u n d when the switch is closed. E x a m i n e for broken wires, loose connections, or burned segments on the commutator at the point of loose or broken connections. Inspect the commutator for a foreign metallic substance which might cause a short between the commutator segments.
(d) Incorrect voltage or frequency	(d) Requires new motor built for operation on local power supply. DC motors will not operate on AC circuit, or vice versa.
(e) Worn or sticking brushes	(e) When brushes are not making p r o p e r contact with the commutator, the motor will have a weak starting torque. This can be c a u s e d by w o r n brushes, brushes sticking in holders, weak brush springs, or a dirty commutator. The commutator should be polished with fine sandpaper (never use emery). The commutator should never be oiled or greased.

Symptom and Possible Cause	*Possible Remedy*
(f) Improper brush setting	(f) Unless a new armature has been installed, the brush holder or rocker arm should be opposite the index and locked in position. If a new armature has been installed, the position may be slightly off the original marking.
(g) Improper line connection	(g) See that the connections are made according to the connection diagram sent with the motor. The motor may, t h r o u g h error, be wired for a higher voltage.
(h) Excessive load	(h) If the motor starts with no load, and if all the foregoing conditions are satisfactory, then failure to start is most likely due to an excessive load.
(i) Shorted field	(i) Take separate current readings on each of the two halves of the stator winding. Unequal readings indicate a short. Shorted coil may also feel much hotter than the normal coil. An increase in hum may also be caused by a shorted winding.
(j) Shorted rotor	(j) Remove the brushes from the commutator and impress full voltage on the stator. If there is one or

77

Symptom and Possible Cause   *Possible Remedy*

more points at which the rotor "hangs" (fails to revolve easily when turned), the rotor is shorted. By forcing the rotor to the position where it is most difficult to hold, the shorted coil will become hot. Do not hold in position too long or the coil will burn out.

Motor Operates Without Lifting Brushes

(a) Dirty commutator

(a) Clean with fine sandpaper. (Do not use emery.)

(b) Governor mechanism or brushes sticking, or brushes worn too short for good contact

(b) See that brushes move freely in slots and that governor mechanism operates freely by hand. Replace worn brushes.

(c) Frequency of supply circuit incorrect

(c) Run motor idle. After brushes throw off, speed should be slightly in excess of full-load speed shown on nameplate. An idle speed varying more than 10% from nameplate speed indicates that motor is being used on an incorrect supply frequency. A different motor will be required.

(d) Low voltage

(d) See that voltage is within 10% of nameplate voltage with the switch closed.

Symptom and Possible Cause	*Possible Remedy*
(e) Line connection improperly or poorly made	(e) See that contacts are good and that connections correspond with diagram sent with motor.
(f) Incorrect brush setting	(f) Check to see that rocker-arm setting corresponds with index mark.
(g) Incorrect adjustment of governor	(g) The governor should operate and lift brushes at approximately 75% of speed stamped on nameplate. Below 65% or over 85% indicates incorrect spring tension.
(h) Excessive load	(h) An excessive load may be started but not carried to and held at full-load speed, which is beyond where the brushes lift. Tight motor bearings may contribute to overload. This is sometimes indicated by brushes lifting and returning to the commutator.
(i) Shorted field	(i) See Paragraph (i) under **Failure to Start.**

Excessive Bearing Wear

(a) Belt too tight, or an unbalanced line coupling	(a) Correct the mechanical condition.
(b) Improper, dirty, or insufficient oil	(b) The lubrication system of most small motors provides for supplying the right amount of filtered oil to the bearings. It is necessary only for the

Symptom and Possible Cause   *Possible Remedy*

user to keep the wool yarn saturated with a good grade of machine oil.

(c) Dirty bearings

(c) When bearings become clogged with dirt, the motor may need protection from excessive dust. The application may be such that a specially constructed motor should be used.

Motor Runs Hot

(a) Bearing trouble

(a) See Parapraghs (a), (b), and (c) under **Excessive Bearing Wear.**

(b) Short-circuited coils in stator

(b) Make separate wattmeter reading on each of the two halves of the stator winding. Sometimes the shorted coil may be located by the fact that one coil feels much hotter than the other. An increase over normal in the magnetic noise (hum) may also indicate a shorted stator.

(c) Rotor rubbing stator

(c) Extraneous matter may be between the rotor and the stator, or the bearings may be badly worn.

(d) Excessive load

(d) Be sure proper pulleys are on the motor and the machine. Driving the load at higher speeds requires

Symptom and Possible Cause   *Possible Remedy*

more horsepower. Take an ammeter reading. If current draw exceeds the nameplate amperes for full load, the answer is evident.

(e) Low voltage

(e) Measure the voltage at the motor terminals with the switch closed. The reading should not vary more than 10% from the value stamped on the nameplate.

(f) High voltage

(f) See (e) above.

(g) Incorrect line connection to the motor leads

(g) Check with the connection diagram sent with the motor.

Motor Burns Out

(a) Frozen bearing

(a) See Paragraphs (a), (b), and (c) under **Excessive Bearing Wear.**

(b) Some condition of prolonged excessive overload

(b) Before replacing the burned-out motor, locate and remove the cause of the overload. Certain jobs which present a heavy load will, under unusual conditions of operation, apply prolonged overloads which may destroy a motor and be difficult to locate unless examined carefully. On jobs where it is assumed somewhat intermittent service will

Symptom and Possible Cause   *Possible Remedy*

normally prevail, a n d which consequently are closely motored, the load cycle should be especially checked as a change in this feature will easily produce excessive overload on the motor.

Motor Is Noisy

(a) Unbalanced rotor

(a) When transportation handling has been so rough as to damage the heavy shipping case, it is well to test the motor for unbalanced conditions at once. I t is e v e n possible (though it rarely happens) that a shaft may be bent. In any event, the rotor should be rebalanced dynamically.

(b) Worn bearings

(b) See Paragraphs (a), (b), and (c) under **Excessive Bearing Wear.**

(c) Rough commutator, or brushes not seating properly

(c) Noise from this cause occurs only during the starting period, but conditions should be corrected to avoid consequent trouble.

(d) Excessive end play

(d) Proper end play is as follows: 1/3 HP and smaller −0.005″ to 0.030″; 1/2 to 1 HP −0.010″ to 0.075″. Washers supplied

Symptom and Possible Cause	*Possible Remedy*
	by the factory should be used. Be sure to tell factory all figures involved. Remember that too little end play is as bad as too much.
(e) Motor not properly aligned with the driven machine	(e) Correct the mechanical condition.
(f) Motor not firmly fastened to mounting base	(f) All small motors have steel bases so they can be firmly bolted to their mounting without fear of breaking. It is, of course, not to be expected that the base should be strained out of shape in order to make up for roughness in the mounting base.
(g) Loose accessories in motor	(g) Such parts as oil covers, guards (if any), end plates, etc., should be checked, especially if they have been removed for inspection of any sort. The conduit box should be tightened when the top is fitted after the connections are made.
(h) Air gap not uniform	(h) This results from a bent shaft or an unbalanced rotor. See Paragraph (a).
(i) Amplified motor noises	(i) When this condition is suspected, set the motor on a firm floor. If the motor is now quiet, then

Symptom and Possible Cause   *Possible Remedy*

the mounting is acting as an amplifier to bring about certain noises in the motor. Frequently the correction of slight details in the mounting will eliminate this, but rubber mounts are the surest cure.

Excessive Brush Wear

(a) Dirty commutator

(a) Clean with fine sandpaper (never use emery).

(b) Poor contact with commutator

(b) See that the brushes are long enough to reach the commutator, that they move freely in the slots, and that the spring tension gives firm but not excessive pressure.

(c) Excessive load

(c) If the brush wear is due to overload, it can usually be checked by noting the time required for the brushes to lift from the commutator. The proper time is less than 10 seconds.

(d) Failure to lift promptly and stay off during the running period

(d) Examine for conditions listed under **Motor Operates Without Lifting Brushes.**

(e) High mica

(e) Examination will show this condition. Take a very light cut off the com-

Symptom and Possible Cause　　　*Possible Remedy*

mutator face and polish with fine sandpaper. Undercut the mica.

(f) Rough commutator　　　(f) True up on lathe.

Brush-Holder or Rocker-Arm Wear

(a) Failure to lift properly and stay off during the running period

(a) No noticeable wear of this part should occur during the life of the motor. Troublesome wear indicates faulty operation. See under **M o t o r Operates Without Lifting Brushes.**

Radio Interference

(a) Faulty ground

(a) Check for poor ground connections, and repair. Static electricity generated by the belts may cause radio noises if the motor frame is not thoroughly grounded. Check f o r loose connections or contacts in the switch, fuses, or starter.

FRACTIONAL-HORSEPOWER MOTOR TROUBLE CHART
(Capacitor-Start Induction Motors)

Failure to Start

(a) Blown fuses or overload device tripped

(a) Examine motor bearings. Be sure that they are in good condition and properly lubricated. Be sure the motor and driven machine both turn freely.

Symptom and Possible Cause	*Possible Remedy*
	Check the circuit voltage at the motor terminals against the voltage stamped on the motor nameplate. Examine the overload protection of the motor. Overload relays operating on either magnetic or thermal principles (or a combination of the two) offer adequate protection to the motor. Ordinary fuses of sufficient size to permit the motor to start do not protect against burnout. A combination fuse and thermal relay, such as *Buss Fusetron,* protects the motor and is inexpensive. If the motor does not have overload protection, the fuses should be replaced with overload relays or *Buss Fusetrons.* After installing suitable fuses and resetting the overload relays, allow the machine to go through its operating cycle. If the protective devices again operate, check the load. If the motor is excessively overloaded, take the matter up with the manufacturer.
(b) No voltage or low voltage	(b) Measure the voltage at the motor terminals with

Symptom and Possible Cause	*Possible Remedy*
	the switch closed. See that it is within 10% of the voltage stamped on the motor nameplate.
(c) Open-circuited field	(c) Indicated by a humming sound when the switch is closed. Examine for broken wires and connections.
(d) Incorrect voltage or frequency	(d) Requires motor built for operation on power supply available. AC motors will not operate on DC circuit, or vice versa.
(e) Cutout switch faulty	(e) The operation of the cutout switch may be observed by removing the inspection plate in the end bracket. If the governor disc does not hold the switch closed, the motor cannot start. This may call for additional end-play washers between the shaft shoulder and the bearing. Dirty or corroded contact points may also keep the motor from starting. See that the contacts are clean. With the load disconnected from the motor, close the starting switch. If the motor does not start, start it by hand and listen for the characteristic click of the governor as the motor

Symptom and Possible Cause   *Possible Remedy*

speeds up and also when the switch has been opened and the motor slows down. Absence of this click may indicate that the governor weights have become clogged, or that the spring is too strong. Continued operation under this condition may cause the phase winding to burn up. Remove the motor to the service shop for adjustment.

(f) Open field

(f) These motors have a main and phase winding in the stator. With the leads disconnected from the capacitor, apply current to the motor. If the main winding is all right, the motor will hum. If the main winding tests satisfactorily, connect a test lamp between the phase lead (the black lead) from the capacitor and the other capacitor lead. Close the starting switch. If the phase winding is all right, the lamp will glow and the motor may attempt to start. If either winding is open, remove the motor to the service shop for repairs.

Symptom and Possible Cause
  (g) Faulty capacitor

Possible Remedy

  (g) If the starting capacitor (electrolytic) is faulty, the motor starting torque will be weak and the motor may not start at all, but may run if started by hand. A capacitor can be tested for open circuit or short circuit as follows: Charge it with DC (if available), preferably through a resistance or test lamp. If no discharge is evident on immediate short circuit, an open or a short is indicated. If no DC is available, charge with AC. Try charging on AC several times to make certain that the capacitor has had a chance to become charged. If the capacitor is open, short-circuited, or weak, replace it. Replacement capacitors should not be of a lower capacity or voltage than the original. In soldering the connections, *do not use acid flux.*

Note 1. — Electrolytic capacitors, if exposed to temperatures of 20° F and lower, may temporarily lose enough capacity so that the motor will not start, and may cause the

89

Symptom and Possible Cause	*Possible Remedy*
	windings to burn up. The temperature of the capacitor should be raised by running the motor idle, or by other means. Capacitors should not be operated in temperatures exceeding 165° F. *Note 2.*—The frequency of operation of electrolytic capacitors should not exceed two starts per minute of three seconds acceleration each, or three to four starts per minute at less than two seconds acceleration, provided the total accelerating time (i.e., the time before the switch opens) does not exceed one to two minutes per hour. This may be approximately determined by checking the ampere input with the nameplate marking. Excessive load may prevent the motor from accelerating to the speed at which the governor acts, and thus cause the phase winding to burn up.

Radio Interference

    (a) Faulty ground   (a) Check for poor ground connections. Static elec-

Symptom and Possible Cause	*Possible Remedy*
	tricity generated by the belts may cause radio noises if the motor frame is not thoroughly grounded.
(b) Loose connections	(b) Check for loose connections or contacts in the switch, fuses, or starter. Capacitor motors ordinarily will not cause radio interference. Sometimes vibration may cause the capacitor to move so that it touches the metal container. This may cause radio interference. Open the container, move the capacitor, and replace the paper packing so that the capacitor cannot shift.

CHAPTER 5

Appliance Testing and Troubleshooting

Every home appliance serviceman is required to test and trouble-shoot electrical circuits by means of simple meters and other test equipment. In order to properly analyze and diagnose the faults that may be found in electrical appliances, it is necessary to possess a fundamental knowledge not only of the meters and instruments themselves and their proper use, but also of the circuit arrangements employed in typical commercial forms of these instruments.

THE MILLIAMMETER

Perhaps the simplest instrument most commonly used in appliance service work is the milliammeter. The "movement" of this instrument, Fig. 1, forms the basic part of all ammeters, voltmeters, ohmmeters, circuit testers, etc., that are used in all types of electrical circuit testing.

This type of meter consists of a horseshoe magnet between the two poles of which is suspended an armature. Attached to the armature is a pointer and spring arrangement, which hold the pointer to its zero position when no current is being passed through the meter coil. When a current is passed through the armature coil, the coil becomes an electromagnet with two poles of opposite polarity. A reaction between the energized coil and the permanent magnet causes the coil to rotate on its axis so as to facilitate the attraction of the unlike poles and the repulsion of the like poles of the two magnets.

The amount of movement is determined by the balance attained between the resiliency of the spring mechanism and the strength

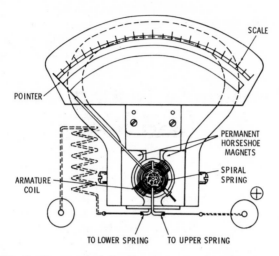

Fig. 1. Construction details of a typical milliammeter.

of the magnetic field set up around the coil. Since the strength of the magnetic field set up around the coil is determined by the amount of current flowing through it, the movement may be calibrated in units of current, or in any other unit, such as volts, ohms, or microfarads, all of which possess a definite relationship to the unit of current.

Shunts and Their Use

All ammeters for use in direct-current measurements may be designed to pass a similar amount of amperes, although the actual amount of current in the circuit may differ greatly. The main difference between the various ammeters is in the types of shunts employed. The function of a shunt, Fig. 2, is to pass only a certain definite amount of the circuit current through the ammeter. If the full amount of current were allowed to pass through the ammeter, its coil would of necessity have to be of heavier wire, thus materially increasing its cost and size, which in addition would decrease the sensitivity of the entire moving element.

A shunt will carry a certain ratio of the total current, depending on the ratio of its resistance to that of the resistance of the ammeter coil. This makes it possible to use the same sensitive ammeter for

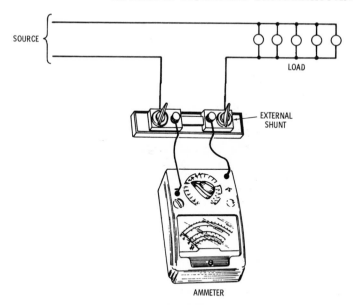

Fig. 2. Shunts allow only a prescribed amount of current to pass through the ammeter, thereby permitting the meter to be used over a wide range of currents.

different current ranges by merely shunting or by-passing a portion of the total current flowing in the circuit. The required shunt size is designed from a knowledge of the current to be measured and of the existing resistance of the ammeter coil. To enable the appliance serviceman to calculate shunt resistances, the following example is given.

Example—If a milliammeter giving full-scale deflection on 500 milliamperes is required to be changed so as to enable the measurement of current up to 5 amperes, what size shunt should be used? Assume the meter coil has a resistance of 0.2 ohm.

Solution—The increase in current for full-scale deflection is 5/0.5 or 10 times; hence, each scale reading would have to be multiplied by 10 for each actual current indication. The resistance of the coil and shunt combined, in order to permit 10 times the current to flow, would have to be designed so that the ammeter coil would carry 0.1 of the current, and the shunt would carry the remaining 0.9 of the total current. This may be written as a formula to express the fact that the shunt resistance is equal to the meter resistance divided by the multiplication factor less one, or,

$$R = \frac{r}{n-1}$$

where,

R = the resistance of the shunt,
r = the internal resistance of the meter coil,
n = the multiplication factor (or the number indicating how many times the meter range is to be extended or multiplied).

If the meter coil has a resistance of 0.2 ohm, the shunt resistance, according to the formula, would have to be,

$$R = \frac{0.2}{10-1} = \frac{2}{90} = 0.022 \text{ ohm}$$

Hence, a shunt having a resistance of 0.022 ohm must be connected across the meter. This resistance should be of a size sufficient to carry the current without overheating.

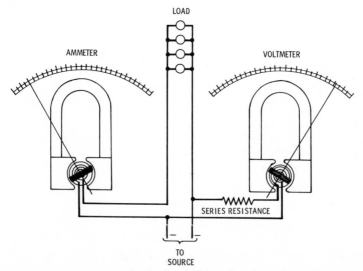

Fig. 3. *The ammeter is connected in series with the source to measure current, whereas the voltmeter is connected in parallel with the source to measure the applied voltage.*

CONNECTION OF METERS

A meter calibrated for current measurement in terms of amperes, or fractions thereof, usually has a comparatively low resistance and is always connected in series with the circuit in which the current is to be measured.

An instrument for the measurement of voltage, or a voltmeter, has a high resistance and is always connected across the circuit or source whose value is to be measured.

Meters for the measurement of power or energy, such as watt-meters and watt-hour meters, usually have two coils; that is, one current coil and one voltage coil. The current coil is connected in series, and the potential coil is connected across the circuit whose measurement is to be made; these connections are illustrated in Fig. 3.

CONVERSION OF METERS

The primary difference between a voltmeter and a milliammeter is that a voltmeter has a high resistance connected in series with the moving coil. By connecting accurate resistances in series with a milliammeter, it is possible to make a voltmeter, which may then be employed when reading voltages or voltage drops across all types of resistance heating appliances, testing of motors, checking source voltages, etc. It is, of course, self-evident that the accuracy of such a converted meter depends on the accuracy of the milliammeter and the fixed resistance used.

Table 1 gives the values of resistance required with different milliammeters to read voltages from one to 1000 volts. The accuracy of Table 1 may conveniently be checked by the following example.

Example—If a five-milliampere meter is to be employed to read voltages up to 50 volts, what resistance should be connected in series with it?

Solution—From Table 1, the resistance required is 10,000 ohms. According to Ohm's law, $E = IR$, or $R = E/I = 50/0.005 = 10,000$ ohms, as already obtained from the table. If the value of resistance required to read the voltage is not found in the table, the resistance may be obtained by calculation in the same manner as that already shown.

Resistors with a wattage rating of one watt will be satisfactory for all those values given in the table. However, it is advisable to use resistors with a rating of approximately two watts, so that there will be little possibility of the resistance value changing due

Table 1. Voltage Multipliers for Milliammeters

Milliamperes	1000 ohms	10,000 ohms	100,000 ohms	1,000,000 ohms
1.0	1.0 volt	10 volts	100 volts	1000 volts
1.5	1.5 volts	15 volts	150 volts	
2.0	2.0 volts	20 volts	200 volts	
3.0	3.0 volts	30 volts	300 volts	
5.0	5.0 volts	50 volts	500 volts	
8.0	8.0 volts			
10.0	10.0 volts			

to the heating effect (I^2R). Also, resistors with a two-watt rating, operating considerably below their rated dissipation, will be likely to hold their calibration longer than resistors of lower wattage.

DIRECT-CURRENT VOLTMETERS

Since the current through a meter is proportional to the voltage applied at its terminals, any ammeter, as previously described, may be used as a voltmeter. In this case, however, a resistor of high value must be connected in series with the movable coil, because if an ammeter were connected directly across the line, it would immediately burn out due to the low resistance of its coil. The high fixed resistance connected in series with the moving coil is considered as part of the meter.

Example—Assume that the moving-coil milliammeter, as used in the previous example, is to be utilized for a voltage measurement of 110 volts at full-scale deflection, with the allowable current drain to be one milliampere. What will be the value of the series resistance?

Solution—The resistance unit must be of a value such that when the voltage across the terminal is 110 volts, exactly one milliampere will flow through the resistance and meter coil at full-scale deflection of the pointer. By Ohm's law,

$$R = \frac{E}{I} = \frac{110}{0.001} = 110,000 \; ohms$$

Since the moving-coil resistance is small compared to the series resistance, it may readily be omitted in most practical problems. The series, or multiplier, resistance may be tapped at various places to obtain more than one voltage range; it is usually placed inside the voltmeter case and connected in series with the coil. A typical connection of a two-range voltmeter is illustrated in Fig. 4. If the 110,000-ohm series resistance is tapped at its center, the voltage range for the same current drain would be $E = 0.001 \times 55,000 = 55$ volts.

In order to obtain proper needle deflection, the binding posts of the meters are marked + (plus) and − (minus). The post marked + should always be connected to the positive side of the line, with either of the other posts connected to the negative side.

RESISTOR ARRANGEMENT IN MULTIRANGE VOLTMETERS

Resistors for multirange voltmeters may be arranged in various ways, as illustrated in Fig. 5. Each resistor will give a certain

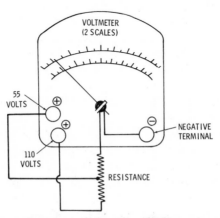

Fig. 4. The schematic representation of multiplier-resistance connections to a two-scale voltmeter.

definite voltage drop, and should be of the percision type, un-affected by normal temperature changes. Voltmeters suitable for electrical appliance work usually have a resistance of 20,000 ohms per volt but can be as low as 1000 ohms per volt. Inspecting the resistance arrangements in Fig. 5, it is found that when using the 0-100 volt scale, the circuit resistance is 100,000 ohms, and when using the 0-250 volt scale, the resistance is 250,000 ohms.

COMBINATION VOLT-AMMETERS

The construction features of voltmeters and ammeters are basi-cally the same; the difference is that in an ammeter, resistors (shunts) are placed *parallel* to the moving coil, while in a voltmeter, resistors are placed in *series* with the moving coil. It is thus possible

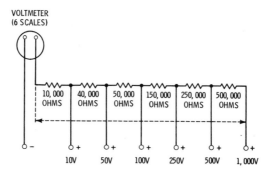

A. One resistor tapped at various points, or separate resistors in series, to obtain the proper multiplier values for each scale.

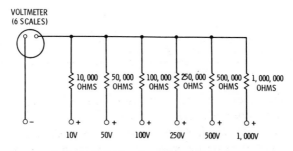

B. Individual resistors for each scale.

Fig. 5. Two methods of connecting multiplier resistances to a voltmeter.

100

to use a single instrument for the measurement of both voltages and currents by employing a proper switching arrangement. Typical circuits of this kind are shown in Figs. 6 and 7.

COMBINATION VOLT-OHM-MILLIAMMETERS

Meters of this type may, in addition to the voltage and current scales, also have a resistance (or ohmmeter) scale, which makes it convenient to check the value of resistances. An ohmmeter is simply a low-current DC voltmeter that is provided with a source of voltage, usually consisting of dry cells, which are connected in series with the unknown resistance. Meters of this type are commonly used in testing and repair work on electrical appliances of

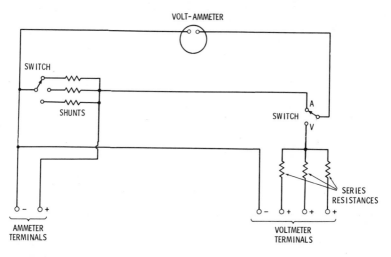

Fig. 6. Principal method of connecting a combination voltmeter-ammeter. If current is to be measured, the selector switch is moved to A, after the proper shunt has been selected. For voltage measurements, the meter is connected across the load, after selection of the proper resistor, and the selector switch is moved to V.

all types, since they combine the measurements of current, voltage, and resistance within a single instrument.

These meters are available in a large variety of shapes and sizes, their cost depending on the size of the indicating meter itself and its accuracy of calibration. A meter of the two- or three-inch face size will be entirely adequate for the practical troubleshooting performed by the home appliance serviceman.

It cannot be too strongly emphasized that a high degree of accuracy is not at all necessary in meters used in this class of work, since relative rather than absolute readings are all that are required for most practical purposes. Most inexpensive meters of this type are much more accurately calibrated than are the circuits or

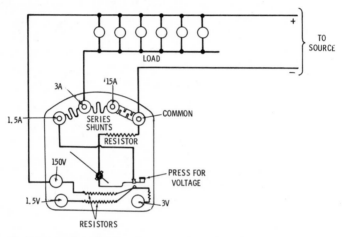

Fig. 7. Construction details of a typical combination voltmeter-ammeter.

devices they test; they are widely used for measurements of voltage and resistance and for checking circuit continuity in electrical appliances. They are not suitable for measurement of the relatively heavy alternating current consumed by the average appliance.

The current-measuring capacity of most inexpensive volt-ohm-milliammeter combinations is limited to about 250 milliamperes, which is only one-fourth of one ampere. Since the average electric toaster or pressing iron takes from five to ten amperes, no attempt at current measurement should be made with such an instrument unless the current-measurement range is increased by a suitable shunt or current transformer, or unless the meter has a high current range.

The voltage ranges usually are applicable to either direct or alternating current, although in some cases the current ranges are limited to direct current. When the instrument is limited to measurement of direct current only, it is useless for checking current on appliances, since alternating current is used almost exclusively in homes throughout the country. Therefore, when current measurements need to be made, the electrical-appliance serviceman should make use of an AC ammeter of sufficient range for the purpose.

A combination instrument of the foregoing type is illustrated in Fig. 8. Before using a multipurpose meter, a precautionary examination should be made to ascertain that the respective controls are properly adjusted to prevent the instrument from being damaged. When measuring unknown values of currents, begin with the highest range, thus identifying the proper range for accurate measurement. When using the instrument as an ohmmeter, the instrument should never be connected across a circuit in which current is flowing. In other words, the appliance power cord should be disconnected when resistance measurements are obtained.

CALCULATION OF CURRENT–RESISTANCE OF APPLIANCES

The amount of current drawn, as well as the internal resistance of an electrical heating appliance, may readily be calculated in most cases by a simple application of Ohm's law.

The "cold" resistance of an appliance will differ considerably from that of the working, or "hot," resistance. In a previous discussion, it was shown that the resistance of all metals increases considerably with an increase in temperature. A few simple examples will suffice to show the commonly employed methods used to obtain the approximate values of resistance and current flow through an electrical heating appliance.

Example—Find the ohmic value of a ribbon-type electric toaster element, which takes 550 watts when connected to a 110-volt outlet. What is the current flow?

Solution—The resistance of the heating element may be calculated from the formula

$$R = E^2/P = 110 \times 110/550 = 22 \ ohms$$

The current flow through the heating element is ,

$$I = E/R = 110/22 = 5 \ amperes$$

Example—A two-heat broiler measures 40 ohms on "low" and 20 ohms on "high" heat. What is the power and current consumption when connected to a 110-volt source?

Solution—The current requirements on "low" and "high" heat are, respectively, 110/40 and 110/20, or 2.75 and 5.5 amperes. The power consumption of the broiler is

$$2.75^2 \times 40 = 302.5 \text{ watts for "low" heat and}$$
$$5.5^2 \times 20 = 605 \text{ watts for "high" heat}$$

Example—The nameplate of an electric kitchen clock is marked "10 watts at 110 volts." What is the approximate resistance of the small synchronous motor connected between its source terminals?

Solution—The ohmic resistance of this circuit is rather high; thus, a typical electric clock resistance may measure in the neighborhood

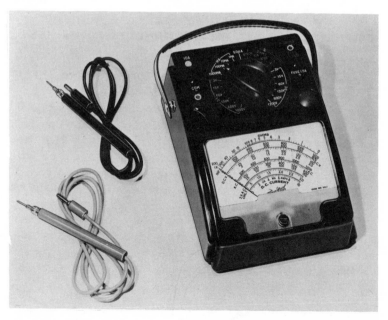

Courtesy The Hickok Electrical Instrument Company

Fig. 8. Exterior view of a volt-ohm-milliammeter with probes. This instrument is equipped with numerous ranges to cover the majority of both AC and DC measurements required in the electric appliance field.

of 1000 ohms. In the example under consideration, the resistance is,

$$110 \times 110/10 = 1210 \ ohms$$

POWER MEASUREMENT

The home appliance serviceman will often find it convenient to ascertain whether the actual *power* being consumed by an appliance is the same as that for which it is rated. The instrument used in such tests is known as a wattmeter and should not be confused with a watt-hour meter, which registers the total amount of *energy* consumed in a circuit.

A single-phase wattmeter contains two coils, one of which must be connected across the line to measure voltage, and one which must be connected in series with the load to measure current. The moving coil is the voltage (or potential) coil, and the fixed coil is the current (or field) coil. The deflection of the wattmeter is approximately proportional to power. Therefore, the scale of the instrument is substantially uniform.

TROUBLESHOOTING

It should be clearly noted that it is not necessary to employ a large number of instruments for successful appliance troubleshooting. An accurate multimeter for testing voltage, current, and resistance, such as shown in Fig. 8, is all that is required in most instances for detection of circuit faults. A multimeter is not only a timesaver but is an invaluable aid in discovering troubles when other methods have failed.

Advantages of Meter Checks

A voltmeter is the only practical device that can be used to test for a variation in voltage, which will impair proper performance of an appliance, but a voltmeter alone has its limitations. Line voltage may be correct, and the appliance may be mechanically perfect, but still it may not function properly. Here, both an ammeter and an ohmmeter may be essential in discovering and locating the electrical difficulty.

Continuity checks for shorts and grounds can be made efficiently and effectively by using an ohmmeter between points in a circuit. High current, which can blow fuses, destroy wiring, and start electrical fires, can be located by using an ammeter at representative sections of a circuit. One of these three meters alone cannot make a complete and correct electrical diagnosis, but a combination of all three can be an invaluable instrument and an immeasurable aid to the appliance serviceman.

Uses of the Ohmmeter

The ohmmeter, Fig. 9, can be used for making continuity checks on any electrical appliance. It is best to have the manufacturer's wiring diagram, which is on the back or bottom of most units. If this is not available, some visual circuit tracing should be done to make certain which part of the circuit is being checked.

If a certain item of electrical equipment (such as a motor, relay, or solenoid) is being checked, first make sure that this item is isolated from other circuits. The connections to this item may have to be removed to isolate it completely. Connect the ohmmeter leads to the connections of the item being checked. These could be two terminal screws or two wires. The reading obtained on the ohmmeter must be interpreted correctly before the condition of the item can be judged. The approximate electrical resistance of the item should be known beforehand, or it can be estimated.

For an item that has only resistance wire, the resistance value can be found by squaring the applied voltage and dividing by the wattage rating. For an item that has wire-wound coils (that is, the item possesses electrical inductance), the resistance cannot be found in this way. Look at the wiring diagram—the resistance values of the different items are often included. As a last resort, remember that the resistance of devices designed to operate on 110 volts AC can range from 0.01 ohm to approximately 100 ohms.

An extremely small reading, such as 0.01 ohm, can be hard to read on most ohmmeters, unless the meter has a low-reading resistance scale. With a reading this low, the item may be in good condition, or it may be faulty. A final check may be made by applying 110 volts AC to the item with an appropriate fuse or

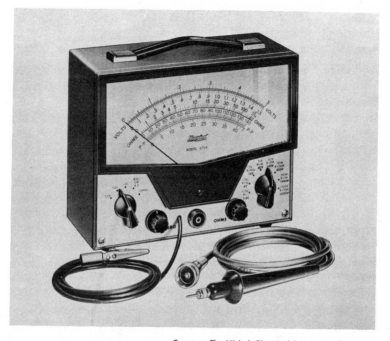

Courtesy The Hickok Electrical Instrument Company

Fig. 9. A typical multimeter suitable for use in testing electrical appliances. This meter can be used to measure resistances, DC voltages, and rms voltages of alternating current.

circuit breaker in series with the applied voltage. If the fuse blows or the circuit breaker trips, the item is faulty. If the item operates properly, it can be assumed to be good.

A resistance reading that is much higher than the actual or calculated resistance value of the device being checked can indicate faulty (corroded) connections within the device. Either disassemble and repair the device, or replace it. In all cases, be absolutely sure that no power is being supplied to the appliance. The plug must be disconnected from the outlet socket, or the circuit must be isolated.

The ohmmeter can also be used on other electrical appliances to check for opens, shorts, and grounds. This instrument is especially effective in making bench checks. In order to test for grounds, touch one test probe to one lead wire and one probe to the appliance unit frame. There should be an infinite-resistance reading on the resistance scale, thus showing that no circuit exists between the wiring system and the unit housing.

Timer Continuity Checks—The ohmmeter can also be used to make continuity checks on automatic washing-machine timers. Faulty timers are a source of constant irritation to the serviceman, and this is one way to check their effectiveness. Place one ohmmeter probe on the common timer terminal. Touch the other probe to each timer terminal in turn, thus checking each individual switch circuit. The timer must be rotated manually to each succeeding part of the cycle in order to test all terminals. It is not necessary to remove the timer from the washing-machine assembly for testing, but it is absolutely essential that all wires are removed from the timer terminals, thereby completely removing the timer from the circuit. Again, it should be remembered that when making continuity checks on any electrical appliance with the ohmmeter, the appliance must be completely disconnected from its power source.

Testing Hermetic Units—Frequently, it is difficult to identify the leads attached to the motor windings on a hermetically sealed refrigeration unit. By using the ohmmeter, this problem can be resolved with a minimum of time and effort. There are three terminals at the side of the sealed unit, as shown in Fig. 10. Attached to these terminals are two motor windings—running and starting. The windings are connected by three leads; the common

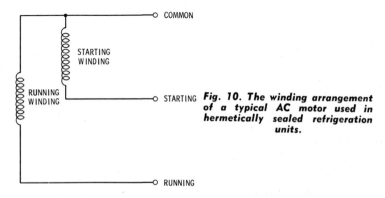

Fig. 10. The winding arrangement of a typical AC motor used in hermetically sealed refrigeration units.

lead is connected to both the running and starting windings, and the running and starting windings are each connected by a separate lead. By using an ohmmeter, the serviceman can determine which is the running winding and which is the starting winding. Attach the two ohmmeter probes to two of the terminals, designating these two terminals as "one" and "two." Note the ohmmeter reading. Then, note the readings with the probe attached to "one" and "three" and "two" and "three."

The two terminals with the greatest amount of resistance between them are connected to the running winding and the starting winding. This is due to the fact that they are in series with each other. If the highest ohmic rating is obtained between "one" and "three," then "two" must be the common terminal or connection. A check from the common terminal to the starting winding will give the second largest resistance reading, so it will be necessary to check from "two" to "three" and from "two" to "one." If the combination of "two" to "three" gives the second highest reading, it is obvious that terminal "three" is connected to the starting winding. By deduction again, the remaining winding is the running winding and is connected to terminal "one." On some units, an overload-protection device with two terminals will be found directly below the motor terminals. To check this device, disconnect it, and apply the test-meter leads across the terminals, thus performing a continuity check. If the device is to be effective, there must be a reading of zero on the ohmmeter scale. By using the multimeter now as an ammeter, the actual current for any appliance can be determined as long as its voltage or current does not exceed that on the meter scales.

Voltage Tests

Frequently, an excessively high or low heat temperature in automatic clothes dryers is found; this condition can be due to an improperly functioning heating element. It may be caused by a current that does not match the element rating. This, in turn, is due to high or low voltages or the wrong resistance.

To determine the power entering or consumed by the heating element, attach the two voltmeter test probes directly across the line to the two terminals on the heating element; then take the reading from the voltmeter scale. By using the correct formula—*power equals voltage multiplied by current*—the actual wattage may be determined and compared to the rated wattage of the element. The comparison should be quite close, depending on how close the applied voltage is to the element rating, and whether the element is heated to the proper temperature.

Both electric range surface units and dryer heating elements are dependent, for proper functioning, on whether or not the wattage supplied by the power source is equivalent to the rated wattage of the element and is sufficient to make them function correctly. Since the components of wattage are voltage and current, it will first be necessary to determine these quantities. In these cases, if an excessively low current reading were obtained, it would account for the unusually slow rate of heating.

In order to make a voltage check, attach the voltmeter test probes to both sides of the 220-volt line connected to the switch or heat control. Read the voltage from the correct scale, and compute the actual wattage as previously described. A comparison can then be made to the rated watts, and any necessary adjustments can then be effected. In cases where the line voltage is at fault, the only recourse is to replace the surface unit with another having a voltage rating that more closely matches the actual voltage. In special cases, it may be possible to have the power company set up a special transformer to correct the voltage.

Use the ohmmeter to make a continuity check with the range disconnected from its power supply. Since this will be a resistance reading, the battery accessory will be attached to the test probes. By measuring across the total element with the test probes, a resistance reading can be taken. If there is a full-scale needle deflection, the element is shorted and should be replaced.

The fill-valve solenoid on automatic dishwashers is another trouble source to appliance servicemen. To run a continuity check on this solenoid, disconnect the machine from its power supply, and remove the wires from the solenoid terminals. Use the battery accessory, because the reading obtained will be in ohms. A voltage check on the fill-valve solenoid may be made by first reconnecting the appliance and then attaching the wires to the solenoid. Then, by connecting the meter probes across the terminals, the load voltage can be read. This procedure also acts as a line voltage check, because the solenoid is in parallel with all the other components in the dishwasher. Therefore, by making this test, a check is also made to see that the timer is functioning on the *fill* cycle.

In some cases, it may be difficult to take a current reading from only one wire of a two-wire power cord; however, this is absolutely necessary. If both wires are used, their magnetic fields would counteract each other, thus producing a cancelling effect on any scale reading. As an example, in using one line of a power cord, the reading may be 6-amperes; if both lines were used, the reading would be zero. This is especially important to note when working with the total heating element on a combination washer-dryer unit. In this case, it is advisable to take the reading from one lead-in wire. Do not use the neutral wire if, for example, tests are to be made on a 220-volt dryer. The voltage can be determined on the combination washer-dryer by using the same procedure as has been described for any 220-volt appliance. Take the test probes from the meter, and attach them from the center post, or *common*, to either of the two outside posts. Connect the power cord from the appliance to the outlet. This procedure should give a 110-volt reading. Attach the probes to the two outside posts; this should give a 220-volt reading.

The multimeter is a versatile piece of test equipment, possibly the most complete test instrument a serviceman can own. The possibilities for testing appliance components are infinite, and the tests themselves are absolutely correct, if the device is used properly.

OTHER APPLIANCE TEST EQUIPMENT

In addition to a set of test instruments, the electrical-appliance serviceman finds numerous uses for such tools as a bell-ringer tester, a series-circuit lamp, and a "hot-circuit" lamp, with accompanying test leads.

The Bell-Ringer Tester

A bell-ringer tester, as shown in Fig. 11, consists of two dry cells and a bell (or buzzer) connected as illustrated. It is used as a continuity tester, and, since the bell will ring when the two test leads are brought together, it is evident that the continuity of a

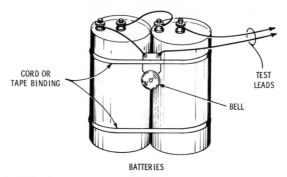

CORD OR
TAPE BINDING

TEST
LEADS

BELL

BATTERIES

Fig. 11. A bell-ringer test set, used as a continuity tester for electrical appliances.

circuit may be tested by simultaneously touching the adjacent heating-coil ends of an appliance. Similarly, a ground test can be made by touching the metal frame of the appliance with one test lead and one of the insulated terminals with the other. If the bell rings, there is a ground, and this condition should be corrected before the appliance is operated.

The Test Lamp

Another useful testing device employed to test for grounds, opens, or short circuits is the series test lamp, consisting of a lamp, a set of test leads, and a wall plug, as shown in Fig. 12A. As in the previously discussed bell-ringer tester, the lamp, when connected to its source, will light when the two test probes are brought together. When the two terminals of an appliance are touched simultaneously with the test probes, the lamp will light if the circuit is complete. If the lamp does not light, the circuit is open. A grounded circuit is indicated if the lamp lights when the metal frame of the appliance and one of its terminals are connected in the circuit.

A hot-circuit test lamp, as shown in Fig. 12B, is employed to test any "hot" or directly connected circuit. With the leads inserted in any "live" wall outlet or appliance-cord terminal, the lamp will light if there is no fuse failure or other breaks in the circuit.

112

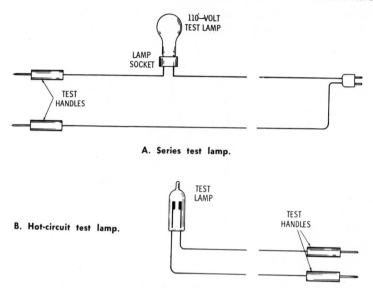

A. Series test lamp.

B. Hot-circuit test lamp.

Fig. 12. Two types of test lamps commonly used to check electrical appliances.

Appliance Test Board

A simple test board, such as that shown in Fig. 13, will prove its value in a short time for circuit testing numerous electrical household appliances. It may also be used as an appliance short-circuit tester by inserting test leads in the plug receptacle and closing the snap switch shown in the "off" position.

To test for an open circuit in the appliance, proceed as follows:

1. Connect wires to a proper source of current by inserting the plug in any 110-volt AC fused outlet.
2. Insert plug of appliance to be tested in the receptacle.
3. Flip the toggle switch to the "off" position.
4. If there is a continuous circuit through the appliance, the 7.5-watt lamp will light.
5. Test leads may be used by inserting the plug of the leads into the receptacle.

The test board may also be used to signal the opening or closing of thermostat contacts in the following manner:

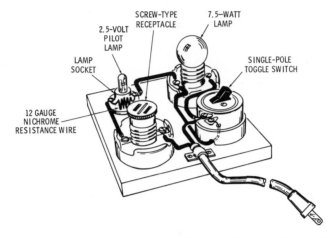

Fig. 13. A typical test board for electrical resistance-heating appliances with a maximum rating of 1500 watts.

1. Turn toggle switch to "on" position.
2. Insert plug of appliance to be tested in the receptacle.
3. When the switch is in the "on" position, there will be no circuit through the 7.5-watt lamp. The pilot lamp will glow but only faintly, since it is connected in series with the appliance being tested.
4. When the pilot lamp goes out, it is a signal that the thermostat contacts have opened, thus breaking the circuit.

Other commonly used testing devices are a growler for testing motor armatures, temperature-indicating devices (such as thermocouples or special thermometers) for accurate determination of oven temperatures in electric ranges, etc.

CHAPTER 6

Shop Technique

When dealing with the repairs of electrical appliances, it is often necessary to replace frayed or worn-out connection cords, attach various types of plugs, and make satisfactory soldered joints.

REMOVING INSULATION

To prepare insulated conductors when making a splice, the insulation must first be removed from the end of each conductor for a suitable distance, depending on the type of splice to be made.

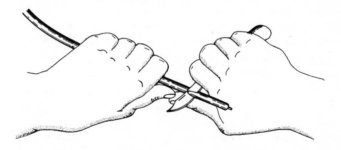

Fig. 1. Using a knife to strip insulation from a wire. Hold the blade so that it lies flat with the wire to avoid damaging the wire.

This process is sometimes called "skinning" or "stripping" and is usually performed by the use of an ordinary knife blade, as shown in Fig. 1. The insulation can also be removed by crushing it with a pair of pliers or by using a wire scraper, such as shown in Fig. 2. When using a knife, Fig. 1, care should be taken so as not to nick the wire, since this may damage it and cause a break under normal service conditions.

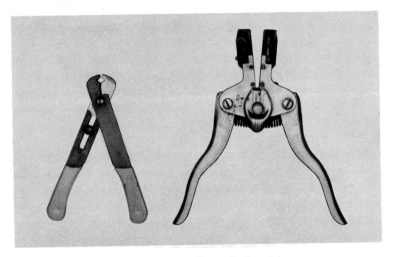

Fig. 2. Two commonly used wire strippers.

CLEANING OF CONDUCTORS

After removing the insulation, the wire must be thoroughly cleaned to insure a good electrical contact between the ends of the wires, so that the solder will adhere properly. The wires may be cleaned by scraping.

SOLDERING

The essential conditions for successful soldering are that the wire surfaces to be joined are clean, that the soldering iron is the correct temperature, and that a suitable flux is used. There are several ways of performing the essential operations of soldering, depending on the type of work to be done. In electrical-appliance work, a small electrical soldering iron of the type shown in Fig. 3 will be entirely satisfactory.

Tinning

Before soldering, the iron must be coated with solder; this operation is known as *tinning*. To tin a soldering iron, heat it until it is

Fig. 3. A commonly used type of electrical soldering iron.

hot enough to melt a piece of solder rapidly when the solder is lightly pressed against the soldering iron. When the iron is at the right temperature, its copper surface tarnishes slowly; if it is too hot, the surface will tarnish immediately.

Put some solder and flux on a piece of tin plate; then rub the soldering iron on the plate. The molten solder should be spread over the entire surface of the soldering-iron tip. The extra solder should then be wiped off the tip with a clean, damp rag. The surface of the soldering iron should now appear bright and silvery; this condition indicates proper tinning.

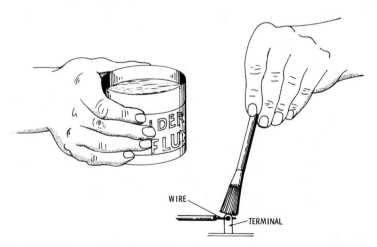

Fig. 4. Apply a coating of rosin flux to the mechanical connection of the wire and the terminal before soldering.

Application of Flux

Prior to applying the solder to the joint, a light coating of flux should be applied to the wire connection. There are several types of flux on the market, although for electrical work, rosin flux is universally used. The function of the flux coating is to clean and free the copper wire from copper oxide, thus promoting a better union between the wire and the terminal when soldered. An approved method of flux application is shown in Fig. 4.

Solder

Solder used in electrical work is a metallic alloy of lead and tin, and is available in bars, in ribbons, or as solid or hollow wire filled with rosin flux. The latter type is quite convenient for soldering light wire to terminal lugs.

Soldering Procedure

Solder alone cannot be relied on for mechanical strength. A good mechanical connection must, therefore, be made before the

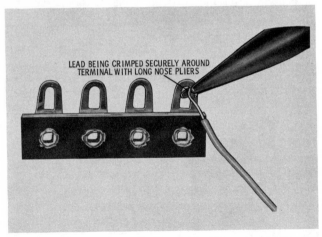

LEAD BEING CRIMPED SECURELY AROUND TERMINAL WITH LONG NOSE PLIERS

Fig. 5. Crimp the wire to the terminal with a pair of plier before soldering the connection; this procedure increases the mechanical strength of the connection.

wire is soldered in place. This is accomplished by securely crimping the wire to the terminal with a pair of pliers, as shown in

Fig. 5. Apply the soldering iron to the joint, and hold the iron there until the joint becomes hot enough to melt the solder. Then, apply solder to both the iron and the joint simultaneously. Allow the solder to flow freely into and around the connections as shown in Fig. 6.

Fig. 6. The right way and the wrong way to make solder connections.

WIRE ATTACHMENTS

The appliance serviceman is often required to connect wire to appliance plugs, switches, and other devices equipped with terminals, as shown in Fig. 7. To attach the wire, the insulation is re-

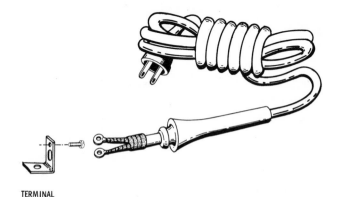

TERMINAL

Fig. 7. The method used to attach an appliance cord to a screw terminal.

moved, and the wire strands are twisted together, leaving no loose strands; a loop is then made in the bare wire. After the terminal screw is opened to admit the wire loop, the binding is tightened to make a good connection between the terminal and the wire.

When attaching the wire to the screw terminals, it is important to insert the loop under the binding-screw head in such a manner that tightening of the screw will close the wire loop. If this precaution is not observed, the loop will be enlarged due to binding-screw pressure, thus resulting in an unsatisfactory connection.

APPLIANCE CORDS

Appliance cords are subject to considerable wear. They consist essentially of two wires embedded in rubber insulation, with a plug connected at one end and a connection to the appliance at the other. A faulty appliance cord is a common cause of trouble. This is mostly due to rough handling or incorrect repairs often made by the novice who has little, if any, concept of electricity.

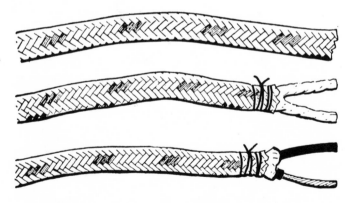

Fig. 8. Step-by-step method of insulation removal from wire.

When an appliance cord shows signs of wear, it should be replaced. To install a new appliance cord, cut the new wire to the correct length, and remove the insulation from the end of the new cord, thereby baring the copper wire. All of the insulation must be carefully removed, as shown in Fig. 8, from both wires before they can be put under the binding screws in the attachment plug cap, or the appliance terminal.

Whenever possible, a knot, such as shown in Fig. 9, should be tied in the cord after it has been passed through the plug, to remove the strain from the binding screws produced by pulling the wire when the plug is removed from its outlet receptacle. Be sure that all small wires in the conductor are properly twisted

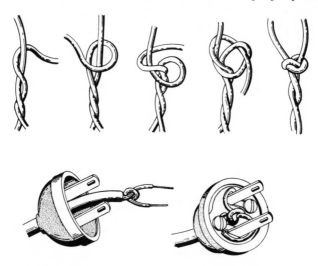

Fig. 9. A knot should be securely tied in the end of the wire before it is attached to the plug binding screws; this procedure will prevent the wire from being pulled off the screws when the plug is withdrawn from the receptacle.

together, and that they are all properly fastened underneath the binding screws. None of the strands from either conductor should be allowed to come in contact with strands of the other conductor: this precaution will prevent a short circuit.

Fig. 10 illustrates how connections are to be made on a standard type of detachable appliance heater plug (used a few years ago), which is usually furnished with electric pressing irons, toasters, waffle irons, etc. When replacing a cord in a plug of this type, a disassembly of the plug will reveal the connection method used. It is an easy matter to make the new connections in the same way; be careful not to damage the plug in the repair process.

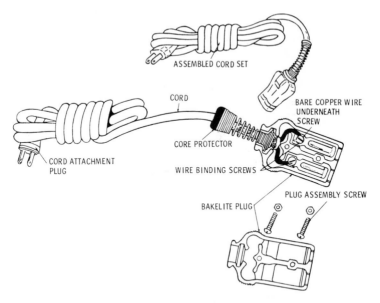

Fig. 10. The method used to connect wire to the binding screws in a detachable heater plug. The assembly screws should be carefully tightened to prevent the plug parts from vibrating.

RESISTANCE WIRES

The function of resistance wire, as used in the heating elements of household appliances, is to reduce the current to a controllable value, in addition to producing the necessary heating effect. It is a well-known fact that when electricity flows through a conductor, some of the energy is used to overcome the electrical resistance of the conductor. This energy appears in the form of heat. Certain alloys having a high resistance transform all this energy into heat, and thus are inefficient conductors. Therefore, the more inefficient a conductor is for transmitting electricity, the better it is for producing heat electrically. It then follows that copper wire would not be suitable for resistance purposes, since its conductivity is too high, and it deteriorates at high temperatures. To overcome this difficulty, considerable research has been made, and several types of resistance wire that are suitable for use in appliance-type heating elements have been developed.

When selecting a resistance wire for a heating element, the first consideration must be the operating temperature of the element. Other factors are the wattage, voltage, and space allowed for mounting the coil. It should also be remembered that a resistance wire has a higher resistance when hot than when cold. The current value may easily be obtained when the wattage and voltage are known by a simple application of Ohm's law.

When designing heating coils, the available space must receive close consideration. In close-wound coils, a space equal to the diameter of the wire is usually allowed between adjacent turns. To determine the length of one turn in a coil, the diameter of the wire selected must be subtracted from the coil diameter; the remainder is multiplied by π (3.14159). The number of turns for each inch of coil is found by dividing 1000 by the diameter of the wire in mils. The length of the wire in one turn multiplied by the number of turns for each inch gives the length of wire in one inch of close-wound coil; this length divided by 12 gives the length of wire in feet in one inch of coil. The number of ohms per inch of close-wound coil is found by multiplying the length of wire per inch of coil of the wire selected by the standard resistance per

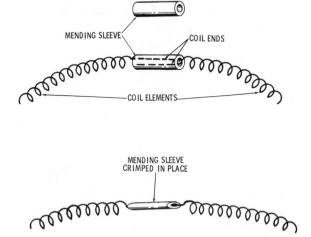

Fig. 11. Broken elements of resistance wire are repaired by means of a mending sleeve. The wires must be thoroughly cleaned before their ends are inserted into the hollow mending sleeve.

foot. The length of the close-wound coil is then determined by dividing the cold resistance by the number of ohms per inch of close-wound coil.

Manufacturers frequently supply resistance wires on spools, which are available in various gauge sizes. Tables furnished with these spools usually supply information about the length of wire necessary for a certain wattage. Thus, when exchanging elements in an appliance, caution should be taken to note the wattage and voltage on its nameplate. In this manner, it is a comparatively simple matter to obtain the correct heating element.

MENDING SLEEVES

Mending sleeves are used in certain instances to connect broken resistance wires in open-coil-type heating elements. A mending sleeve consists of a small hollow metal tube of nichrome or manganese nickel and is fitted on the broken wire ends, as illustrated in Fig. 11.

Connecting resistance wires by means of mending sleeves should only be resorted to as a temporary expedient, since broken wires are a sign of rapid deterioration of the wire coil as a whole. Hence, such coils should be replaced as a unit at the first opportunity.

CHAPTER 7

Electric Irons and Ironers

All electric irons used for hand pressing or ironing of clothing and other materials, such as the two shown in Fig. 1, are equipped with a special resistance-heating grid, or element, whose resistance wire is designed to furnish the necessary heat required for the ironing process when properly connected to an electric circuit. The sole plate and heating-element assembly form the base of the iron and include the ironing surface. Heating elements in electric irons consist of nichrome wire, which may be wound on special mica insulating sheet material, or in a spiral and assembled on a protective metal tubing; this tubing must be properly insulated from the surrounding conducting surface.

Electric irons, depending on their construction, may conveniently be divided into two classes, namely, the automatic iron and the steam iron.

AUTOMATIC IRONS

Automatic irons incorporate a special heat-control device, or thermostat. The function of the thermostat is to regulate the heat to a specific predetermined value, thus disconnecting the heating element when the iron becomes sufficiently hot and reconnecting it to its source when the iron reaches a lower temperature. In this manner, the temperature of the iron is held nearly constant; the actual temperature cycles back and forth between certain set values. Most irons of this type are also equipped with a heat-control knob, which regulates the heat between certain specified limits, depending on the type of fabric being ironed. In addition to these features, some automatic irons are furnished with a special indicat-

Courtesy Dominion Electric Corporation

Fig. 1. Typical automatic electric irons.

ing lamp, which is usually fitted into the front of the handle; this lamp shows whether the iron is at the proper temperature.

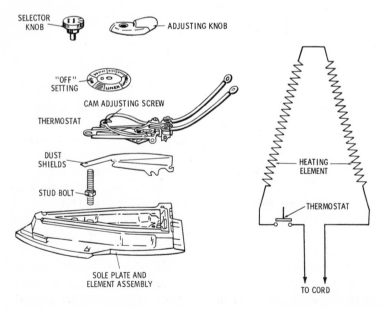

Fig. 2. Exploded view and heating-element diagram of a typical automatic iron.

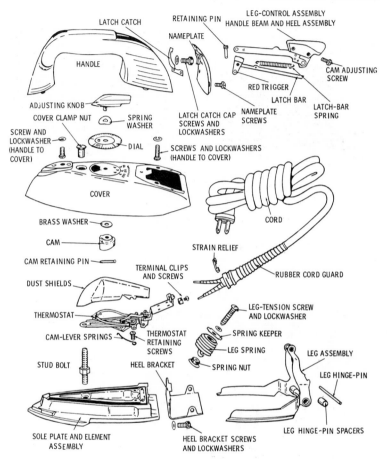

LEG-CONTROL ASSEMBLY
HANDLE BEAM AND HEEL ASSEMBLY
RETAINING PIN
LATCH CATCH
NAMEPLATE
HANDLE
CAM ADJUSTING SCREW
RED TRIGGER
LATCH BAR
LATCH-BAR SPRING
ADJUSTING KNOB
COVER CLAMP NUT
SPRING WASHER
LATCH CATCH CAP SCREWS AND LOCKWASHERS
NAMEPLATE SCREWS
SCREW AND LOCKWASHER (HANDLE TO COVER)
DIAL
SCREWS AND LOCKWASHERS (HANDLE TO COVER)
COVER
CORD
BRASS WASHER
CAM
STRAIN RELIEF
CAM RETAINING PIN
TERMINAL CLIPS AND SCREWS
RUBBER CORD GUARD
DUST SHIELDS
THERMOSTAT
LEG-TENSION SCREW AND LOCKWASHER
CAM-LEVER SPRINGS
THERMOSTAT RETAINING SCREWS
SPRING KEEPER
LEG SPRING
LEG ASSEMBLY
STUD BOLT
HEEL BRACKET
SPRING NUT
LEG HINGE-PIN
SOLE PLATE AND ELEMENT ASSEMBLY
HEEL BRACKET SCREWS AND LOCKWASHERS
LEG HINGE-PIN SPACERS

Courtesy Proctor Electric Company

Fig. 3. Completely disassembled automatic iron.

STEAM IRONS

Steam irons differ from automatic irons in that they contain a steam chamber and water reservoir, as noted in Figs. 4 and 5. Here, the heat from the heating element raises the water temperature to boiling, thereby causing steam to be emitted from a set of slots in the sole plate. In this manner, moisture-laden steam issues from the steam chamber in the iron through the slots in the sole plate, and finally through the material being ironed.

The use of steam irons is preferred in a great many ironing processes. For example, in the ironing of wool and similar heavy material, the use of a steam iron does away with the necessity of using a moistened press cloth. Certain makes of electric irons, generally known as the steam-dry type, have a separate water reservoir and sole-plate attachment, which converts the iron from a dry iron to a steam iron, as shown in Fig. 6.

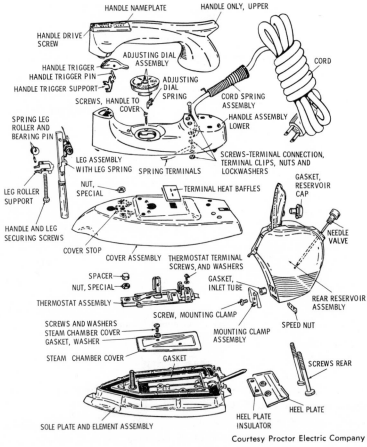

Courtesy Proctor Electric Company

Fig. 4. Completely disassembled steam iron.

Courtesy General Electric Company

Fig. 5. A typical modern steam iron with water-reservoir attachment.

AUTOMATIC IRON OPERATION

Although there is a great variety of automatic electric irons available, the operating principles are quite similar. To understand the principles of operation, refer to Fig. 3, which illustrates a typical thermostat, or thermostatic switch.

The heart of the thermostat assembly consists of two bimetal strips, or blades, equipped with contact points at one end and firmly anchored at the other. With the iron connected to its source, the current consumed by the heating element in the sole plate passes through the series-connected bimetallic strips, which, when heated sufficiently, will bend, thus opening the circuit. After a brief period of time, the iron will again have cooled adequately, and the bimetallic blades will unbend and close the circuit. This action is repeated over and over again, and, in this manner, a nearly constant iron temperature is maintained.

A variation in the iron temperature is obtained by means of a temperature-control switch and dial arrangement, which is located in the iron handle, or hood. By turning this dial, as shown in

129

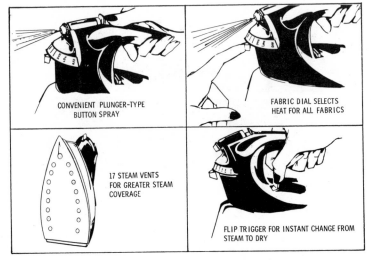

CONVENIENT PLUNGER-TYPE BUTTON SPRAY

FABRIC DIAL SELECTS HEAT FOR ALL FABRICS

17 STEAM VENTS FOR GREATER STEAM COVERAGE

FLIP TRIGGER FOR INSTANT CHANGE FROM STEAM TO DRY

Courtesy Dominion Electric Corporation

Fig. 6. A typical spray-steam and dry iron with conversion arrangements.

Fig. 7, the thermostat can be set to actuate at different temperatures. In this manner, a suitable temperature for the particular fabric to be ironed can be secured. To raise the operating temperature of the iron, the control switch is turned in a clockwise direction; turning it counterclockwise will lower the temperature.

An automatic iron requires an alternating-current supply for its operation. If an automatic iron is used on direct current, the silver thermostat contacts, which open relatively slowly, will cause an electric arc to be formed across the contact gap, and the contacts will weld together. This can completely ruin the iron.

STEAM IRON OPERATION

The steam iron, Fig. 7, is much like the automatic iron; the only noticeable difference is that the steam iron has facilities for the generation of steam. There are two common methods used to generate steam by the majority of iron manufacturers—the boiler-type method and the flash-type method.

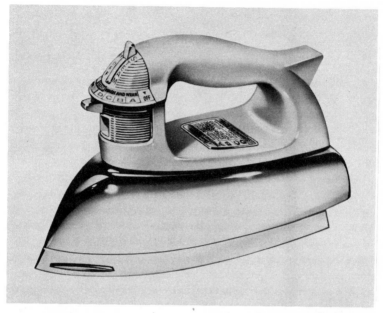

Courtesy Sunbeam Corporation

Fig. 7. A typical steam or dry iron with an easy-to-set heat dial. By turning the thermostat setting, different temperatures may be obtained for the various types of fabrics to be ironed.

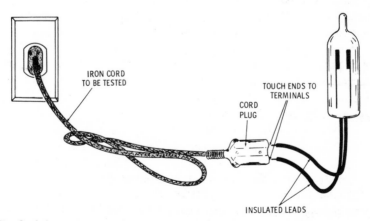

IRON CORD
TO BE TESTED

TOUCH ENDS TO
TERMINALS

CORD
PLUG

INSULATED LEADS

Fig. 8. A hot-circuit test lamp can be used to check the condition of the appliance cord.

131

The boiler-type method consists of the addition of a water tank resting directly on the heating elements. The temperature of the heating elements causes the water to boil, thereby producing steam. This steam then flows through a steam tube to the outlets on the sole plate.

The flash-type steam generator employs a water tank so mounted that the tank does not receive much heat directly from the heating elements but, instead, serves as a reservoir. The water from this supply is metered, by means of an adjustable valve, into a steam chamber, which is part of the sole plate. As the water strikes this plate, it is immediately vaporized and is forced out under pressure through the outlet holes in the sole plate in the form of steam.

SERVICING AND REPAIRS

Although an electric iron, when properly used, will perform satisfactorily over a long period of time without failure, most service calls relate to defective cords, broken handles, or internal circuit openings.

Defective Cords

If an electric iron will not heat, first test the cord. A cord test may easily be accomplished by means of a hot-circuit test lamp, connected as shown in Fig. 8. Insert the skinned end of the test leads into the detachable iron plug, as illustrated. If the lamp lights, the cord is free from defects. If, on the other hand, the lamp does not light, squeeze the cord along its length. This procedure will usually cause the lamp to light temporarily when the broken conductor completes the circuit. In this manner, the fault may easily be located.

If this test does not locate the fault, examine carefully the cord connections at both the wall and heater plug. Disassemble the heater plug by unscrewing the screws that hold the two parts together. Check for broken or burned conductors. See that insulation is perfect up to the terminal binding screws, and bend individual conductors to make sure the wires are not broken.

In many instances where a faulty cord is causing ironing trouble, sparks around the wall plug or heater plug may be observed; in

132

other instances, a short may have formed due to excessive wear or improper wire attachment at the terminal-binding screws, causing fuses to blow. In all such cases, the trouble and methods of repair are self-evident. The failure of the cord conductor indicates a considerable amount of wear, and only in rare instances should repairs be attempted. A new cord will last for several years.

Iron Fails to Heat Up

When an electric iron fails to heat up to its normal operating temperature, the cause may be a faulty wall receptacle or improper contacts between plug and receptacle. Sometimes the posts may be in such a burned and pitted condition that a high-resistance circuit is established. An experienced appliance serviceman can usually locate cord troubles in a short time merely by careful inspection.

Another cause of heat failure or low heat is insufficient voltage. A voltmeter can be used to ascertain whether the voltage is approximately the same as that given on the appliance nameplate. Resistance-heating appliances, such as electric irons, are manufactured to operate properly on a voltage variation of approximately plus or minus 10%. The standard 60-cycle AC voltage in most sections of the country is usually 110 to 120 volts, and most irons are rated at 105-125 volts AC.

Since the heating effect, in Btu, is directly proportional to the consumed power (E^2/R), it will be found that an electric iron designed to function properly on a 115-volt AC source delivers only about two-thirds of its rated heat when connected to a 100-volt circuit. Thus, a difference in voltage of only 15 volts will cause a heating loss of nearly 33%. This is due to the fact that the heating effect varies directly as the *square* of the voltage.

Sometimes the failure of an electric iron to heat up is caused by a worn-out heating element or one or more open internal connections. The element should be tested for an open circuit with a series test lamp arrangement, as shown previously. Touch the test prods together to see if the lamp lights, and then apply them to the terminal posts of the iron. The lamp should light if the circuit is continuous; otherwise there is an open circuit.

A ground test is accomplished in a similar manner, except that in this test, one of the test prods is applied to the side of the iron, while the other is applied to one of the terminal posts. The lamp should *not* light in this test; if it does, the iron is grounded and is unsafe to use.

Repairing Automatic Irons

As previously noted, an automatic-iron assembly includes a thermostat and certain other heat-regulating features. To test the heating element and thermostat circuit, first place the heat-regulating lever on any point except "off," and apply the test prods of a series test lamp to the terminals. Be sure to remove the wall plug from the power source. If the lamp fails to light, disassemble the iron. Loosen the setscrew in the regulating lever, and remove this piece. Under its bottom flange will usually be found two screws, which must be removed. Note the exact position of the heat-regulating lever, so that it will have the same registration for different fabrics on reassembly. The top cover can now be removed, thereby exposing the thermostat and inner parts. The thermostat with its connections, including the bimetal blades connected in series with the heating element, is located in the center of the assembly.

In order to determine whether the heating element or the thermostat is at fault, place a jumper across the thermostat to shunt out the thermostat circuit. Touch the element terminals with the test

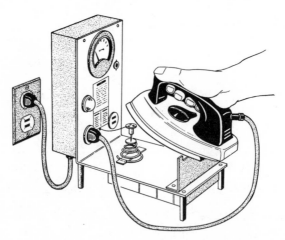

Fig. 9. An instrument-test-stand combination can be used to measure the sole-plate temperature of an automatic iron.

prods. If the lamp now lights with the thermostat shunted out, the trouble is not in the heating element but in the thermostat. If the test lamp fails to light with the shunt in place, the element is probably open, and a new heating element must be installed.

While it may be possible to make some adjustments on the thermostat, such as cleaning and realigning the contacts, it will often be necessary to replace it. Thermostat units are readily available and should cause no difficulty in replacement. It is never advisable to disturb adjustments on thermostats that function properly. As a general rule, thermostat-protected irons are long lived, with respect to the element, because the thermostat cuts off the current whenever the iron has reached its preset temperature, thus preventing the iron from overheating. Thermostats in modern irons are built quite ruggedly and should give little, if any, operational trouble.

Repairing Steam Irons

Steam irons differ from automatic irons mainly in that they require special care and maintenance because of their water and steam reservoirs. In irons of this type, it is of the utmost importance for the iron to reach its operating temperature before use. For example, if the temperature were too low, the water from the reservoir would not turn into steam but would drip out through the holes in the bottom of the sole plate. If, on the other hand, the temperature were too high, the steam would emerge dry and would lose its required wetness for proper ironing.

For proper functioning, it is desirable not to use ordinary tap water containing lime and other mineral deposits, which, when boiled, leaves a deposit inside the iron and eventually plugs up the steam passage. Chlorine, used in many locations to make water safe to drink, reacts with aluminum to form aluminum chloride. This is a flaky gray substance that is often mistaken for mineral or other deposits. It does not stick to the steam passages and is often carried along by the steam into a garment. When ironed into the garment, it leaves a black smear that is difficult to remove.

Distilled or demineralized water should be used whenever possible. If this is not available, rain water, filtered through several layers of cloth to remove all foreign matter, may be used. After each use, the iron reservoir should be carefully cleaned and dried by plugging it into a wall receptacle for a brief time.

Iron Temperature Test

Manufacturers of automatic electric irons usually specify a certain maximum and minimum sole-plate temperature, with the iron operating at a standard voltage on 60-cycle alternating current. For example, a typical universal automatic flatiron has its sole-plate temperature adjusted to operate at 475° to 500°F., with the temperature control set at maximum "high" heat. In other words, the 475°-500°F. setting is the cut-in point of the thermostat control at "high." After operating for a few minutes, the minimum value of the temperature cycle should fall between 475° and 500°F. This temperature may be checked with any suitable temperature-recording instrument, with the thermocouple placed under the center of the sole plate. A typical instrument-test-stand combination for the measurement of sole-plate temperatures is illustrated in Fig. 9. To make a temperature test, proceed as follows:

1. Plug the iron cord into the proper socket of the tester, and adjust the heat-regulator knob for the highest setting.
2. Place the iron on the stand so that its toe rests flat on the button, otherwise a true temperature reading cannot be obtained.
3. With the iron properly connected, the small white bulb on the tester will light each time the thermostat cuts in and will remain lighted until it cuts out.
4. After the light goes out, wait until the temperature-recording needle stops climbing before taking a reading. Always disregard the first reading because of the tendency of the thermostat to overshoot. The second and following readings will be correct.

If the iron under test operates correctly, the readings obtained should compare with the manufacturer's sole-plate temperature specifications. Failure to perform within the range specified will require thermostat adjustment. Thermostat adjustments should not be attempted prior to a careful reading of the manufacturer's service instructions, since they determine the exact adjustment procedure.

IRONER

Although ironers have lost their popularity in the last few years, and it is even doubtful if you could purchase one in your neighborhood appliance store in most areas, they usually (at one time) served as a companion piece to the electric washer and automatic dryer, thus completing the home laundry. They are furnished in two general types, the rotary type and the flatplate type. In the rotary type, the ironing process is performed by placing the material to be ironed on a rotating roll that passes the clothes under an electrically heated shoe to which considerable pressure is applied. In the flatplate type of ironer, the electrically heated shoe is brought to bear on the material under pressure, thus accomplishing the ironing, or pressing, operation.

ROTARY IRONER

The rotary-type ironer, shown in Figs. 10 and 11, consists essentially of an electrically heated concave ironing shoe, as shown in Fig. 12, against which is pressed a rotating drum that holds the material to be ironed on its surface. The curved ironing shoe, which contains the heating elements, has the same curvature shape as that of the rotating roll, shown in Fig. 13, thus providing a close fit between the two surfaces.

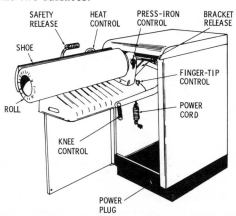

Fig. 10. A typical rotary-type electric ironer. The ironer mechanism is mounted on counterbalanced brackets, which permit it to swing into the operating position when the hinged door is opened.

137

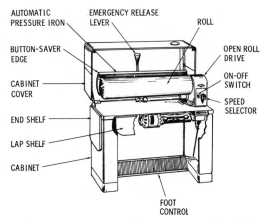

AUTOMATIC
PRESSURE IRON
EMERGENCY RELEASE
LEVER
ROLL
BUTTON-SAVER
EDGE
OPEN ROLL
DRIVE
CABINET
COVER
ON-OFF
SWITCH
SPEED
SELECTOR
END SHELF
LAP SHELF
CABINET
FOOT
CONTROL

Fig. 11. The component parts of a rotary-type electric ironer.

The operative action is similar to that obtained with a hand iron, except that here the iron is stationary, and the material to be pressed is moved under the heated iron by means of the rotating drum, without any effort on the part of the operator. When pressing large articles of clothing, such as trousers, etc., the drum may be pressed against the shoe without the customary revolving action.

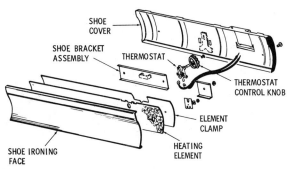

SHOE
COVER
SHOE BRACKET
ASSEMBLY
THERMOSTAT
THERMOSTAT
CONTROL KNOB
ELEMENT
CLAMP
SHOE IRONING
FACE
HEATING
ELEMENT

Fig. 12. The component parts of an ironing shoe, as used on the rotary ironer.

The electrical system is comparatively simple and consists of an electric motor, which provides the motivating power for the roll, and a thermostatically controlled heating element mounted in the shoe, together with a speed selector and a group of disconnecting switches. Fig. 14 shows a typical wiring diagram of an electric ironer.

The iron and the ironing roll are usually controlled by a foot or knee-action switch; the movement of the switch causes the roll to rotate and the iron to press against the roll. A speed-selector switch equipped with a pointer, which indicates the position at which the switch is set, is commonly located on the front control

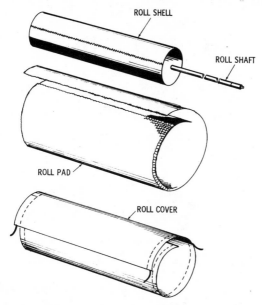

Fig. 13. The disassembled view of a rotary-ironer roll.

panel of the machine. This switch enables the operator to select any convenient roll speed by a movement of the switch to any position, while the ironer is in operation. The roll is usually powered by a ¼- or a ½-horsepower, split-phase, 115-volt, 60-cycle AC motor. The motor is usually mounted under the ironer table and is connected to the drive mechanism by a flexible coupling.

The heating element, Fig.14, is usually divided into two sections, right and left, with a separate thermostat to provide heat control for each element. Since the thermostats are adjustable to provide different ironing temperatures, it is therefore possible to select the desired heat intensity for either end of the shoe. For example, if the operator wishes to iron small pieces of clothing, one element may be disconnected while the other element is set at the desired temperature by means of the thermostat. For large articles, however, both thermostats are set at the same temperature, thereby providing an even heat distribution throughout the entire length of the shoe.

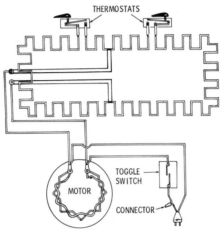

Fig. 14. The diagrammatic representation of the electric circuit in a rotary-type ironer. Two thermostatically controlled switches permit the user to select a different heat intensity for each end of the shoe.

The thermostats normally employed function in the same manner as those used on electric hand irons. When the thermostat is set at a certain preselected temperature, the contact points remain closed until the selected temperature is reached, at which point the thermostat contacts open, thus disconnecting the heating element. After the temperature falls below the selected value, the thermostatic contact points again close the circuit to the heating element. This cycling process prevents the ironer shoe from becoming overheated and provides a desirable safety factor on all types of thermostatically controlled heating elements.

THE FLATPLATE IRONER

The flatplate-type electric ironer differs from the rotary type mainly in the action employed to obtain the ironing results. In the flatplate ironer, the heated shoe is brought down against the ironer board, and the material to be pressed is placed in between. The shoe is mounted on swinging brackets, thereby permitting it to be moved up or down as required for the proper placing or removal of material. The shoe containing the electric heating element is covered by layers of heavy insulation, which are held in place by an outer shell. The proper ironing temperature is obtained by thermostatic control in the same manner as that employed for the rotary-type electric ironers. The heating element consists of two separately controlled units; each section is equipped with a thermostat to provide heat control for each unit.

Modern ironers of this type employ a special motor-oil pump or thrustor to provide the necessary ironing pressure. The pressure obtained by the oil-pump motor arrangement may reach approximately 400 pounds. Because of its simplicity and efficiency in operation, many commercial dry-cleaning establishments employ heavy-duty ironers of this type, since they are particularly well adapted for the pressing of outer garments, such as men's coats, trousers, etc.

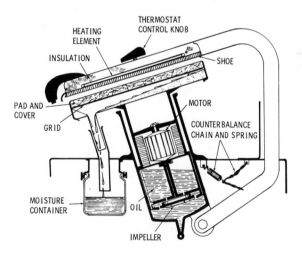

Fig. 15. The cross-sectional view of a typical flatplate-type electric ironer. The thruster motor is the principal part of this mechanism; when energized, it rotates the impeller and, through a spinning action, lifts the motor, thruster assembly, and ironer board against the ironer shoe, thus providing the necessary ironing pressure.

A typical automatic electric ironer of flatplate construction is shown in Figs. 15 and 16. The ironing mechanism, Fig. 15, consists essentially of an electrically driven oil pump, or thrustor; an ironing surface, or buck; and a moveable iron shoe whose position is guided by means of a hinged bracket, which is fastened to the lower part of the thrustor assembly. The electrical circuit in Fig. 16 contains an electric motor, two individually controlled heating elements, and contact and handle switches for opening or closing the motor circuit.

In operation, the ironer mechanism comes into motion when the shoe is pulled forward. This action closes the handle and contact switches, thus energizing the oil-pump, or thrustor, motor. Rotation of the motor shaft and impeller causes a pumping action in the oil chamber, which, in turn, forces the pump assembly upward and raises the buck against the shoe, thereby providing the ironing pressure. The shoe is released by merely removing the pressure from the handle switch. This action opens the motor circuit. The motor then stops, and the thrustor reverses its motion, thus completing the pressing cycle.

During the ironing process, moisture from the material being pressed turns into steam. The condensate is drained into a special

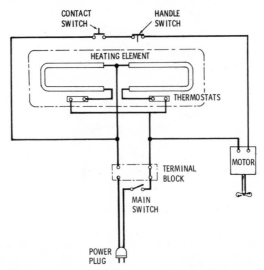

Fig. 16. The internal wiring circuit of a typical flatplate-type ironer.

moisture container or glass water trap, thus preventing a water leak from the ironer.

Figs. 17 and 18 illustrate a flatplate ironer of slightly different construction. As in the previously described type of ironer, the ironer pressure is provided by the thrustor assembly, in which the electric motor supplies the motivating power. With the motor operating, the rotating impeller in the oil-pump cylinder provides an upward motion, thereby raising the mechanism and buck against the shoe. The entire heating-element assembly is mounted on sturdily hinged U-arms, which facilitate the protection of the heating element and thermostat wiring.

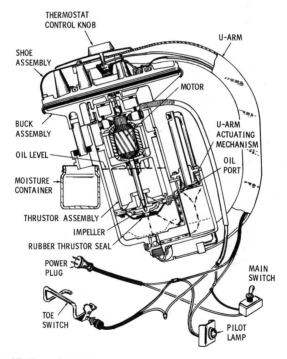

Fig. 17. The principal component parts of a typical flatplate ironer.

The electric circuit, shown schematically in Fig. 18, is controlled by a main switch, a temperature-control switch (the thermostat), and a toe switch. When the ironer is plugged into the wall outlet, and the main switch is closed, current is available at both the temperature-control switch and the toe switch. The temperature-control switch can be set to whatever position is desired, and this temperature will be maintained automatically. When the toe switch is depressed, the circuit is closed, and the shoe moves forward in the ironing position. Then the thrustor and buck assembly rises to contact the shoe, thereby developing an ironing pressure of approximately 400 pounds. The dotted lines shown at the bottom

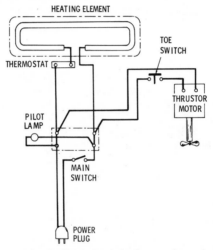

Fig. 18. The circuit diagram of the ironer shown in Fig. 17.

of the thrustor cylinder and of the U-arm actuating mechanism in Fig. 17 indicate the thrustor and pinion positions when the shoe is in the "off" pressure position.

The sequence of ironing operations in the automatic flatplate-type ironer, shown in Fig. 15, is fundamentally as follows:

With the thrustor in the down position and the toe control depressed, the thrustor motor begins to pump, thus causing oil pressure to be developed. The oil is forced down through the hole in the bottom of the thrustor cylinder, then through a small discharge tube into the U-arm actuating mechanism, thereby forcing the piston up. When the piston in the U-arm actuating mechanism rises to its uppermost position, the oil runs through a by-pass port in the side of the thrustor cylinder and flows directly under the outer portion of the thrustor. The oil then causes the thrustor to rise and develop the ironing pressure. The thrustor does not rise when the motor begins to run, because a rubber seal on the bottom of the thrustor keeps the fluid pressure from expanding and getting under the thrustor itself. Therefore, the thrustor cannot rise until after the U-arm actuating mechanism has reached its uppermost position; the oil is then allowed to flow to the by-pass port and create the pressure necessary to raise the thrustor assembly.

A moisture jar is provided that collects the water forced out of the damp clothes during the pressing operation. There is also a red pilot lamp on most models that glows as long as the main switch is in the "on" position.

ROTARY IRONER WITH AUTOMATICALLY OPERATED SHOE

An ironer in which the shoe automatically rises to the roll when it begins to rotate is illustrated in Figs. 19 and 20. Convenient control provides either hand or knee operation to start and stop the roll and to bring the shoe into position for ironing. There is also an emergency-release bar that, when pressed, opens a solenoid circuit, which in turn disconnects the motor and heating element and also moves the shoe away from the roll. Pressing controls are usually designed to be operated by the left knee.

Controls

In order to understand the operation of the type of ironer under consideration, a knowledge of the electrical controls is quite helpful. The following controls are usually found on an ironer of this type:

1. Left knee control,
2. Right knee control,
3. Emergency-release control,
4. Thermostat control,

145

5. Finger-tip control,
6. Motor-control switch,
7. Heat-control switch,
8. Speed-control shifter.

In addition to these controls, the ironer may have various pilot lamps, which indicate the positions of the various switches, such as a motor pilot lamp, a heat pilot lamp, and two thermostat dial lamps.

As shown in the ironer circuit diagram of Fig. 19, the various controls perform the following functions:

Left Knee Control—The left knee lever is applied for the purpose of stopping the roll while the shoe remains pressed against it. To operate this control, apply a slight knee pressure to the left; when held in this position, the control causes the roll to remain stationary with the shoe in the "press" position. Releasing the left knee control causes the roll to resume its rotation.

Right Knee Control—The right knee lever performs the same function as the finger-tip control. When moved toward the right and immediately released, it causes the mechanism to move the

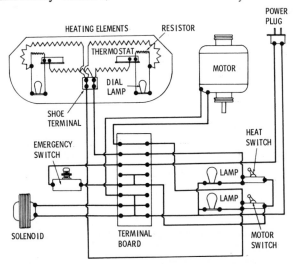

Fig. 19. The circuit arrangement for a rotary ironer with an automatically operated shoe.

146

shoe against or away from the roll. If the right knee control lever is held in its operating position for any length of time, it will cause the shoe to oscillate back and forth toward or away from the ironer roll.

Emergency-Release Control—This control, also called the "safety release," is provided as a means of quickly removing the shoe from the roll in the event of power failure or other circumstances that may make the shoe-operating mechanism inoperative. By operating this switch, all electrical circuits are immediately broken, and the ironer shoe is released from the roll. Before operation of the ironer can be resumed, it will be necessary to actuate the finger-tip and emergency-release controls, which will again restore the circuit to its normal operation condition.

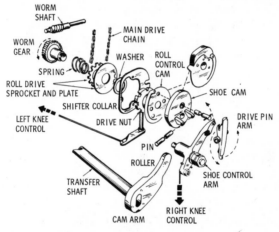

Fig. 20. The disassembled view of the transmission assembly for a rotary ironer with an automatically operated shoe.

Thermostat Control—The heat of the ironer shoe is controlled by two thermostats that are equipped with graduated temperature-control dials, one at each end of the shoe. Their function is to turn the thermostats to predetermined settings, thereby providing heat selectivity in the shoe. Each thermostat controls one half of the shoe-heating element, thus making it possible to independently control the heat on each end of the shoe.

Finger-tip Control—The finger-tip control duplicates the function of the right knee control, thus enabling the operator to control the ironing action with the fingers. By pressing the control lever down firmly and immediately releasing it, the shoe is moved against or away from the roll. If the finger-tip control is held down for any length of time, the shoe will move back and forth.

Motor Control—The motor-control switch opens or closes the motor circuit. On some ironers, this control is illuminated by means of a lamp, which is connected in parallel with the switch circuit. This lamp lights when the motor is energized.

Heat Control—The heat-control switch opens or closes the circuit to the shoe heating element. Certain ironers are furnished with an indicating pilot lamp that lights when the heat switch is in the "on" position.

Speed Control—A two-position motor-control switch provides roll-speed selectivity and permits the use of either of two speed ranges. Although the speed ranges differ in the various ironer types, the average roll speed is about 7 revolutions per minute in "fast" speed and 3½ revolutions per minute in "slow" speed. Speed ratios of 3 to 2 are common in several ironers having single-speed motors and speed gear shifts.

Transmission

To obtain a working knowledge of the ironer transmission assembly, reference is made to Fig. 20, which shows an exploded view of typical ironer transmission parts. The ironer roll is supported by two bronze bearings, which are incorporated in the transmission housing. Power transfer from the motor to the worm shaft is made through a V-belt and two pulleys, one on the motor shaft and one on the worm shaft.

The worm shaft drives the worm gear, and the worm gear, through a coupling arrangement, actuates the roll in addition to supplying the motivating power for the movement of the shoe toward or away from the roll. The function of the worm gear can best be explained by a study of its construction. This worm gear bears a distinct resemblance to a compound gear; that is, it has two gears of different diameters mounted on a common hub. In this case, the smaller gear has been omitted and a steel disc with numerous slots cut in its circumference, called the drive nut, has been used. Instead of being a permanent part of the worm gear, the drive nut is threaded on the hub and moves up against a shoulder. The main drive sprocket floats between the drive nut and the worm gear and turns on the outside diameter of the same hub; it is welded to its component part, the main drive plate. Two lugs on the drive plate are spaced so as to engage opposite slots in the drive nut. A light coil spring exerts pressure on the sprocket and drive-plate assembly to maintain engagement. The main drive chain transfers the power from the main drive sprocket to the idler sprockets, which in turn drive the roll-drive gears through chains. When the roll-drive gears revolve, the roll rotates.

A flat ring, called the shifter collar, pivots on a vertical shaft through the transmission housing. Its design and position are such that the drive nut can rotate within it. Two raised "dimples" bear on the main drive plate constantly. The lower end of the shifter-collar shaft, which extends below the transmission housing, is linked to the left knee control by a lever and rod assembly. Pressure exerted on the left knee control is transferred through the shifter collar to the main drive plate, and the resultant motion disengages the lugs from the drive nut, thereby stopping the rotation of the roll. The "press" position is attained in this manner.

Shoe Operation—Mounted on the same shaft that supports the worm gear and drive nut is an eccentric cam. A drive pin is mounted through the cam at the proper distance from the shaft center to align the cam with the slots in the drive nut. When the drive pin is extended, one end enters one of the drive-nut slots, and the cam is revolved with the worm gear. The other end of the drive pin has a pin through it, forming a cross arm, which rests in the forked end of the drive arm. The drive arm is actually a lever with its fulcrum crossing the center line of the cam pivot point.

Tongues on each end of the drive arm contact a double-faced chisel on the end of the drive-arm stop lever. The tongue on the pin end of the drive arm contacts the outer side of the chisel and pulls the lever away from the cam, thereby disengaging the drive pin from the drive nut and stopping the cam. When the drive-arm stop lever is lifted (by the right knee or finger-tip control) and released, spring tension forces the drive pin to engage the drive nut and turn the cam 180 degrees, or until the tongue on the opposite end of the drive arm meets the inside face of the drive-arm stop lever. The lateral motion imparted to the drive arm forces that end of the lever down and lifts the pin end, once more stopping the rotation of the cam by pulling out the drive pin, which is mounted through the cam.

An arm rides on the circumference of the eccentric cam and is fastened to the transfer shaft; this is called the cam arm and follower. As the cam assumes either of its two stationary positions, the cam arm and follower cause the transfer shaft to rotate and raise or lower the shoe through the release mechanism and the rear shoe lever.

The necessary coordination between the movement of the roll and the extension and retraction of the shoe is obtained by a side-faced cam that is mounted on the inside of the shoe cam. This cam is so constructed as to bear on a third "dimple" on the opposite side of the shifter collar. Its riser is positioned so that, as the shoe cam revolves and raises the shoe to the desired ironing position, the shifter collar is released to permit the drive-plate lugs to engage the drive nut and start the rotation of the roll. Conversely, when the shoe is retracted by further rotation of the shoe cam, the riser on the roll cam forces the shifter collar to disengage the drive-plate lugs, and the roll becomes motionless.

The same drive nut is the driving factor in each section of the transmission. Neither the drive-plate lugs nor the drive pin will occupy more than one half the depth of the drive-nut slot when fully engaged, so it is entirely possible for both members to occupy the same slot.

Speed Shifting—Roll-speed variation is accomplished by the use of three concentric idler sprockets that are welded together and driven by the main drive chain. Each of the outer sprockets drives a chain, which, in turn, drives a floating sprocket on the roll-drive shaft. On the inside face of each of the floating sprockets are lugs that engage similar lugs on either end of the speed-shift collar. The roll-drive shaft is milled flat on opposite sides, and it drives the speed-shifter collar, which has matching flats on its inside diameter. Since the roll-drive sprockets are not fastened to the shaft, motion can be induced only when the speed-shifter collar lugs are engaged with one of the sprockets. By varying the ratio between the idler sprockets and their corresponding roll-drive sprocket, roll-speed change can be obtained.

PORTABLE IRONER WITH MANUALLY OPERATED SHOE

Rotary ironers of this type are particularly adaptable for use in small or cramped quarters. They can be used on any convenient table or stand and require only small storage space in a cupboard or closet. A typical portable ironer is shown in Fig. 21.

In operation, the heated shoe is brought down against the rotating roll by the forward movement of a centrally located control lever, which is also called the shoe-operating lever or control pedal. This action automatically starts the roll, and the material to be processed may now be fed in between the two surfaces for ironing while the roll rotates. To release the shoe, the operating lever is pushed back. This action moves the shoe to an "off-roll" position and simultaneously stops the rotation of the roll. In most ironers of this type, the roll can be stopped without moving the shoe away from the roll by a movement of the operating lever in a horizontal direction.

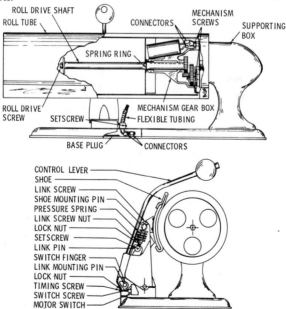

Fig. 21. The operating mechanism of a portable ironer with a manually operated shoe.

151

The construction of the ironer shoe, roll, and heating element is similar to that employed in other types of rotary ironers. The roll is driven by a fractional-horsepower electric motor through a series of gears, which constitute the drive mechanism; the roll speed is approximately 6 revolutions per minute. Heat control is provided by a centrally located knob-and-dial assembly by means of which the thermostat may be set for any desired ironing temperature.

ROTARY IRONER WITH AUTOMATICALLY LIFTING SHOE

A rotary ironer with a somewhat different transmission and shoe control is illustrated in Figs. 22 to 24. The operating mechan-

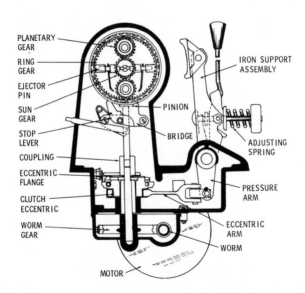

Fig. 22. The operating mechanism of a rotary ironer with an automatically operating shoe.

ism is shown in Fig. 22. As noted, the roll-driving motor is located underneath the ironer base and is connected to the operating mechanism by means of a worm and worm gear assembly. The gear assembly is capable of two speeds; these are usually a high speed at 6.5 revolutions per minute and a low speed at 1.7 revolutions per minute.

When the ejectors of the roll driver mesh with the ring gear, the entire roll assembly is caused to turn as a complete unit. When the ejectors are not meshed, the stop lever positions itself and causes the ring gear to become stationary. The planetary gears, which are driven by the sun gear, roll on the teeth of the ring gear, thereby acting to reduce the roll speed. (The driving mechanism is

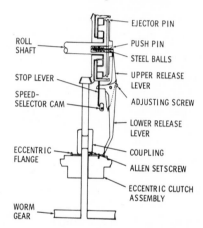

Fig. 23. The driving mechanism of a rotary ironer with an automatically operating shoe.

Labels: EJECTOR PIN, ROLL SHAFT, PUSH PIN, STEEL BALLS, STOP LEVER, UPPER RELEASE LEVER, SPEED-SELECTOR CAM, ADJUSTING SCREW, LOWER RELEASE LEVER, ECCENTRIC FLANGE, COUPLING, ALLEN SETSCREW, ECCENTRIC CLUTCH ASSEMBLY, WORM GEAR

shown in Fig. 23.) When the ejectors are not meshed, and the stop lever is not engaged, the planetary gears are freed from the roll, and the roll stops.

The shoe heating element is divided into two sections—right and left—with a separate thermostatic control for each half of the element. This makes possible a selection of heat intensity for either end. An emergency-release lever is usually provided behind the ironer assembly to permit manual release of the shoe in case of power failure while the shoe is pressed against the roll. Shoe pressure can be adjusted by the use of an adjusting spring, which is located at the base of the emergency lever.

SERVICING AND REPAIRS

Service operations on electric ironers do not differ appreciably from those of other types of laundry equipment. Any ironer

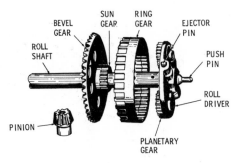

Fig. 24. The planetary-gear assembly of a rotary ironer with an automatically operating shoe.

mechanism, irrespective of model or manufacture, consists of a series of gears and clutches that are driven by an electric motor, which in turn is controlled by a number of hand-, knee-, or foot-operated switches. It then follows that the most common troubles may be either mechanical or electrical or occasionally a combination of both.

Installation

When installing an electric ironer, it is advisable to locate the ironer in dry surroundings. Over a period of time, dampness can cause slight shorts, which may result in shocks at contact and are quite annoying to the operator. Rubber mats should be employed where excessive dampness prevails. Excessive dampness may also cause undue rusting of the internal parts. The possibility of connections being jarred loose during shipment should also be considered. Accordingly, each newly delivered ironer should be operated through several complete ironing cycles to make certain it is operating satisfactorily.

Electrical Requirements

The power-consuming components of an electric ironer consist of the electric motor and the heating elements. The electrical rating may vary from approximately 1200 to 2000 watts, depending on the particular type used. Most ironers require a 115-volt, 60-cycle alternating current for their operation. Extension cords should not be used. The ironer should be plugged into a wall outlet of sufficient wiring capacity to supply the required power without over-heating or blowing the fuses.

Troubleshooting Flatplate Ironers

Insufficient Heat or No Heat—If the shoe fails to heat when the heat switch is in the "on" position, and all thermostats are turned to a position other than "off," check first to be certain that current is supplied to the shoe. This can easily be done by applying the prods of a test lamp across the two leads to the shoe. If there is a current flow to the shoe, shunt the thermostats with a suitable piece of jumper wire. If the shoe still fails to heat, the heating element is faulty and should be replaced. In certain ironer models of this type, the heating element consists of two sections. If either side of the shoe fails to heat, apply the previously mentioned checks. If either circuit of the heating element is faulty, the entire element should be replaced. If the shoe heats but fails to get hot enough, adjust the thermostats. If the thermostats cannot be adjusted, they should be replaced. If the ironer does not operate, although the shoe heats, the motor and motor circuit should be inspected for loose connections. The motor switches should be tested for open circuits, and defective switches should be replaced with the substitute specified by the manufacturer.

Lack of Pressure—Lack of sufficient pressure may be due to one of the following:

1. Insufficient thrustor oil,
2. Cracked U-arm,
3. Cracked buck.

Testing for the proper amount of oil can be accomplished by disassembling the thrustor and measuring the oil level. The fluid surface should be approximately 5½ inches below the front edge of the cylinder bottom, depending on the model. If the oil level is less than this, oil should be added, preferably the type recommended by the manufacturer of the ironer. If the oil has been used for a considerable length of time, the old oil should be drained out, and new oil should be supplied. To remove the oil, remove the pipe plug usually found in the lower part of the thrustor. After the oil has been thoroughly drained out, the pipe plug threads should be painted with white lead before replacement to prevent oil leakage. Oil may now be poured into the cylinder to the proper level.

Another cause of insufficient pressure may be due to a cracked U-arm or cracked buck. To replace a cracked U-arm, proceed as follows:

1. Remove thermostat knobs.
2. Remove screw on edge of cover.
3. Remove handle by removing two screws.
4. Remove cover and insulation.
5. Remove nut holding thrustor motor lead spring.
6. Remove nut holding thrustor switch in place.
7. Pull out thrustor switch, and remove wires; cut or unsolder.
8. Remove snap-ring, cotter-pin, or hair-pin clip to remove rod or pivot pins holding U-arm to thrustor housing.
9. Disconnect spring and chain, or tie link holding U-arm to channel reinforcement plate.
10. Remove lead clamps at bottom of thrustor motor housing and under table.
11. Remove U-arm and shoe.
12. Disconnect wires through clamp at lower end of U-arm.
13. Remove screws holding U-arm to shoe.
14. Replace with new U-arm, using original lead wires if possible.
15. Reassemble by reversing this procedure.

To replace a cracked buck, proceed as follows:

1. Remove padding and screen.
2. Remove setscrew in side of thrustor cylinder.
3. Pull buck and thrustor motor out of cylinder.
4. Remove retaining wires from buck retaining screws, and remove screws.
5. Remove buck, and replace; reassemble.

Buck Does Not Come Up to Meet the Shoe—This may be due to an inoperative main or handle switch, a faulty motor, or a collapsed piston head. In all of these cases, replacement of components may be necessary.

Troubleshooting Rotary Ironers

Shoe Does Not Heat—This may be due to several factors, such as a loose connection at the terminal block or switches or an inoperative thermostat or heating element. A circuit-continuity check with a test lamp will conveniently establish the cause of the trouble.

The thermostat is replaced by removing the thermostat knob and cover plate. Disconnect the heater lead from the thermostat; the thermostat can now can be removed by removing the screws that hold the thermostat to the shoe. The new thermostat is then installed by reversing the dismantling procedure.

To replace the heating element, it is necessary to first remove the thermostat knob and cover. Disconnect and remove the wiring behind the shoe. Remove the two holding studs that secure the shoe to the pressure mechanism; the shoe may now be removed from the ironer. Disassemble the shoe by removing the nuts from the back. This allows the pressure plates, insulation, and heating element to be removed. To assemble the new element into the shoe, make sure that the shoe surface on which the element is to be assembled is clean and free from all pieces of the old element. Place the heating element in place on the shoe. Be sure that none of the wires in the element are out of place so that they touch any part of the shoe. Add the insulation strips and pressure plates. Assemble the washers and holding studs to the shoe, and tighten firmly. Connect the heating-element leads to the thermostat, and string the two covered leads through the shoe cover. Assemble the shoe cover, shoe ends, and thermostat knobs. Reassemble the shoe to the pressure arm. The holding studs should be tightened firmly. Reconnect the electrical leads.

Shoe Does Not Heat at One End—This is a common occurrence in rotary ironers equipped with two thermostats in which the heating element may be adjusted for different heats. The most common cause is a loose wire connection on the thermostat, an inoperative thermostat, or a burned-out element. If any one of these conditions exist, repair or replace as necessary.

Ironer Will Not Operate—If the ironer will not operate, even though the shoe element heats normally, the cause may most likely be found in the motor or the motor circuit. If tests show that power is available at the motor terminals, the motor itself should be checked for possible electrical trouble and repaired or replaced as necessary.

With the capacitor in the circuit and a test cord attached to the motor, check to determine if the motor will run; if not, disconnect the motor. With a series test lamp, check the motor leads to the frame of the motor to determine whether the motor windings are shorted. If a short is not indicated, replace the capacitor with a new one, and try the motor again. The motor shaft must be free and must not be bound by congealed oil. Gummed oil can be resolved by washing out the gear case with naphtha.

Roll Does Not Turn—If the motor runs but the roll does not turn and the shoe does not move, the drive mechanism or the motor belt may be inoperative. The drive arrangements differ for the various types of ironers; some ironers employ gears or belts, while others use chains and ratchets. Whatever the system arrangement, a thorough inspection should be made to determine the cause of the failure of the motor to transfer its rotary motion to that of the roll and shoe.

Failure of the roll to turn is also commonly caused by a hung-up roll clutch, which in turn is due to a stuck shifter fork, a worn or binding roll-clutch collar pin, or a broken clutch collar spring. In this case, it will be necessary to remove the roll shaft and roll-clutch assembly to determine the inoperative part that needs replacement. When disassembling the drive mechanism, carefuly note the location of parts as they are removed, so that they may be reassembled in their proper positions. When reassembling the mechanism, all parts should be thoroughly washed with naphtha, new gaskets should be installed, and the required gear-case oil should be put in the mechanism.

Lack of Pressure—After an extended period of use, the roll pad may become hard and may set to the roll, thereby causing insufficient pressure and making satisfactory ironing results difficult. To restore the pad to its original flexibility, it can be removed for washing, airing, and fluffing.

Insufficient pressure may also be caused by weak pressure-spring tension. There are usually pressure-adjustment points available on the ironer assembly to increase the pressure-spring tension;

these simply require the turning of a screw or a knob. After making the adjustment, the ironer should be checked for proper pressure, since too much pressure can cause the motor to stall and can also put additional strain on the ironer mechanism.

CHAPTER 8

Electric Toasters

The electric toaster, Fig. 1, is one of the oldest and most widely used of electrical appliances. In its simplest form, a toaster consists

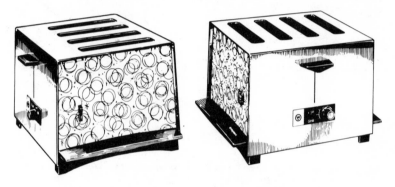

Courtesy Westinghouse Electric Corporation

Fig. 1. A modern automatic toaster, capable of toasting four pieces of bread at one time.

of a line cord attached to the heating elements or grids, which are arranged so that bread will be toasted when brought in close proximity to the heated elements or grids. The heating elements commonly consist of nichrome ribbon wound on mica strips, although certain types of toasters employ nichrome-wire coils for their resistance-heating elements.

AUTOMATIC TOASTERS

There is a great variety of automatic toasters with somewhat different construction principles on the market. In all cases, however, the heat from the electrical current in the heater element

161

Fig. 2. An automatic toaster with a spring-actuated clock mechanism, which controls the degree of brownness by regulating the heating time.

toasts the bread. The main difference in the various types is in the method employed to control the toasting cycle.

A great number of automatic toasters are classified as the "pop-up" type, as shown in Figs. 2 and 3, and include, in addition to the heating element, a thermostatic switch or timing device that automatically controls the toasting process. When the operating lever is pressed down, the switch contacts are closed, and they remain closed until the bread carriages are released. Thus, the units are energized from the time the operating lever handle is depressed until the bread pops up.

One type of automatic toaster, as illustrated in Fig. 2, contains a spring-actuated clock mechanism; the lowering of the operating handle with its bread carriage energizes the spring. This action also completes the electrical circuit through the heating elements. At the completion of the toasting cycle, the clock automatically

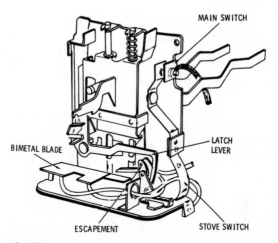

MAIN SWITCH

BIMETAL BLADE

LATCH
LEVER

ESCAPEMENT

STOVE SWITCH

Fig. 3. The internal mechanism of an automatic toaster.

trips the carriage, which returns to its original position, thereby opening the circuit through the heating element.

SERVICING AND REPAIRS

Automatic toasters, which are dependent on a thermostat switch or other timing device for their operation, are made to operate on 105 to 125 volts AC only. Since all toasters employ the same method of connections—they are all equipped with a double-conductor cord set—a common no-heating condition may be caused by:

1. Loose wall plug in outlet,
2. Blown fuse in house circuit,
3. Defective wall outlet,
4. Open circuit in supply cord,
5. Loose connections at terminals.

The remedy in each of these cases is obvious. Detailed repair methods have been fully covered in previous chapters.

Operation of Automatic Toasters

In order to obtain a comprehensive knowledge about automatic toasters, it will be necessary to understand the method of operation.

ELECTRIC TOASTERS

It should be observed that, due to various design modifications, their operations will vary somewhat. However, the treatment of various basic makes should enable the serviceman to repair any toaster needing repair, since, in principle, they all function similarly. A pop-up-type automatic toaster, as shown in Fig. 3, operates as follows:

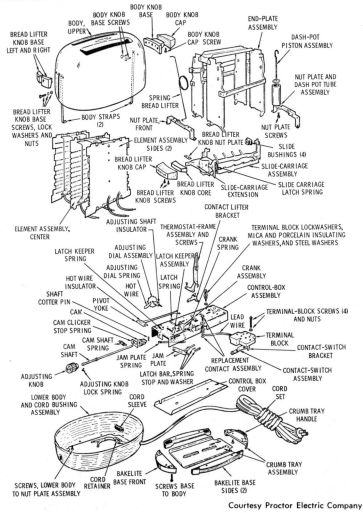

Courtesy Proctor Electric Company

Fig. 4. The disassembled view of a modern automatic pop-up toaster.

When the starting knob is pushed down, the latch lever engages the lower roller on the escapement. At the same time, the master switch is closed, and the stove switch is opened. The bimetal blade is then heated by the stove unit; the free end of the blade rises until it has raised the escapement lever high enough to disengage the latch lever from the bottom roller on the escapement. When this occurs, the latch lever engages the top roller. At the same time, the stove switch contacts close, thereby cutting the stove unit out of the circuit. When the stove is cut out of the circuit, the bimental blade begins to cool, which causes a lowering of the free end of the blade. The escapement lever follows the bimetal blade in its return to normal. Finally, the latch lever disengages from the top roller, and the starting lever, toast rack, etc., return to their starting positions. The main switch then opens, thus completing the toasting cycle.

The exploded view of another popular "pop-up" toaster is shown in Fig. 4; its operation is as follows:

The bread-carrier knob (starting knob) is attached to the carriage. When this knob is pushed down, the carriage descends, depresses the controls in the control box assembly, and latches. The switch-control arm moves the contact-lifter bracket and lowers the gravity-switch arm until its contact rests on the stationary contact and completes the electrical circuit. The upper part of the contact-lifter bracket, which protrudes from the top of the control box, falls toward the hooked end of the thermostat assembly. In addition, movement of the crank assembly permits a spring to slide the jam plate along the latch bar. With the circuit closed, the heater element commences to toast the bread. The continued current flow expands the hot wire and allows the jam plate to slide farther along the latch bar. The main bimetal strip, which is mounted close to the bread, bends in response to the surface temperature of the bread. The result is that when the bread is fully toasted, the hooked end of the bimetal-wire extension is pushed against the top of the contact-lifter bracket, thereby separating the contacts and interrupting the circuit. With the circuit open, the hot wire cools, contracts, and pulls back on the pivot yoke and jam plate, which are now locked or jammed on

165

the latch bar. As the latch bar moves, it disengages the latch keeper, which, with the aid of the bread-lifter spring, actuates the carriage, thus removing the bread from the toasting position. The controls are now ready for the next toasting. If this operation takes place immediately after the first toasting cycle, that is, while the toaster is still hot, the bimetal compensator, due to its slight bend, will work in the opposite direction to that of the main bimetal. In this way, a somewhat shorter toasting period will be obtained, thus preventing overtoasting of the bread at all times.

A third type of automatic "pop-up" toaster is equipped with a clock mechanism to control the toasting cycle. In a toaster of this type, the operating lever, which performs the function of lowering the bread into the oven, closes the electrical circuit and winds the timing mechanism, which is automatically locked, until it is released at the end of the correct time interval. To obtain variations in timing, the speed of the clock is varied by means of a timing button. Turning the button to the right increases the speed of the clock; turning it to the left decreases the speed. The position at which the timing button is set determines the color of the toast.

Certain late-model units of this type are equipped with a variable clock, which consists of double-wound elements and a shunting switch in the base of the toaster. The purpose of this construction is to increase or decrease the resistance of the heating units, thereby obtaining the desired wattage in localities where the voltages differ greatly. For low-voltage ranges, the shunting switch in the bottom of the toaster is set to shunt out the extra winding, thus decreasing the resistance of the heating units. If the elements operate at too high a temperature with the shunting switch set in this position, the switch should be opened, which increases the resistance, to obtain satisfactory operation on the higher voltage ranges. With the switch closed (high), the toaster operates satisfactorily at an applied voltage of 105 to 115 volts AC. With the switch open (low), it operates satisfactorily on a voltage range of 110 to 120 volts AC.

In this type of toaster, the bimetal strip in the timing mechanism is heated by an auxiliary element that is connected in series with the main toaster elements. The flexing of the bimetal strip controls the toasting cycle. It is voltage-compensating and automatically increases or decreases the toasting time when used within the limits of 105 to 125 volts. A heat-up–cool-off cycle controls the timing operation. During the heat-up part of the cycle, the auxiliary element heats the bimetal strip, thus causing it to flex. This action forces the operating arm slowly forward until it hits the timing shaft. The bimetal strip, still heating, then flexes in the opposite direction until it has moved off the shunt-lever trigger. This causes the auxiliary switch to close and shunt out the auxiliary element. The release link then drops in the path of the operating arm. At this point, the cool-off part of the cycle begins. The operating lever moves back toward the starting position until it comes in contact with the release link, moving the release lever until the toaster trips. The toast automatically pops up, opening up the main switch and shutting off the current. The entire timing mechanism is now reset for the next operation.

Toaster Fails to Heat Up

When an automatic toaster fails to heat up after the bread carriage lever has been properly depressed, it is evident that an open circuit exists, either within the toaster or in the cord or its attachments. Check the complete circuit until the trouble is located; repair or replace inoperative parts. A circuit test on a toaster is usually performed by means of a common circuit tester. If the circuit is complete to the toaster terminals, it is obvious that the circuit within the toaster is open.

Disconnect the wall plug, and remove the shell in order to trace the internal toaster circuit. Plug the cord into the circuit tester, and observe the action of the mechanism in the control box while latching the carriage.

If the electrical contacts need cleaning, disconnect the cord, and clean the contact points with a very fine ignition file and fine emery or crocus cloth. Plug the cord into the circuit tester; if the lamp does not light, disconnect the plug, and check the entire circuit with the test leads.

Lack of Uniformity

In modern toasters, the individual elements are measured for resistance and are graded and matched at the factory to obtain substantially uniform toasting on all four sides facing the two slices of bread in the toaster at any one time. To accomplish uniform toasting, the center element has a somewhat higher heating capacity than the outside element, so that when the toaster is cold, the inside surfaces (those facing the center element) may come out slightly lighter than the other two surfaces. With correctly-matched elements and uniform conditions of bread surfaces, there should not be any difference in bread color as the toaster heats up. It should, therefore, not be assumed that elements are mismatched unless several tests are made, and judgment is passed on the average result. After such a test, if the inside surfaces consistently toast lighter or darker than the outside surfaces, replace the elements with a set of matched elements designed for the particular toaster model.

Degree of Browning

As previously noted, toasters have, in addition to the carriage-mechanism lever, a small control knob near the bottom of the front face, which may be set for different degrees of brownness of the bread to be toasted. The control knob moves by means of a linkage to affect the time setting, and thus it directly controls the duration of the toasting period. If the toasting test shows that the desired degree of toast darkness cannot be obtained by adjusting the time-control knob, a secondary adjustment can be made by removing the crumb tray and turning the adjusting screw toward the rear of the toaster for dark toast and toward the front for lighter toast. This adjustment is usually critical. A fraction of a turn will normally change the length of the toasting cycle by several seconds. Therefore, proceed with caution, and actually test the toaster with bread after each adjustment. It is further suggested that when resetting the toasting time, set the control knob at the midpoint, thereby giving an average setting. This will allow the customer the widest possible range from which to select his favorite color. Fig. 5 illustrates another method of adjusting the time period.

Toaster Fails to Pop Up

If the toast fails to pop up after the toasting cycle has been completed, it is usually due to a binding pop-up lever or knob, a defective thermostat or timing clock, welded silver contact points on the thermostat (due to the use of the toaster on direct current), bent or defective pop-up wire, broken carriage-elevator springs, or a binding dash-pot piston.

The contact-switch assembly and contact points must always be clean. Be sure that on any toaster with a pull-type slide-carriage spring, this spring does not rest against the beam. The trip-link spring must not be overstretched or distorted. The trip-lever as-

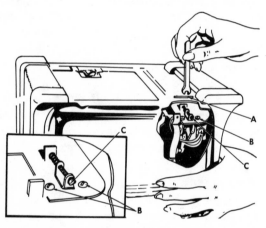

Fig. 5. The time adjustment for this toaster is made by loosening two small screws (B) by one-half turn. Then, insert a 3/16-inch wrench into the opening (A), and engage the adjusting screw (C). Turn this screw toward the rear of the toaster for darker toast, toward the front for lighter toast. After the desired adjustment has been made, tighten screws (B), which will hold the adjustment-screw setting.

sembly must be at right angles to the base, so that it will not bind in the slot of the trip link. If it is bent, align it to the trip-link slot with pliers.

Length of Toasting Cycle

The toasting time will vary somewhat with different types of toasters and is dependent on voltage changes, dryness and size of bread, condition of toaster (whether cold or warm), etc. Under normal conditions (operating with moderately fresh bread, an average moisture content, on 115 volts AC, and with the toaster set at "medium"), the total toasting cycle will vary from approximately 90 seconds when starting with a cold toaster to less than one minute as the toaster warms up.

Disassembly

Since modern toasters include a large number of designs, the method used to take them apart will, in each case, depend on the method of assembly. An inspection will reveal the disassembly method to follow in each instance. The usual procedure in most designs includes the removal of the Bakelite operating handle and timing button, after which the under cover, which is held in place by a couple of thumb nuts, is removed. Remove the screws holding the element and switch assembly to the shell. Turn the toaster bottom down, and press the ends of the shell together for removal. Care must be taken to see that the shell clears the terminal nuts on the sides of the element assembly. If removal of the elements is unnecessary, the guard wires can be taped to the top of the cage to keep them from falling out when the toaster is turned over. The reassembly process is simply a reversal of this procedure.

Element Replacement

Individual elements can be replaced, when necessary, by removing the guard wires above the particular element (tape the remainder to hold them in place) and disconnecting the terminal screws or nuts. The defective element can now be slipped out through the top, and the new element can be installed in the reverse manner. The heating elements are usually not interchangeable; therefore, it is essential when ordering elements for replacement to specify the proper element by name and, if possible, by number. After replacement of any functional part in the toaster, a test of several toasting cycles should be made to be certain of satisfactory operation.

Timing-Mechanism Assembly Replacement

The timing-mechanism assembly necessitates disconnection of the lead wire from the front unit terminal; disconnect this wire from the terminal at the top of the main switch. The timing-mechanism assembly is held in place by several screws, which must be removed for replacement. The new timing mechanism can then be installed by reversing the disassembly procedure.

Thermostat and Stove-Unit Assembly Replacement

The thermostat and stove-unit assembly can be removed by disconnecting the line-terminal wire from the switch and removing the mounting screws. The new assembly can then be mounted in

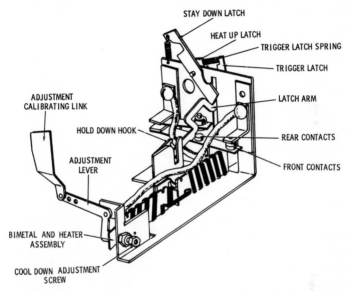

STAY DOWN LATCH

HEAT UP LATCH

TRIGGER LATCH SPRING

TRIGGER LATCH

ADJUSTMENT CALIBRATING LINK

LATCH ARM

HOLD DOWN HOOK

REAR CONTACTS

ADJUSTMENT LEVER

FRONT CONTACTS

BIMETAL AND HEATER ASSEMBLY

COOL DOWN ADJUSTMENT SCREW

Courtesy General Electric Company

Fig. 6. A typical thermostatic operating mechanism, which consists, essentially, of a bimetal strip surrounded by a coil of resistance wire. This heater coil is connected in series with the heating elements and causes the bimetal strip to bend. The bending of this bimetal strip produces the mechanical force to control the automatic operation of the toaster.

place in the usual manner. A common type of bimetallic thermostat is shown in Fig. 6.

Caution: Under no circumstances should the bimetal strip be bent or tampered with. This would ruin the toaster and render it useless, since the metal content of the bimetal strip is calculated in relationship to the amount of heat supplied by the stove to allow for the exact distance of travel. However, if the limits are not sufficient to allow for proper adjustment, carefully bend the stove unit and thermostat assembly mounting bracket. This procedure should only be used in extreme instances.

Electric Toasters

Under no conditions should a repaired toaster be tested unless the shell and handles are fully assembled, with the crumb tray in place.

When ordering replacement parts, always be sure to supply the correct toaster model number. This will assure that the right part is used, which will result in satisfactory operation.

CHAPTER 9

Electric Waffle Irons

Today practically all modern waffle irons, as shown in Fig. 1, are electrically heated and are also of the automatic type, which permits the housewife to attend to other matters while the thermostat or clock mechanism serves as the appointed guardian. The automatic device disconnects the heating element at the exact moment that the waffle is correctly baked, thus assuring an evenly prepared product throughout.

Many waffle irons include an indicating lamp that automatically signals when the baking cycle is completed. All that is required is to properly preheat and grease the griddles and set the thermostat or heat regulator for any desired degree of brownness, prior to putting in the batter. Some waffle irons are, in addition, equipped with replaceable grids, as shown in Fig. 2, thus permitting them to be used for toasting various products, such as bread, sandwiches, fried ham and eggs, etc.

OPERATION

The modern waffle iron, in common with numerous other automatic electric appliances, is thermostatically operated. Fig. 3 illustrates the wiring diagram and baking cycle for a typical waffle iron. Note that the waffle iron contains two equally spaced grid heating elements (one for each half of the iron). Since the heating elements, thermostat, and lamp shunt are connected in series, it is evident that the only disconnecting means in the circuit is the thermostatic switch. The heat-adjustment rod, located at the thermostat, permits various degrees of brownness by turning the rod for the different settings of "light," "medium," and "dark." These settings differ with the various models of waffle irons.

Fig. 1. A typical automatic electric waffle iron, with interchangeable grids.

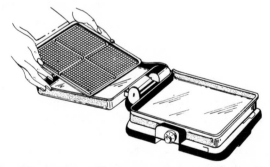

Fig. 2. The grids of most waffle irons can be reversed to grill sandwiches and other foods.

When the waffle iron is connected in an alternating-current circuit of the correct voltage, with the heat-adjustment dial set at "medium," current will flow through the heating elements and commence the heating cycle. The signal lamp will then light, since the circuit is held closed by the action of the thermostat. As the

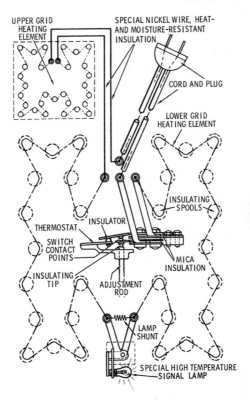

Courtesy Sunbeam Corporation

Fig. 3. The wiring diagram of a typical automatic waffle iron. The thermostat, which is connected in series with the lamp shunt and the heating elements, acts as the disconnecting means for the circuit.

temperature increases, the bimetallic blades in the thermostat will flex, or warp; the maximum amount of bending will occur at the point of highest temperature. When the predetermined temperature, which is dependent on the setting of the temperature-adjustment rod, is reached, the thermostat contacts open the circuit. The indicating lamp then signals that the heating cycle is completed. The waffle iron is now ready for more batter and the next heating cycle. If the "medium" setting has not produced the desired degree of brownness, the control knob should be reset until the exact shading is obtained.

When the circuit opens at the termination of each heating cycle, the grid heating elements and the bimetal blades will cool off slightly. This cooling permits the bimetal blades to straighten out, thereby closing the circuit and lighting the indicating lamp. This type of waffle-iron construction is only one of several designs that are available for home use. One type of waffle iron uses the grid casting itself for thermostatic control, instead of the ordinary bimetal thermostat. In this type of iron, thermostatic switching is obtained by inserting a Pyrex glass rod into a hole in the aluminum grid. Since aluminum has a higher coefficient of expansion than glass, the difference in expansion of the two materials, when exposed to temperature changes, causes the switch contacts to move, thereby opening or closing the heating-element circuit and controlling the heating cycle.

SERVICING AND REPAIRS

If the thermostat gets out of adjustment or is replaced by a new unit, the temperature settings must be readjusted to assure the desired temperature. Use a pyrometer or a glass-bulb thermometer that can indicate temperatures up to 600°F. to make the proper temperature settings. If a thermometer is used, pour a sufficient amount of vegetable oil into the lower grid to just cover the ribs. By placing the thermometer bulb into this oil near the center of the grid, without touching it to the metal, a fairly accurate temperature can be obtained. The temperature adjustment is made as follows:

1. While the waffle iron is at room temperature, turn the heating-control rod until the thermostat contacts are opened.
2. Plug in the waffle iron. Turn the control rod until the signal lamp goes on.
3. The amount of pressure on the control rod will affect the cutout point, so care must be exercised. From the point when the lamp first goes on, turn the control screw approximately one turn.
4. Without disturbing the control setting, push the Bakelite control tightly on to the rod with the pointer in the vertical position ("medium"). The thermostat should cut off at ap-

proximately 400°F. After heating for several minutes, the thermostat will cycle (turn off and on) at a slightly lower temperature, usually between 360° and 380°F.

5. If it is necessary to increase or decrease the temperature setting beyond the stop position, the Bakelite control knob must be pulled off, the shaft must be turned in the desired direction, and the control knob must then be replaced for the new setting; the adjustment can then be repeated with the control knob in the "medium," or vertical position.

In order to protect and support the thermometer during the test, it will be necessary to construct a suitable fixture that will permit temperature readings while the upper grid is closed.

Insufficient Heat

If the waffle iron heats up too slowly, or is insufficient for satisfactory service, check all terminals and switch-contact points to ascertain whether the waffle iron is drawing its rated power (usually about 1000 watts) from the voltage source.

Most waffle irons are designed for operation on 110 to 120 volts (alternating current only); when the voltage is insufficient (100 volts or below), the waffle iron will heat up and bake slowly. House wiring, including extension cords in use, may sometimes be incapable of carrying the load imposed on the circuit by the waffle iron operating at its rated voltage. In such cases, a different electrical outlet should be selected, or correction of the house wiring system should be arranged with the electric utility company. In some cases, if other appliances are used on the same circuit at the same time as the waffle iron, the electric circuit could be overloaded, thereby resulting in a temporary drop in voltage. Under these conditions, no adjustment of the waffle iron is possible, but the use of other appliances should be avoided while the waffle iron is in operation.

Too Much Heat

If the waffle iron becomes too hot, the cause will usually be due to either an incorrectly adjusted thermostat or stuck or welded thermostat contact points (a condition caused by using the waffle iron on direct current). Overheating may also be caused by embedded metal chips on the contact points, thereby impairing thermostat operation. The thermostat must be removed, and its contact points must be inspected. The thermostat unit should be replaced if it is found to be defective.

No Heat

In the event that the waffle iron does not heat at all, an open circuit is usually the cause. If a test indicates that power is available at the wall outlet, check the cord assembly for a broken or loose wire at the wall plug or at the heating-element terminals. After ascertaining that power is available at the wall outlet, test the heating-element terminals by means of a test lamp. If power is available at the element terminals, an open circuit in one of the heating elements, a faulty lamp shunt, or a defective thermostat will be the most likely no-heat cause. In order to locate and eliminate the trouble, the waffle iron must be disassembled, as shown in Fig. 4, by loosening and removing the nuts and screws holding the assembled parts together. Use a soft cloth or pad under the inverted baker to prevent scratching or marring of its surfaces.

If it is found necessary to replace the thermostat, disconnect the flexible leads of the switch assembly. Remove the heat-control rod, the control-rod disc, the movable arm, the contact assembly, and the screws holding the heat-control-rod bracket assembly. Replace the thermostat, and adjust the heat-control rod until the switch points make contact. Adjust for the proper temperature setting. To reassemble, reverse the disassembly procedure, and check the assembly for an improper ground with a conventional test lamp. In a great many modern waffle irons, the signal lamp shunt wire is part of the heating-element circuit. If the shunt wire is broken, or if its terminal connections are loose, the heating element will not heat, and the signal lamp will be burned out.

Signal Lamp Does Not Light

In most wiring arrangements, the connections are such that the waffle iron will operate normally even though the signal lamp is burned out. The wiring diagram, Fig. 3, indicates the lamp shunt wire whose resistance is calculated to supply a specific voltage

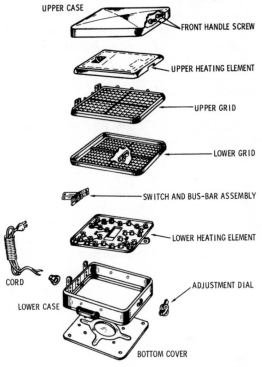

UPPER CASE

FRONT HANDLE SCREW

UPPER HEATING ELEMENT

UPPER GRID

LOWER GRID

SWITCH AND BUS-BAR ASSEMBLY

LOWER HEATING ELEMENT

CORD

ADJUSTMENT DIAL

LOWER CASE

BOTTOM COVER

Courtesy Sunbeam Corporation

Fig. 4. The disassembled view of an automatic waffle iron. Note the similarity between the upper and lower sections.

across the lamp terminals. It is of the utmost importance that the correct lamp is substituted in the event of failure. Should the lamp shunt wire need replacement, use only those resistors supplied by the manufacturer, since resistors of a different ohmic value can cause lamp failure. These resistors usually have a value of only a fraction of an ohm, which makes them somewhat difficult to measure accurately without the proper equipment. Ordinary ohmmeters are not sufficiently accurate for this purpose. To replace the signal lamp, unscrew the bottom cover, which permits access to the lamp; remove the defective lamp, and insert a new lamp. Be careful not to scratch the upper case during the operation.

In certain types of waffle irons, heat indication is obtained by employing a part of the heating element as a signal lamp. Because of the series-connected circuit, any failure (opening) in the heating elements immediately causes a signal breakdown, in which case a new element may be required.

Replacement of Heating Elements

While there may be some difference in the disassembly methods, depending on the type of waffle iron involved, in general the procedure used will be the same for all. If a test indicates an open heating element, loosen and remove the required nuts and screws; remove the top and bottom shells, covers, or cases. Disconnect the element leads from the terminal posts. Detach the heating elements from the grids to permit an inspection of the heating-element resistance wire. In case of element failure, install a new element by reversing the disassembly procedure, connecting the element lead ends to their terminal posts. Be careful to remove any loose foreign material prior to installing the new element. After reassembly, readjust the thermostat control.

CHAPTER 10

Casseroles, Roasters, and Broilers

The casserole, roaster, and broiler, as shown in Fig. 1, are electrically operated portable cooking devices, each having somewhat different characteristics and each suited for a particular use. They are alike in that they operate on the oven principle, with one or more electric heating units as an integral part of a closed container.

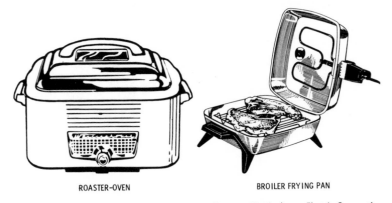

ROASTER-OVEN   BROILER FRYING PAN

Courtesy Westinghouse Electric Corporation

Fig. 1. The automatic electric roaster and broiler represent two of the many time-saving devices available for today's modern housewife.

Automatic cooking devices of this type contain a thermostat, which controls the temperature of the food being cooked by regulating the amount of current flowing through the heating element. Others may, in addition to the thermostat, be equipped with an automatic clock or timer, which turns the device on or off at a predetermined time.

181

CASSEROLES

Electrically operated casseroles are manufactured in numerous sizes and shapes. Fig. 2 shows one type of electric casserole. It consists essentially of a cooking well around which a heating element is wrapped. A layer of insulating material prevents the heat generated by the element from penetrating to the outer shell. During the cooking process, the outside of the casserole remains cool, and can be used on the table as a serving dish. The casserole is an ideal waterless cooker, since it retains all the savory flavors of the food without using water. Since the heating element completely

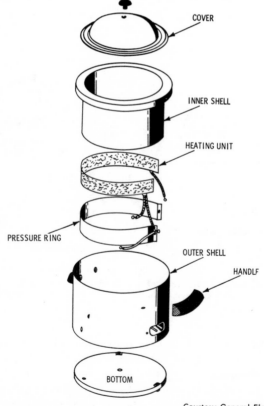

Courtesy General Electric Company

Fig. 2. The disassembled view of a typical electric casserole.

encircles the sides of the casserole, the food is less likely to be burned. Provision for various heat values make for additional convenience and economy.

Casseroles are usually furnished with a porcelain, enamel, or chromium finish, and are comparatively easy to keep clean. They are also available in pairs, or in combination with a hot plate, thus

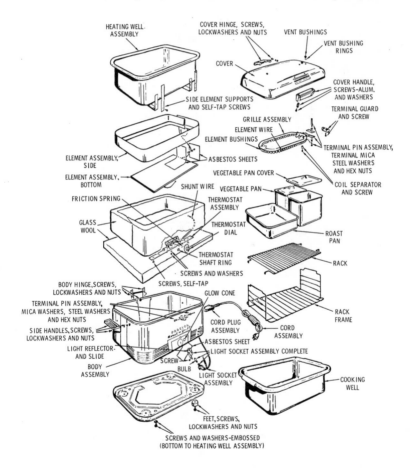

Courtesy Proctor Electric Company

Fig. 3. The disassembled view of a typical automatic roaster-broiler combination. This appliance not only contains side and bottom heating elements for roasting but also incorporates a broiler element in the cover.

increasing their range of service. Casseroles used in a combination can be removed from their heating units for both serving and cleaning.

ROASTERS

Roasters consist essentially of an inner and outer shell, a cover, heating elements, and heating-control components. Fig. 3 shows a typical appliance of this sort; it contains two heating units, one in the bottom and the other encircling the sides, arranged for low and high heats. The inner and outer shells are, in addition, separated by a layer of glass-wool or rock-wool insulation, which prevents heat absorption by the body assembly.

Roasters are manufactured with a variety of special features; thus, for example, the roaster shown in Fig. 3 is a roaster-broiler combination, incorporating a special broiler element in its cover. The circuit diagram for this type of roaster is shown in Fig. 4.

In operation, the foods are placed in the cooking well, or in separate removable cooking pans of porcelain enamel that nest

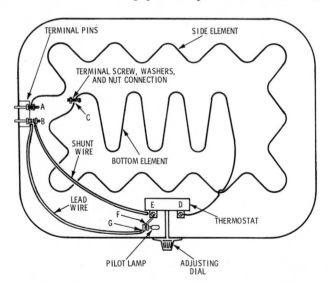

Courtesy Proctor Electric Company

Fig. 4. The circuit diagram for the roaster-broiler combination.

in the oval well, so that three or more foods can be cooked at one time. In some roasters, a baking set is included for baking bread, cakes, pies, etc. Roasters usually have a porcelain lining, and, like the casserole, can be used on the table as a serving dish.

Automatic roasters are provided with thermostats or clock-timing devices, which control the cooking process. Thermostat-equipped roasters, in some instances, permit the user to select the desired cooking temperature by turning a dial on the appliance. This dial setting provides various spring tensions on the bimetallic blade and in this manner prevents the bimetallic blade from operating until the preselected temperature has been reached. Roasters equipped with timing devices operate similarly to range-oven timers and permit the user to select a specific temperature and time limit.

BROILERS

An electric broiler consists essentially of a hinged container with a heating element embedded in the top part or cover; the lower part is equipped with a grill or wire rack for support of the food to be broiled. Depending on construction, a broiler may have one or two heating elements. The double-element type provides either high or low heat. A typical motor-equipped broiler of the rotisserie type is shown in Fig. 5. The heating element (or elements) consists of nichrome resistance wire supported by insulators or attached to a slab of heat-resistant insulating material with grooves into which the wire coil is fitted. Temperature control is obtained by means of a timing device, which regulates the length of the heating cycle.

SERVICING AND REPAIRS

Electrically operated cooking vessels, such as casseroles, roasters, and broilers, are comparatively easy to repair. The most common complaints are that the appliance will not heat, that the user receives a shock when the device is touched, and, in certain instances, that the thermostat is inoperative or causes underheating or overheating because of faulty adjustment. In addition, pilot lamps or fuses may burn out and require replacement.

Courtesy General Electric Company

Fig. 5. A typical automatic broiler-rotisserie combination, which consists essentially of a motor and a heat-control switch mounted on the broiler enclosure with a conventional heating element.

Appliance Will Not Heat

This may be due to a defective cord set, a loose connection in the appliance, or an open heating element. A defective cord set is checked for continuity with a test lamp in the conventional manner and should be replaced when necessary.

To test for a loose connection or an open heating element, circuit continuity is easily verified by touching one prod of the series test lamp to the single lead terminal and the other prod to the terminal of each of the remaining leads. If the test lamp lights, the circuit is continuous, although one of the paralleled bottom elements may be open. As shown in Fig. 6, an open circuit in the side elements will make the roaster inoperative, whereas an open circuit in one of the bottom elements will allow circuit operation but will impair the thermostat regulations because of a change in the circuit heating resistance. A loose internal connection, which may not be found with a continuity test, or an open heating element usually requires the disassembly of the unit in order to locate the trouble.

If a shock is felt when the appliance is touched, there is a "live" wire in contact with the metal enclosure or other conducting part

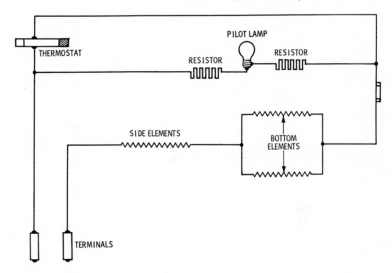

Fig. 6. The circuit wiring diagram of a typical automatic roaster.

of the unit. This dangerous condition can be remedied only by disassembly; if an old element lacks sufficient insulation properties, it should be replaced.

When it is necessary to replace an open-coil resistance element from any cooking appliance of this type, it is important that the new element to be installed has the same length, has been coiled to the same diameter, and is of the same wire gauge as the old element.

In order to obtain a certain wattage, wire of a gauge capable of carrying the required current involved must be employed. This wire has a rating of length with respect to its gauge number, which determines its resistance. Therefore, a certain number of feet and a fraction thereof must be used in a circuit. This wire is coiled on different sized arbors to a closed-coil spring form, as purchased. To determine the length that the coiled wire must be stretched to fit properly in the element space, place a string in the element groove, and measure the distance. Then clamp one end of the coil in a vise, and pull the coil out with a pair of pliers to a little less than the string length. Pull the turns out uniformly so they will be equally spaced when completed. Ends of wires can be looped to properly fit the terminal screws. Coiled wires should be stretched slightly to make connections, so that a small amount of tension will keep the element in place.

Cooking appliances with heating elements embedded in ceramic insulated bricks or other insulating material of a type that requires replacement as a unit are usually available at appliance service centers or may be obtained directly by request from the manufacturer of the appliance in question. In any event, it is important to supply the manufacturer with sufficient information, giving the nameplate data in addition to any catalog or number type of the appliance, in order to facilitate correct heating element replacement.

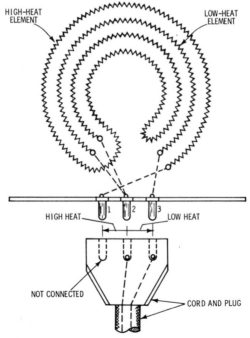

Fig. 7. The connection method for a two-heat (high and low) cooking device. If the cord plug is inserted in the position indicated, thereby energizing only terminals 2 and 3, the low-heat element will be connected. If this position is reversed, terminals 1 and 2 will be energized, and the high-heat element will be connected in the circuit. This connection design permits heat selection without the aid of an external switch.

Special Two-Heat Connection Methods

Certain types of cooking appliances are equipped with special terminal post connections, as shown in Fig. 7. Three terminal posts provide two heats, one post being common to the other two. As illustrated, two elements are used for high heat and the other is used for low heat. By turning the cord plug over, it is possible to connect either one of the two elements with the power source.

Thermostat Adjustment

Cooking appliances that are furnished with thermostats having provisions for external heat settings by means of knob rotation around a graduated dial, such as shown in Fig. 3, have no further heat adjustments. Customers who complain of inaccurate thermostat adjustments are usually those who have attempted to check the temperature on the control-knob setting against the temperature inside the appliance, using an oven thermometer in an empty unit. It should be clearly understood that the thermostat is not calibrated to conform to the temperature markings unless the unit contains its normal food load; thus, any adjustments made when the unit is empty will be incorrect. Thermostat adjustments should be attempted only after a careful and well-planned test indicates that an adjustment is necessary and should be performed according to the manufacturer's instructions for the appliance in question.

CHAPTER 11

Electric Coffee Makers and Urns

All automatic electric coffee makers have an electric heating element, or stove unit, fitted into the base. Electric coffee makers may be classified, according to the method used to make coffee, as the *percolator* type and the *brewer* type.

In the percolator type, as shown in Fig. 1, the heated water is forced upward repeatedly through a percolating tube, which extends from the center of the base into the coffee basket located in

Fig. 1. A common percolator-type coffee maker. The scale on the handle, an optional feature, indicates the number of cups remaining in the percolator at any time.

Courtesy Dormeyer Corporation

the upper part of the assembly. The percolating tube may or may not have a valve, depending on the type of unit.

In the brewer-type coffee maker, such as in Fig. 2, all the heated water is forced into the upper bowl at one time where it is retained with the coffee grounds until it drains down into the lower bowl to complete the coffee-making cycle.

AUTOMATIC COFFEE MAKERS

An automatic coffee maker is one in which a thermostat and a switch serve to regulate the coffee-making cycle in such a way that

Fig. 2. An automatic brewer-type coffee maker. The dial at the bottom permits two different temperature settings.

Courtesy Sunbeam Corporation

the only work necessary is to supply the proper quantity of ground coffee and water to obtain the finished brew.

Operating Principles—Brewer Type

A typical fully automatic electric coffee maker of the brewer type is illustrated in Fig. 3, which shows the thermostat-switch mechanism and heating element. Note that the heating-element switch and thermostat are fitted on the bottom of the lower bowl inside the Bakelite base. The thermostat is set for high or low heat according to the position of the switch. Proper operation of the coffee maker depends on the accurate setting of the low and high adjustment screws.

One of the features of this automatic coffee maker is that the water rising to the upper bowl reaches a temperature slightly in excess of 200°F., which is supposedly the best temperature for brewing coffee. This delayed rising is achieved by a small opening in the tube near the upper bowl that allows the pressure in the lower bowl to equalize itself during the preliminary heating period. When the water reaches the proper temperature and goes up, all but a small quantity of water in the bottom leaves the lower bowl. This small quantity boils away and the resulting steam agitates the coffee in the upper bowl. When all the water is out of the lower bowl, the heat increases rapidly; the thermostat then automatically shuts off the power and switches the control to low. Then, as the temperature decreases, a vacuum is produced in the lower bowl, and the coffee is forced down through the filter into the lower bowl, where it is automatically kept at a temperature of between 165° and 185°F. by the low-heat setting of the thermostat.

The following is an explanation of the thermostat and switch operation for the mechanism shown in Fig. 3:

Alternating current enters at the terminal posts, passes through the heating element into the low-heat contact spring, through the silver contact points, into the terminal contact spring, and back to the terminal post. This circuit is exactly the same on a low or high heat setting. When the switch is set at "low," as shown in Fig. 3, and the current is turned on, the thermostat blade is heated. The blade deflects until it moves the low-heat adjustment screw and contact spring, thereby separating the silver contact points. The contacts remain open until the loss

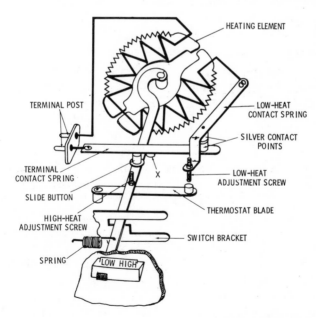

Courtesy Sunbeam Corporation

Fig. 3. The internal component arrangement for a brewer-type coffee maker.

of heat straightens the thermostat blade, closes the points, and starts another cycle. If the low-heat adjustment-screw setting is correct, the heating element will keep the coffee in the lower bowl hot but not hot enough to rise to the upper bowl. To make coffee, set the switch in the "high" position. The lever will go down into the lower step in the switch bracket, bringing the high-heat adjustment screw closer to the thermostat. At the same time, the slide button will come under projection X on the diagram, thus pushing both the springs and the adjustment screw away from the thermostat. When the current is switched on, the temperature rises until the thermostat is deflected enough to raise the high-heat adjustment screw; the switch lever is then forced out of the lower step in the switch bracket, and the spring pulls it back to the "low" position. When this happens, the low-heat adjustment screw is immediately moved by the thermostat, and the circuit remains open until enough heat is lost to permit closing at the low-heat setting, as previously described.

Operating Principles—Percolator Type

The assembled view of a typical automatic percolator-type coffee maker is shown in Fig. 4. The opearting unit, including temperature controls, is sealed in the base of the coffee maker. Prior to commencing the coffee brewing cycle, the coffee is supplied to the basket in the top of the unit. A quantity of cold water is then added to the coffee maker. When the coffee is put into the basket, care should be taken to see that none of the grounds fall into the inset tube. The control lever is then set at the desired position, usually between "mild" and "strong," and the percolator is connected to the wall outlet.

The pump is the heart of the percolator and consists of a pump chamber, a valve washer, and a valve seat. When the inset tube is in position, the base of the pump rests on the heater-plate nose. When the pump is placed on the heater-plate nose (with water already in the percolator), the water enters the pump chamber and rises in the inset tube to the level of the water in the percolator. As soon as the unit is energized, the small amount of water in the heater-plate nose is heated quickly, and a small amount of steam is formed. The valve washer, which is seated tightly against the valve seat, prevents the steam and water from escaping into the main body of the unit. Therefore, the steam pressure pushes some of the water in the heater-plate nose and pump chamber up the tube. This reduces the pressure in the pump chamber, thereby allowing the valve washer to be raised by the pressure of the water in the main body of the unit. Water then enters the pump chamber and heater nose. This cycle is continuously repeated; it is in this manner that the heated water is forced up through the inset tube and out onto the spreader plate, as illustrated in Fig. 4.

The percolation time is controlled by the operation of the control in the base of the unit. The movement of the control lever positions the control-switch assembly at a given distance from the roller on the end of the bimetal blade. As the liquid in the body of the percolator becomes heated, the heat is transmitted through the bimetal mounting bracket to the bimetal blade, which flexes,

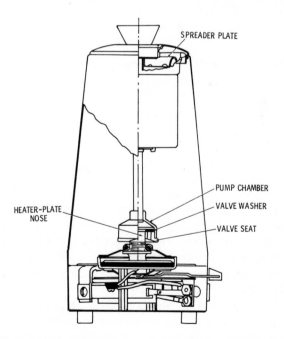

Fig. 4. The operating mechanism of an automatic percolator-type coffee maker.

thereby resulting in a movement toward the end of the contact spring. As the bimetal blade continues its movement in this direction, the roller comes into contact with the end of the contact spring and breaks the circuit by opening the contacts. During the part of the cycle in which the contacts are closed, only the operating unit is in the circuit. The warming unit, bimetal heater, and

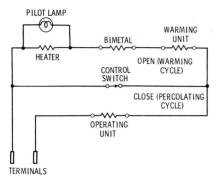

Fig. 5. The schematic circuit diagram of an automatic percolator-type coffee maker.

pilot lamp are shorted out by the control-switch assembly, as shown in Fig. 5. With the switch contacts open, the warming unit, main operating unit, and bimetal heater are all in series. The warming unit will remain in the circuit until the percolator is disconnected from its power source.

SERVICING AND REPAIRS

All automatic coffee makers are dependent on a thermostatic switch for their operating cycle. Since these switches are of the slow make-and-break type, proper operation necessitates their use on alternating current only. If direct current is used, the sustained arc, resulting when the contacts open, will melt the contacts and weld them together, thus making it necessary to replace the thermostatic unit.

TWO-HEAT AUTOMATIC COFFEE MAKERS

A schematic wiring diagram of a brewer-type coffee maker of the two-heat variety is illustrated in Fig. 6. The circuit provides for two different temperatures by means of two specially designed heat units, or elements—one for boiling the water and the other

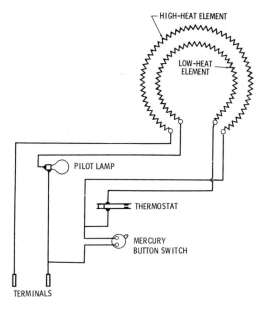

Fig. 6. The schematic circuit arrangement of an automatic two-heat brewer-type coffee maker.

for keeping the coffee warm in the lower bowl. After having placed the coffee and water in the coffee maker, the operating switch is set on "high." At this time, only the high-heat element is connected to the power supply. The thermostat, which is attached to the bottom of the baffle and is in series with the low-heat element, will open, due to the temperature of the high-heat element. After the water has risen to the upper bowl, the switch is moved to the low position, thereby putting the low-heat element and thermostat in series with the high-heat element. The heating elements are not connected to the power supply until the thermostat, and hence the coffee, has cooled enough to close the circuit. When both elements are in series, the heat is sufficient to maintain the brew at approximately 180°F. The thermostat may or may not cycle again, depending on the amount of coffee in the lower bowl.

Servicing Automatic Coffee Makers

Complaints that automatic coffee makers are not functioning properly may be due to inoperative heating units, defective parts (such as faulty stem-cap and lift-disc assemblies), improper supply voltage, use of direct current, or the manner in which the coffee maker is used. In the absence of a specific complaint from a customer, the unit should be carefully checked. If no defect is found, the entire appliance should be put through a complete operational cycle for a further check. On pilot-lamp-equipped units, if the pilot lamp lights, it indicates that the low-heat circuit is operating. When the reset switch is depressed and held down, the pilot lamp should go out, and the high-heat element should glow if the circuit is complete.

Other complaints on brewer-type coffee makers may be that the coffee does not return to the lower bowl or starts to return and then goes back to the upper bowl, oscillating back and forth every few minutes; coffee boils over, the low-heat element does not keep the coffee at the correct temperature, poor coffee flavor, odor from base of coffee maker, etc.

Coffee Does Not Return to Lower Bowl—If the coffee does not return to the lower bowl, or cycles back and forth between the upper and lower bowls, the thermostat is out of adjustment. To adjust the low-temperature control setting on the type of unit illustrated in Figs. 2 and 3, proceed as follows:

1. Have the coffee maker at ordinary room temperature. It is important that there is no heat remaining in the internal parts of the appliance from previous use or testing.
2. Place the lower bowl upside down on a pad or cloth to avoid scratching of surface.
3. Remove the nameplate, and connect a two-watt neon glow lamp in series with the coffee maker, so that the element cannot heat up while the control parts are set.
4. Set the switch lever to the low position. Place a wrench on the low-heat adjustment screw, and loosen the lock nut by turning the wrench in a counterclockwise direction while holding the screwdriver stationary in the screw slot. Loosen the lock nut approximately two turns. Turn the screw in a

clockwise direction while holding the nut stationary until the glow lamp goes out.

5. When the point at which the lamp first goes out is found, hold the screwdriver in a stationary position so that all the slack in the screw slot is held in one direction. This procedure must be followed or erratic results may be obtained.

6. Turn the screw in a counterclockwise direction for one and one-quarter turns. Be careful to apply only slight pressure on the screw; do not press down, since too much pressure can throw the heat adjustment out of the specified temperature range.

7. Finally, hold the screw stationary, and turn the nut in a clockwise direction until it becomes tight.

The high-temperature control setting is made as follows:

1. After making the low-temperature control setting, set the switch lever to the high position. Place the wrench and screwdriver on the high-heat adjustment screw, and loosen the nut by turning the wrench in a counterclockwise direction while holding the screwdriver in a stationary position. Back the nut off approximately two turns. Hold the nut stationary, and turn the screw in a clockwise direction. Be careful to apply only slight pressure to the screw; do not press down until the switch lever snaps to the low position.

2. Try relatching the switch lever in the high position. The lever should not latch now, but you should feel a tendency to latch, if no downward pressure is exerted on the switch knob. After this condition is obtained, the high-heat adjustment screw should be turned in a counterclockwise direction about one and one-half turns. Then, with the screw held stationary, turn the nut in a clockwise direction until it becomes tight. Now check the temperature-control settings.

Checking Low-Temperature Control Setting—The low-temperature control setting should be checked as follows:

1. Fill the lower bowl with water at about 150°F. (cold water

may be used, but warm water will speed up the operation) to the lower edge of the handle screw.

2. Set the control lever to the low position, and connect the unit to the wall outlet for about one-half hour.

3. Insert a glass thermometer in the water. After a steady condition is reached, a temperature of 165° to 185°F. may be expected. If the temperature is less than 165° or more than 185°, the low-heat adjustment screw should be turned in the required direction (clockwise to decrease the temperature and counterclockwise to increase the temperature) and retested until the correct temperature is obtained.

4. One complete turn of the low-heat adjustment screw will produce approximately a 60° change in temperature.

Checking High-Temperature Control Setting—The high-temperature control setting should be checked as follows:

1. The lower bowl and its internal mechanism must be thoroughly cooled prior to setting the control.

2. Place one and one-half ounces of water in the lower bowl. Set the switch to the high position, and connect the coffee maker to a wall outlet. Note the exact amount of time (on a stop watch, if available) that it takes for all the water to be boiled away. All the water must be dried up in the bottom of the bowl, including the bubbles of water in the corner between the sides and the bottom.

3. Then, as the switch is snapped to the low position, check the time on the watch again. The time elapsed between the starting point and the shifting of the switch from "high" to "low" should be approximately 15 seconds. If the time is not within these limits, readjustments may be made by turning the high-heat adjustment screw in the required direction (clockwise to decrease the temperature and counterclockwise to increase the temperature).

After setting and checking the low- and high-heat controls, the coffee maker is ready for the final test. This test can be made with or without coffee. Regardless of the method used, always clean the bowls after testing. The coffee maker operates with water as well as with coffee, except that when the temperature decreases in the lower bowl and the liquid in the upper bowl descends, this liquid will descend more quickly if there is no coffee around the filter cloth. If the setting and checking procedures are performed as outlined above, a properly working coffee maker will result.

Coffee Boils Over—One of the most frequent causes of boiling-over is the result of a vacuum leak. Vacuum leaks may be caused by a faulty rubber seat ring, a leak around the handle-holding screw, a hole in the filter cloth, or an incorrect fitting of the filter cloth when placed on the frame. An excessive high-temperature setting can also cause boiling-over trouble.

AUTOMATIC COFFEE URNS

Fig. 9 shows the disassembled view of a typical 12 to 30 cup coffee urn. This urn is a 120-volt AC/1090-watt unit and is fully automatic with an aluminum body.

OPERATION PROCEDURE

While the coffee urn is cold, the temperature control contacts remain closed and lamp is shorted out of the circuit. The current flows through the pump heater and temperature control and in parallel with the "keep hot" unit. At a predetermined brew temperature, the temperature control opens causing the lamp to glow and the pump heater to cut off. The "keep hot" heater remains in the circuit.

SERVICING PROCEDURE

Unit will not work. Check line cord for continuity. Visually check all connections for loose wires and all connections for corrosion.

Coffee tastes bitter. Inspect the interior for signs of accumulated residue or stains. Clean if necessary.

Coffee too weak or too strong. Check pump for damage and make sure the breather hole is clear. Check temperature control.

Unit reperks. Check "keep hot" for open circuit. Check temper-

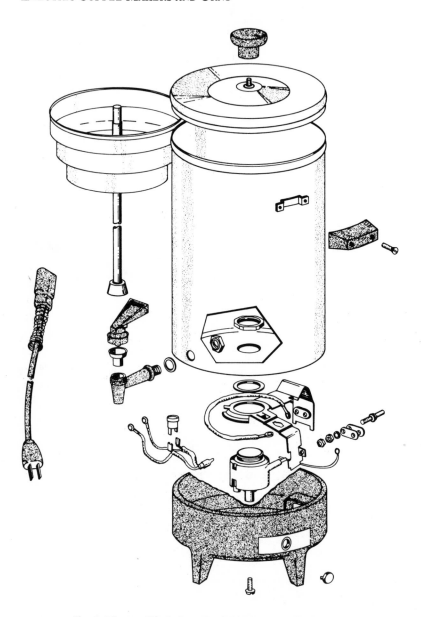

Fig. 9. Disassembled view of a 20 to 30 cup coffee urn.

ature control cut-off temperature. Check bleeder hole in body of pump stem. See Fig. 10 for circuit diagram.

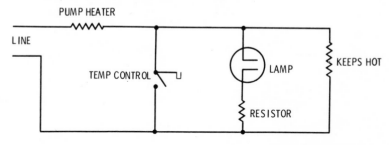

Fig. 10. Schematic diagram of the coffee urn shown in Fig. 9.

Disassembly and Repairs

When replacing any part, carefully position it in the urn exactly in the same position as the old part with particular attention to placement of wire leads. When a defective lamp and broken leads are to be replaced, the complete assembly must be used. Do not attempt to repair the lamp or wiring harness assembly. When testing unit, fill with at least 12 cups of water; insert basket and replace cover. Cut off temperature is generally between 173°F to 200°F.

CHAPTER 12

Electric Space Heaters

Electric heaters, such as shown in Fig. 1, are commonly used in and around the home and are usually called *space heaters*, since their primary function is to heat the air space in the room or area in which they are placed. They are manufactured in various sizes and types to suit different conditions of service and heat require-

Fig. 1. An oscillating-type electric space heater. Heaters of this type are equipped with special oscillating gears, and they rotate back and forth through a certain number of degrees in much the same manner as the conventional oscillating fan.

ments. Irrespective of their construction and size, however, they all work on the electrical resistance principle; that is, a length of resistance wire becomes heated when an electric current passes through it.

While it is true that heating with electricity is an ideal method in that it is available in most sections of the country and is clean and efficient, the cost per kilowatt-hour is too high in most locations for its full utilization in the total heating of homes. Therefore, electricity is used only sparingly as a heat source, and electric heaters are usually employed to provide a supplementary heating means during temporary cold weather spells in homes, summer cottages, camps, etc.

TYPES OF ELECTRIC HEATERS

There are several types of electric space heaters that can be used for general and special heating, as desired. They are:

1. Bowl-type heater (radiant),
2. Convection-type heater,
3. Immersion-type heater.

Depending on the method of air circulation, space heaters are of two general types, namely, natural draft and forced draft. In natural-draft heaters, the air rises by natural draft over electrically heated bars, coils, or wires. The air is heated by contact with the heating element, and by ascending through the natural draft, the heated air distributes itself throughout the room. In forced-draft heaters, the air is blown by an electric fan over electrically heated wires, bars, or coils, which heat the air by contact. The heated air is then distributed throughout the room or area to be heated by the draft of the fan.

Bowl-Type Heaters

This type of electric space heater has obtained its name from the bowl-like shape of its metallic reflector, which is mounted on a sturdy base and is provided with wire guards to prevent accidental contact with the cone-shaped heating element. This is a popular type of heater; it is light in weight and may easily be carried about and connected to any wall outlet as conditions dictate. Modern heaters of this type consist essentially of a screw-in type of heating element. The resistance wire is wound on a cone-shaped insulator and is mounted in the center of the reflector bowl. Because of its parabolic shape, the reflector bowl radiates heat in a cone-shaped wide beam, much in the same manner as light is radiated from a reflector-type lighting fixture.

Convection-Type Heater

This is another type of portable electric space heater in which the heated wire heats the air by convection and radiation. Heaters of this type are built in various sizes and shapes. They consist of perforated sheet-metal cases through which air can circulate over the heating element surface. The warmed air, through convection, is caused to rise, thereby providing circulation of warm air in the room. The heating elements, shown in Fig. 2, may consist of resistance wire wound on cylindrical insulators or resistance wire mounted on special heater strips or bars of suitable wattage, depending on the type of heater in question.

Immersion-Type Heaters

Immersion heaters, often called electric steam radiators, are so designed that the heating units can be placed directly in the water to be heated. They consist essentially of one or more strip heaters that are made of seamless sheets or casings, with the external electrical connections so enclosed that the heater can be placed in the

Courtesy General Electric Company

Fig. 2. A typical automatic space heater with a thermostatic heat control; maximum power consumption is equal to approximately 1650 watts.

water to be heated. When an electric current is passed through the resistance wire, it causes the water to boil. The steam produced by this bodily process supplies the heat that is then radiated through the room to be heated. After the steam has given up its heat, it condenses into water; when this water contacts the heating elements, it is converted into steam once more, thereby completing the conventional heating cycle. All types of immersion heaters may also be equipped with thermostats for automatic temperature control.

Forced-Draft Heaters

Blower-type heaters are used for heating large rooms or locations where warm air must be circulated through greater areas than could be heated by convection and radiation. Heaters of this type consist essentially of one or more heating units and an electric fan, which blows the heated air through the heating units and circulates it in a given area. Forced-draft heaters are available with or without thermostats for room temperature control. Some heaters of this variety incorporate a two-heat switch that permits two different heat-setting selections. An additional feature in most types of forced-draft heaters permits the fan to be operated without actuating the heating elements, thus providing cool air circulation during the summer months.

Operating Voltage

Portable electric space heaters are commonly available for conventional circuits of 115 volts. The amount of heat can be varied by connecting heating elements in series or parallel; however, to do this, each unit must be designed to operate at full-circuit voltage. For example, two similar 115-volt units connected in parallel will produce full heat on a 115-volt circuit, but these same two heaters will only produce one-fourth as much heat when they are connected in series.

SERVICING AND REPAIRS

Electric room heaters of the bowl or convection types are simple in construction. The most prevalent trouble in these two types of heaters consists of an inoperative cord or heating element. In the bowl-type electric heater, the screw-in heater element permits replacement in the same manner as an electric bulb, after the removal of the guard wire. Since the electric current drawn by a room heater

Courtesy Sunbeam Corporation

Fig. 3. A portable convection-type electric space heater.

is several times larger than that consumed by an electric bulb, the screw-in heater element should be carefully tightened to avoid arcing when in use. If this precaution is not taken, the heater element may become welded to its socket, thereby resulting in damage to the heater unit itself. When installing a new element on this type of heater, note the wattage rating of the element in order to obtain the correct heating value. This information is usually available on the heater nameplate.

Replacement of heating elements in convection-type electric heaters, as shown in Figs. 3 and 4, usually requires the removal of the bottom panel. After ascertaining that the element is defective, carefully disconnect the wires from their terminals; be careful not to damage the asbestos or mica insulating washers when removing the element mounting insulators from the heater frame. If any part of the insulating material is damaged, it should be replaced during the reassembly process.

Immersion-type heaters require the periodic addition of water for proper functioning. In locations where distilled water is not readily obtainable, periodic cleaning is necessary to remove scale, which will deposit itself on the inside. Such scale deposits result from the use of impure water and can greatly impair the efficiency of the heater. The manufacturer's instructions with respect to water

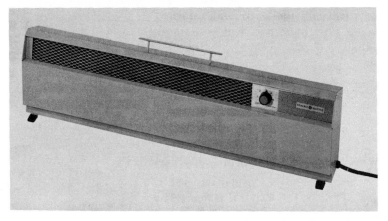

Courtesy General Electric Company

Fig. 4. A baseboard-type automatic electric space heater.

replacement should be carefully followed in each instance. Element replacement in immersion-type heaters differs for various models or types, but an inspection of the assembly will readily reveal the method to be used in the disassembly process. Since the element is assembled within a waterproof jacket or enclosure, it is almost always necessary to use the manufacturer's replacement, which usually includes the complete assembly. Great care should be observed when replacing such an element. All gaskets used must be in good condition; the assembly must also be airtight, since this is one of the fundamental requirements for the production of steam in the heater.

Fan-type, or forced-draft, heaters depend for their proper operation on a motor-operated fan in addition to the heating element.

The fan, motor, or heating elements may easily be replaced in the customary manner. In each case, use identical replacement parts. When adjusting the fan blades, it should be noted that these are carefully aligned at the factory to have a certain predetermined pitch. If it is necessary to remove the fan or install a new fan blade, the track and pitch of the blade should be checked by placing the fan blade face down on a smooth surface and measuring each blade individually to its highest point. After installing the fan on the motor, and before replacing the front shell on the fan, the blades should be checked for their proper track, so that all blades are traveling in the same path. When adjusting the fan track, take care not to get the blades out of pitch.

CHAPTER 13

Electric Water Heaters

Electric water heaters differ from other forms of water heaters mainly in the form of energy used for heating the water. Water heaters are commonly identified by the method of heating. Since the popularity of any water heater depends on the availability and cost of electrical energy used, it is clearly evident that most electric water heaters are found in sections of the country where the cost per kilowatt-hour of electricity is low. The trend toward the use of electric water heaters, however, is constantly increasing, particularly in rural sections where transportation and fuel storage make it convenient to use electricity.

CONSTRUCTION

An automatic electric hot water heater is a comparatively simple appliance and consists of a metal water-storage tank that is heavily insulated to prevent the escape of heat. One or two electric heating elements, with the necessary manual and thermostatic controls, are mounted inside the storage tank. The type of insulation employed may vary, depending on the particular manufacture, although most modern water heaters use Fiberglas or rock wool as the insulating material.

The condition of the available water has a direct bearing on the type of tank used. In general, copper or Monel metal should be used where there is any acid-reactive condition in the water. If the water is hard, however, a heavy galvanized steel tank is preferable and usually provides long and satisfactory service. All tank seams are welded and are usually tested with a hydrostatic pressure of approximately 300 psi (pounds per square inch).

The tank capacity may vary and is directly dependent on the hot water demand; the smaller tanks hold as little as 32 gallons,

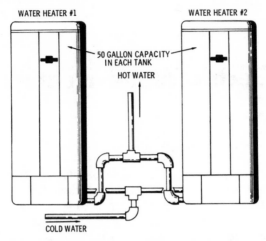

Fig. 1. Two similar electric water heaters connected in parallel. This method is used in locations where one heater is unable to meet the hot water demand.

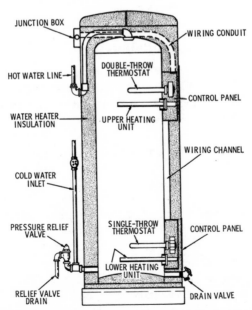

Fig. 2. Construction features of a typical round-shell electric water heater.

and the larger tanks have a capacity of up to 150 gallons. Normally, a 40- to 50-gallon tank will meet the needs of the average-sized home. If hot water demands exceed the normal limit of tank capacity, which is usually about 50 gallons for home application, two such tanks may be connected in parallel, as shown in Fig. 1.

The geometrical form of the external heater cabinet may also differ with the various manufacturers' preferences, varying from a circular to an oval to a rectangular cross section. All water-heater tanks, however, are cylindrical in shape, as shown in Figs. 2 and 3, with a convex bottom that permits complete flushing or draining. Since many water heaters, for economical reasons, are installed in kitchens, the steel cabinets are usually finished with a coating of colored enamel to match any interior coloring scheme desired. The water heater illustrated in Fig. 4 is a typical table-top design for use in the kitchen of a home that does not have a basement or utility room.

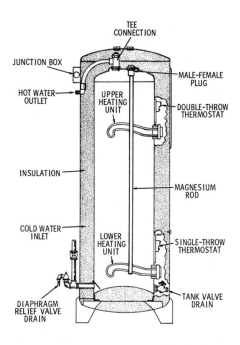

Fig. 3. The principal components of a round-shell electric water heater.

HEATING UNITS

The heating unit, with its interconnected thermostat, is the heart of any automatic electric water heater. All heating units operate on the same principle irrespective of their geometrical form; that is, they all have a heating coil wound on a suitable insulator. The heating coils are shielded from direct contact with the water with a metal (usually copper) watertight enclosure. These metal sheaths are pressed on the heating coils and then brazed on the heating unit flange, as shown in Fig. 5. The fact that the heating units are immersed in water practically assures 100% heat-transfer efficiency.

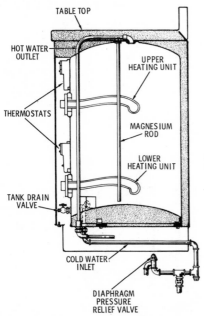

Fig. 4. The construction features of a table-top electric water heater.

Depending on the capacity of the tank, the water heater may be furnished with one or two heating elements. The heating effect, that is, the temperature increase per unit of time for a given volume of water, depends on the wattage of the unit in question. For example, a typical 32-gallon tank may have only one 1500-watt heating element, whereas an 80-gallon tank may have two heating elements with a combined wattage of up to 6000 watts.

Wiring Methods

Electric water heaters may be equipped with either one or two heating units, depending on the particular design of the appliance.

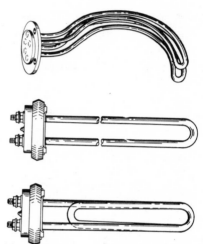

Fig. 5. Typical heating units employed in electric water heaters. Heating elements are commonly named according to their shapes, such as sickle, tubular, life belt, etc.

The circuit shown in Fig. 6A is a schematic representation of a single heating unit and is controlled by one single-throw thermostat. When the temperature of the water falls below a certain predetermined value, the thermostat closes the circuit, thereby energizing the heating element. The heating element then raises the temperature of the water to another preset value and opens the circuit, thus completing the heating cycle.

In electric water heaters equipped with two heating elements, various connection methods are employed. The two common connection methods are known as the nonlimited-demand circuit and the limited-demand circuit. In areas having ample power-generating capacity, the heating elements may be wired to permit the connection of one or both to the line at any time, as shown in Fig. 6B.

Since most power company generating facilities do not permit unlimited use of power for water heating purposes, the amount of current any one heater can draw from the line must be restricted. This restriction usually consists of allowing only one of the heating elements to be connected to the power supply at any one time.

Electric Water Heaters

Since hot water can be stored efficiently, the power companies can use time switches to control the charging periods of the heaters; the water can be kept at a relatively high temperature with only intermittent charging. This type of control reduces the load on the

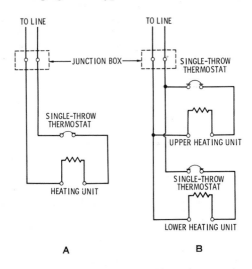

Fig. 6. The wiring connections for a one- (A) and a two-element (B) electric water heater; these connection methods are used only in locations where there is no current limitation in effect.

power lines during the hours when the general current demand is the greatest. For example, in a locality where there is a heavy load on the power lines during the day, service to the water heaters can be disconnected during this period and then turned on at night after the peak demand is reduced. In other localities, the peak demand will occur several times a day, at noon and the early evening hours. Service to the water heaters can be so regulated that heating can be accomplished after these periods are passed. This procedure eliminates the necessity for the power companies to buy and install additional generating equipment in a great many instances. It also permits a lower power-consumption rate per kilowatt-hour for water heaters than for general residential service. The type of service available to the consumer largely determines the size and type of water heater to be installed. Some of the most common types of service offered by power companies are:

1. *Nonlimited-demand service for a single- or double-unit water heater.* In this type of service, the current is available to the heater 24 hours a day. The unit is connected to the line as shown in Fig. 7.

2. *Limited-demand service for a single-unit water heater.* This type of service is illustrated by the wiring diagram of Fig. 8. In this type of service, the water is heated by one heating unit, which is located near the bottom of the tank. A single-throw thermostatic switch is employed to control the water temperature. The flow of electric current to the heater is controlled by a time switch and is available at predetermined intervals only.

3. *Limited-demand service for a double-unit water heater.* A typical wiring diagram representing the connection for this type of service is shown in Fig. 9. The heater tank is filled with cool water; the double-throw thermostat will then complete the circuit to the upper heating element and disconnect

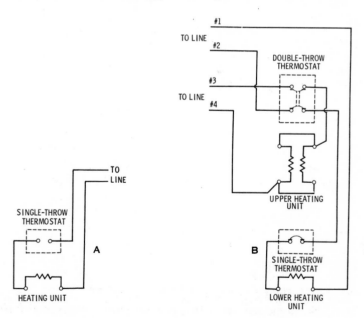

Fig. 7. The wiring connections for a nonlimited-demand service to electric water heaters having one (A) or two (B) heating units.

219

the circuit to the lower heating element. Thus, the upper element provides all the heating during this period. When the water in the uppermost part of the tank has been heated sufficiently, the double-throw thermostat disconnects the upper element and connects the lower element. The lower element then heats the remainder of the water in the tank. When the hot water is withdrawn from the tank and replaced by cool water, the single-throw thermostat connects the lower element, which again starts the heating process. If most of the hot water in the tank is withdrawn, the double-throw thermostat will again connect the upper heating element to the line and disconnect the lower heating element.

It should be observed that the upper heating element has a larger wattage capacity than the lower heating element. For example, in a

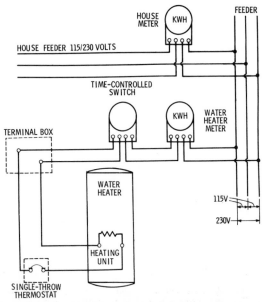

Fig. 8. *The wiring diagram for a limited-demand service on a single-unit water heater. In a circuit of this type, the single-pole single-throw thermostat opens and closes the heating-unit circuit at a specified and preset temperature. The time-controlled switch determines the hours of the day when the circuit is opened or closed, thereby preventing the unlimited use of hot water during the hours of greatest power load.*

typical 40-gallon tank, the wattage rating of the upper unit is 2000 watts, whereas that of the lower unit is only 1000 watts. Because of its larger wattage, the upper element heats its surrounding water much more rapidly and, therefore, acts as a "booster" to serve only when an excessive quantity of hot water is called for.

Voltage and Current

Electric water heating units and controls are usually manufactured for connection to a 115- or a 230-volt single-phase alternat-

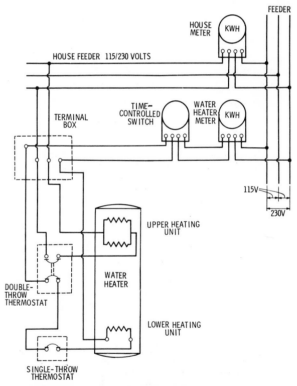

Fig. 9. The wiring diagram for a limited-demand service on a double-unit water heater. The upper heating unit is controlled by a single-pole double-throw thermostatic switch, which has two sets of contacts; one controls the current flow to the upper heating unit, while the other controls the current flow to the lower thermostat. The lower heating unit is controlled by a single-pole single-throw thermostatic switch, which has only one set of contacts that open and close in response to the water temperature in the lower part of the tank.

ing-current line. The amount of current drawn depends on the wattage capacity of the heating unit. On a 115-volt circuit, the current demand may vary from 8.7 amperes for a 1000-watt heating unit to 35 amperes for a 4000-watt unit. It can thus be seen that special wiring is required for all types of electric water heaters, and they may under no conditions be connected in the house wiring circuit, which is used for lighting and small appliance service.

Prior to the installation of an electric water heater, it is recommended that the local power company's representative be consulted with reference to the wiring requirements and other specifications. In some localities, the power company permits the installation of a special watt-hour meter for measurement of the energy consumed by the electric water heater and the electric range combined. This energy is then supplied to the customer at a lower rate than that charged for lighting and small electrical appliances. In any event, all wiring must conform to the requirements of the National Electrical Code and any local requirements in effect at the particular location.

Thermostats

All modern electric water heaters are furnished with one or two thermostats. These function to open or close the circuit at pre-

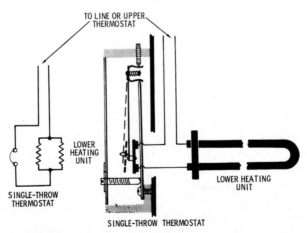

Fig. 10. The schematic arrangement of a single-throw thermostat, with a diagram illustrating the physical connections.

determined water temperatures like any ordinary switch; the only exception is that the switching action is performed automatically. Most thermostats used for water heaters have a range of from 100°F. to 200°F., and they are usually set at the factory to deliver water at a constant temperature of 150°F. at the tank outlet.

Although several types of thermostats are used to control water temperatures, they may all be divided into two general classifications, as single-throw and double-throw thermostats. A single-throw thermostatic switch is illustrated schematically in Fig. 10, with its circuit connections. This switch is connected in series with the lower heating element and has only one set of contacts, which open and close in response to the temperature at the bottom of the water-heater tank.

A double-throw thermostatic switch is shown in Fig. 11. It controls the flow of current to both the upper heating element and the lower thermostat. The double-throw thermostatic switch closes the circuit in the upper heating unit whenever the water temperature in the top part of the tank becomes lower than the thermostatic switch setting, thereby permitting current flow in the upper heating unit only. When this portion of the tank reaches a preset temperature, the switch opens the contacts to the upper unit circuit and, by a toggle action, closes the contacts in the lower unit circuit, thereby permitting current flow in the lower thermostatic switch and heating unit. This is the limited-demand type of hook-up; it means that the greatest current demand at any one time is limited to the wattage of the largest heating unit. At no time with this type of hook-up can both units be simultaneously in the circuit.

HEATING CALCULATION

The cost of electricity in a particular location will largely determine the use of electric water heaters. If the cost per kilowatt-hour of electrical energy is known, it is a comparatively simple matter to calculate the cost of heating a tankful of water. Since the volume of water and its temperature, together with the insulation of the particular installation, may vary, it is evident that only approximate values of the heating cost can be obtained.

Records show, however, that typical families do not use more than 50 gallons of hot water per day. Assuming further that the

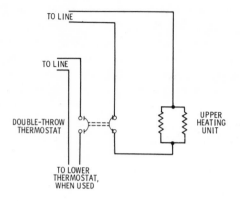

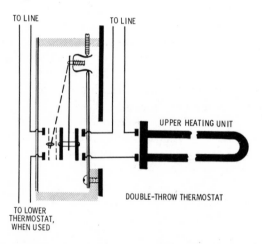

Fig. 11. The schematic arrangement of a double-throw thermostat, with a diagram illustrating the physical connections.

water must be heated from a 50°F. input temperature to a 130°F. outlet temperature, the number of Btu required can be determined by using the following formula:

$$Heat\ (in\ Btu) = W \times 8.34\ (T_o - T_i)$$

where,

    W is the quantity of water, in gallons, to be heated,
    T_o is the output temperature, in degrees Fahrenheit,
    T_i is the input temperature, in degrees Fahrenheit.

The number *8.34* is the factor for converting gallons of water to pounds of water. Therefore, the amount of heat required for this example is

$$Heat = 50 \times 8.34\,(130 - 50) = 33{,}360 \; Btu$$

Now, since one kilowatt-hour of electrical power is equal to 3413 Btu, the number of kilowatt-hours required to raise 50 gallons of water 80° is

$$\frac{33{,}360}{3413} = 9.8 \; kilowatt\text{-}hours$$

If the cost of electrical energy is from one to one and one-half cents per kilowatt-hour, the toal cost of operating the electric water heater for a period of one 30-day month will be from $2.94 to $4.42, depending on the local rate per kilowatt-hour.

CHAPTER 14

Gas Water Heaters

Gas water heaters can be divided into groups or classes, depending on the method used for control of the heat generated by the gas burner, as:

1. Manual.
2. Automatic.
   a. automatic storage.
   b. automatic instantaneous.

MANUAL GAS WATER HEATERS

Although manual gas water heaters are not used presently to any great extent, their operation, installation, and connections are shown in order to be of assistance in locations where such installations are found. Manual gas water heaters (alternately called circulating-tank or side-arm heaters) are of conventional design; the gas burner and accompanying heating coils are mounted on the side of the hot water storage tank. The manual water heater is used mainly because of a comparatively low initial cost and economy in operation. They are, as the name implies, nonautomatic and can supply hot water quickly by igniting the gas shortly before the warm water is required. For this limited service, the operating cost is relatively low. However, when they are used continuously at a rather low temperature setting, much of their economy is lost.

In order to facilitate service when the heater is remotely located (such as in the basement of homes), special clocks, or timing devices, as well as distant control lighters are available. Manual water heaters are generally equipped with copper coils 16.5 to 20 feet in length, with an outside diameter of either ¾ or 1 inch. These coils are usually made of No. 20 Stubbs gauge copper tubing. Other designs are occasionally employed; for example, internal or underfired units are used to overcome liming in hard-water territory. These heaters usually have between a 20,000- and a 30,000-Btu capacity, although the largest size has an 85,000-Btu capacity. The smallest manual water heater can deliver approximately 19 gallons of hot water per hour; this size heater is generally ample for most homes in which the conventional 30-gallon boiler is used for storing the hot water.

A great many installations of the manual type of water heater have given poor service, not because of any fault in the heater but mainly because of the improper method of connecting the water heater to the storage tank. This lack of general good service can largely be overcome, if a few simple installation rules are followed. All boilers of recent manufacture are provided with taps for hot and cold water connections; there are two taps located in the side

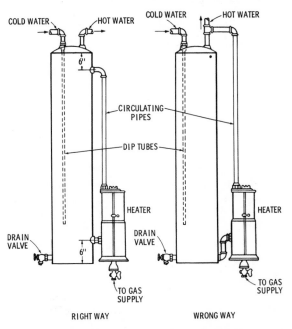

Fig. 1. The right way and the wrong way of connecting hot and cold water taps on a manual water heater.

of the tank six inches from the top and bottom, as shown in Fig. 1, to accommodate the circulating-water connections, which should be made of ¾-inch or larger pipe. The side of the tank near the bottom has a tap for connection to the drain, or blow-off, which is used to drain water or sediment from the boiler. Circulation pipes between the heater and tank should be made as free from fittings and bends as possible.

Brass pipe should be used on the hot water circulating line from the heater to the tank, particularly for high temperature circulation. Unions, when required, should be placed as close to the heater as possible on the hot and cold water circulating lines. The arrangement of hot and cold water taps at six inches from the top and bottom of the tank permits free circulation, which produces a relatively large volume of stored water without overheating. This arrangement also eliminates short circuiting of water through the heater and provides ample sediment storage below the circulating line, thus preventing sediment from entering the heating element.

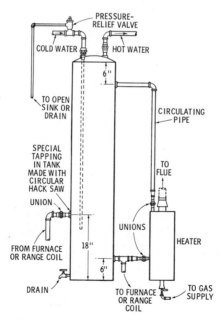

Fig. 2. The external taps on the water heater must be modified when connecting the water heater to a range coil or furnace.

Fig. 2 illustrates the proper method of connecting a water heater to a range coil or furnace.

Manual gas water heaters must be installed in accordance with their listing and the manufacturer's instructions. Unlisted water heaters must be installed with a minimum clearance of twelve inches on all sides, including the rear. Combustible floors under unlisted water heaters must be protected in an approved manner. Gas water heaters must not be installed in bathrooms, bedrooms, or any occupied room that may normally be kept closed. All gas water heaters with an input in excess of 5000 Btu must be vented or provided with some other means for removing the flue gases to the outside atmosphere.

AUTOMATIC GAS WATER HEATERS

Automatic gas water heaters are those in which the hot water tank, the gas burner, the combustion chamber, and the necessary

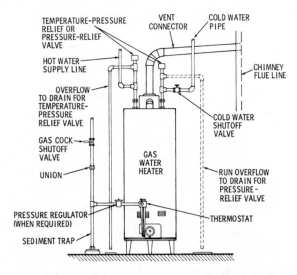

Fig. 3. A typical automatic gas water heater installation.

insulation are combined in a single unit with provisions for an automatic supply of hot water. Depending on such provisions, gas water heaters of this type are usually called *automatic storage* or *automatic instantaneous*. The installation of an automatic gas water heater is illustrated in Fig. 3.

Automatic Storage Gas Water Heaters

As the name implies, these heaters are equipped with provisions for hot water storage. Storage tank sizes vary, depending on their requirements; for single family use, however, a tank of approximately 75 gallons storage capacity is usually sufficient.

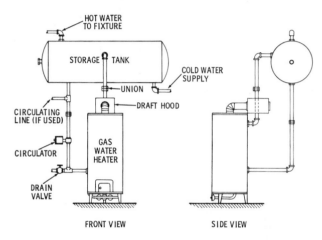

Fig. 4. The installation of a typical automatic storage gas water heater.

The storage tank heating elements and controls are usually delivered separately. The installation involves setting the tank, installing the copper-tube heating coil and thermostatic heat control unit in the tank, connecting the hot water supply line, and insulating the unit. This procedure is shown in Fig. 4. In this type of heater, the heat from the gas flame is transmitted to the water by direct conduction through the tank bottom and flue surfaces. Some heaters have multiple central flues, while in other designs the hot exhaust gases pass between the outer surfaces of the tank and the insulating jacket. In either case, these areas become radiating surfaces and dissipate the heat of the stored hot water to the flue or chimney when the burner is off. This is particularly true if the flue or chimney has a good natural draft.

Automatic gas water heaters can also be connected in tandem for a single storage tank, as shown in Fig. 5. A two-temperature hot water service can be obtained in a system of this type by means of a mixing valve or by using an automatic storage unit in conjunction with the circulating heater and storage tank.

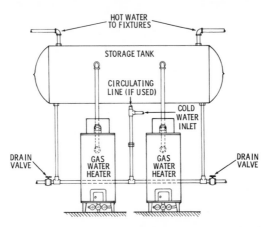

Fig. 5. A tandem gas water heater arrangement for a single storage tank.

Slow-Recovery Heaters—The gas-fired slow-recovery water heater is designed to keep a constant supply of hot water in the storage tank; by means of a constantly burning gas flame, it can continuously deliver hot water to this tank. Thermostats employed in this type of water heater are of the graduating or snap-action type. With the graduating-type thermostat, the burner operates between a low and high flame; the low position of the flame acts as a pilot to keep the heater lighted and also serves as a source of standby heat. The slow-recovery type automatic water heater is extremely economical, since it can never burn more than a certain amount of gas, depending on the regulation offered by the thermostat. The small amount of heating surfaces keeps the standby loss at a minimum, thereby making this type of heater advantageous where economy is the primary consideration.

Quick-Recovery Heaters—This class of heater has the ability to produce hot water at a more rapid rate than the slow-recovery type. It is frequently employed where repeated heavy requirements for hot water make it essential to have a sufficient amount of water available on demand. The thermostats generally used are of the throttling or snap-action type. Primarily, the throttling-type thermostat is one in which the amount of gas-valve opening is directly proportional to the temperature changes of the water in the tank. In the snap-action type of thermostat, the change from a completely open to a completely closed valve position, or vice versa, is accomplished by a snap action, which is produced by a clicker diaphragm that is motivated by the temperature of the water in the water-heater tank.

The essential difference between the slow- and the quick-recovery type of water heater lies in the amount of gas consumed by each unit. For example, a quick-recovery heater with a 25,000-Btu input can deliver 23 gallons of hot water per hour indefinitely, whereas a slow-recovery heater can never burn more than a relatively low and predetermined quantity of gas, with an accompanying reduction in hot water delivery.

The average ratio of hourly gas input to the storage capacity in gallons for gas-fired automatic storage water heaters of the quick-recovery type is such that the recovery capacity in gallons of water raised 100° in one hour, in most instances, approximately equals the storage capacity of the tank in gallons. If the water must be raised by 120°, the recovery capacity in gallons per hour will be approximately 83% of the storage capacity in gallons; where the water must be raised 140°, the recovery capacity in gallons per hour will be approximately 71% of the storage capacity of the water heater.

Automatic Instantaneous Gas Water Heaters

These heaters differ from the previously discussed types mainly in that they do not store any heated water; they must therefore heat the water directly on demand. Instantaneous-type water heaters are actuated by flow valves or water motors that actuate the gas flame whenever water is drawn from the faucet. Gas burners for this type of service are usually built up of several parts, with a hood to protect the burner from condensation and to direct and hold the flames. In addition, burners are usually equipped with a gauge to protect the flame from backflash. Since instantaneous-type water

233

heaters operate by a water motor or flow valve, they have burners whose gas flames go up and down rapidly. Therefore, a careful pilot adjustment is necessary to insure effective burner ignition without extinguishing the pilot flame. In this type of heater, the safety pilot flame must be adjusted so that the flame heats the element sufficiently to keep the valve wide open. For proper functioning, a pressure difference of about 15 psi in the water piping is necessary to hold the flow valve or water motor wide open. The Btu input to the heater can be checked with water flowing freely from a sufficient number of faucets to keep the water temperature down and the thermostat wide open, to assure a wide-open position of the flow valve.

In a properly adjusted burner, the flame should not reach higher than the bottom of the third coil when the proper Btu input is supplied. If the gas continues to burn too long after the hot water faucet is shut off, either low water pressure or excessive friction in the water valve should be suspected. Friction is usually due to a tight fitting plunger or to particles of rust that become lodged in the water chamber. Although the plunger should fit closely, it should not bind. High spots on plungers can be smoothed off with an extra fine grade of emery cloth or a fine file. If the water from the heater is not hot enough when the gas consumption is adjusted to the manufacturer's rating, and the thermostatic control is correctly set, the cold water regulating valve at the heater should be throttled until the water stays hot under a continuous flow.

Proper functioning of an instantaneous water heater does not, however, assure satisfactory performance of the installation, if the actual recovery capacity of the heater (expressed in terms of the rate of hot water flow that the heater can supply per minute based on the number of degrees that the water must be raised in temperature) is inadequate to supply the hot water demand, which may originate in the simultaneous requirement for hot water from two or more fixtures or outlets. The necessity of restricting the hot water flow from the heater to the maximum rate available, based on the number of degrees that the water must be raised in temperature, cannot be too strongly emphasized. An insufficient hot water temperature will inevitably result if the rate of flow is not so restricted. The operating instructions of the manufacturer for this type of heater should carefully be adhered to.

234

GAS WATER HEATER CONTROLS

Thermostat Operation

The thermostat most commonly used in gas water heaters is the snap-action type. A snap-action thermostat is one in which the thermostat valve closes or opens instantly. The snap action is secured by the overcentering movement of a concave disc, or clicker, in conjunction with the movement of the Invar rod or other actuating mechanism. A typical snap-action thermostat operates as follows:

The copper tube, when cold, is in the contracted position, and presses the Invar rod against the clicker. This produces the snap action, which, in turn, holds the valve in the open position and allows the gas to flow to the burner. As the copper tube is heated, it expands, thus withdrawing the Invar rod, which is screwed into the end of the copper tube. Pressure is relieved from the clicker until the clicker passes the snap point and closes the valve. As the copper tube cools, the procedure is reversed; the copper tube contracts, causing the Invar rod to exert pressure against the clicker and open the valve.

For the thermostat to operate properly, the valve gap must be adjusted so that the valve does not open or close while the clicker is being moved by the Invar rod. For example, if the valve stem, which is the adjustment for the gap, is too short, the thermostat graduates off; that is, the gas flow through the burner slowly diminishes until the burner is completely shut off. If the valve stem is too long, the converse occurs; the thermostat graduates on, thereby allowing the gas flow through the burner to gradually increase until the burner reaches its maximum input. This type of operation puts a tremendous burden on the gas burner, because it means that the burner must operate successfully from practically no input to a maximum-input condition. If the thermostat dial is turned to a higher setting, the Invar rod is shortened so that the copper tube must travel farther to close the valve; by turning the thermostat dial to a lower setting, the Invar rod is lengthened so that the copper tube does not have to travel as far to close the valve. Therefore, the farther the copper tube has to travel before closing the valve, the longer period of time the gas must flow to the burner, and the hotter the water becomes. The thermostat has other functions not mentioned here; this explanation covers only that function which the thermostat accomplishes to control the water temperature in the heater.

Pilot-Thermocouple Operation

The function of the pilot in a gas water heater is twofold, to ignite the burner and to heat the thermocouple. The purpose of the thermocouple, in turn, is to supply a low-voltage electrical current to the magnet in a safety switch, valve, or thermostat. The operation of the pilot-thermocouple combination is essentially as follows:

When two dissimilar metals are tightly fastened together, and heat is applied to one end, an electrical current is produced. The thermocouple, usually made of an iron-nickel combination, is bonded together at one end to form a "hot" junction. Two leads are then attached to a coil which is wound around an iron core. When the pilot heats the thermocouple, an electric current is produced; this current then energizes the coil on the core, thereby producing a miniature magnet.

Thermostat Relation

When the thermostat button or dial is actuated, the magnet (located inside the thermostat) is pushed forward and engages a safety catch. The safety catch is then drawn back when the button or dial is released. The spring-loaded safety catch snaps back in position if the pilot-thermocouple combination does not produce enough current to create the necessary magnetic holding power. When the safety catch is in the "off" position, pressure is applied against the main valve. The main valve then closes and stops the flow of gas to the main burner. When the safety catch is held away from the main valve, the valve will open and close as the element tube in the tank contracts and expands in proportion to the temperature changes of the water, thereby allowing gas to flow to the main burner.

INSTALLATION OF WATER HEATERS

When a water heater is delivered to a customer's home, it must be properly assembled in accordance with the instructions furnished by the manufacturer. Care must be taken to see that any valves or parts which must be assembled on the premises are placed in their correct positions, as indicated by the stamped markings on such parts.

Location of Heater

The first factor to consider when planning the layout for a heater installation is the location of the heater itself. This is of paramount importance from the standpoint of economical performance. The heater should be located as close to the point of greatest use as possible; this is usually in the basement, if the house has one, as near as possible to a direct line beneath the kitchen sink; however, convenience of installation is normally the dictating factor for locating the heater. Since the kitchen hot water faucet is opened on the average of twenty times a day, the loss due to these frequent draw-offs can be materially increased in direct proportion to the distance between the heater and the sink.

Pipe Size and Insulation

When considering economy of operation, attention must also be paid to the size of the hot water pipes. In many homes, ¾-inch pipe has been used throughout. Although ½-inch pipe will ordinarily supply sufficient water for two faucets, the larger ¾-inch pipe, particularly when exposed, can be replaced, thereby resulting in lower operating cost. Copper tubing, which usually can be installed more easily than regular piping in existing houses, may also be used. A typical piping arrangement is shown in Fig. 6; individual tube connections (solid lines) to fixtures are compared to a conventional pipe installation (broken lines). Small diameter pipe or tubing will reduce heat losses to a considerable extent, since the amount of water stored in the pipe or tubing will be less than that contained in standard size pipe.

Copper tubing or iron pipe of a ⅜-inch diameter is adequate for hot water lines supplying only one fixture. For wash-basin faucets or runs not exceeding 20 feet to other types of faucets, ¼-inch pipe

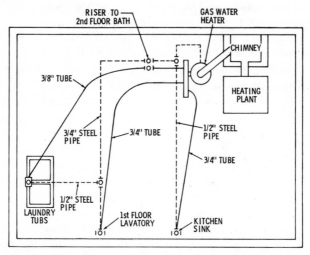

Fig. 6. *A comparison between the standard iron pipe or steel pipe and a copper tubing installation; by using the tubing, the distances between the water heater and the various faucets can be substantially decreased; thereby decreasing the heat loss in the stored water.*

or tubing is sufficient. All exposed hot water pipes should be insulated with a good grade of 1-inch commercial insulation.

Tempering Tank

A tempering tank that is properly installed in the incoming cold water line, as shown in Fig. 7, can aid in reducing operating costs. The purpose of the tempering tank is to preheat the incoming cold water by utilizing the ambient heat in the basement or place of installation. The use of a tempering tank also shortens the time required to raise the water to the preset temperature after some has been used and thereby increases the recovery capacity of the heater. A tempering tank can be any type of *uninsulated* tank installed in the cold water line ahead of the water heater. The tempering tank should have a capacity of at least 30 gallons; it should also be uninsulated, so that the cold water can absorb heat from the surrounding air and in this manner reduce the temperature differences between the hot and cold water. If, for example, the cold water entered the home at 50°F., and after standing in the tempering tank, the water increased its temperature to 70°F., the water heater would only have to raise the temperature of the water in the heater 60°F. to reach a constant temperature of 130°F.

Pressure-Relief Valves

These valves are used to relieve excessive pressure in the heating system. One of the physical characteristics of water is that it is incompressible. For example, when water is heated from 60°F. to 212°F., it will expand to slightly more than 4% of its original volume. As a safety factor, some type of relief valve must be installed to take care of this expansion in case of thermostatic-switch failure. A pressure-relief valve is always installed in the cold water inlet as close to the heater as possible, as shown in Fig. 7. A check, or shutoff, valve should never be installed between the pressure-relief valve and the heater. The diaphragm-type pressure-relief valve gives more positive and satisfactory operation than any other type; local plumbing codes usually specify which type of valve to use. All relief-valve outlets should be piped to the nearest open drain or sink. A physical connection should never be made between this drain and the sewage or waste pipes; this precaution is necessary to prevent pollution of the water system in the house in the event of a build-up in sewer back-pressure.

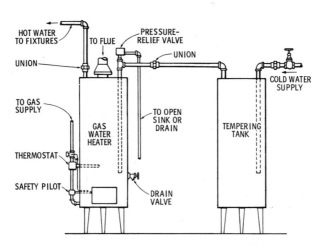

Fig. 7. The connection of a tempering tank to a conventional automatic gas water heater.

Flue Pipe

The size of the flue pipe should not be less than that specified by the manufacturer, or that shown in available tables for the rated gas input. Horizontal runs of the flue pipe should pitch upward toward the chimney connection and should run as directly as possible to avoid unnecessary bends or elbows. The back-draft diverter (usually supplied by the manufacturer) or some other approved draft hood of adequate size should be employed.

Many building codes prohibit the connection of appliances to a common flue or chimney with coal- or oil-fired equipment. If local regulations do not require the venting of gas-fired equipment to a separate flue, and this equipment is vented to a common flue with coal- or oil-fired units, the flue pipe of the gas-fired water heater or other appliance should be connected to the chimney at a point *above* the flue pipe of the coal- or oil-fired equipment.

Meter and Supply Line

The gas meter must be of adequate size or capacity to supply not only the requirements of the water heater but also the requirements of all other gas-fired equipment on the premises. The gas-supply line to the water heater, as illustrated in Fig. 8, should be adequate in size to supply the full rated gas input of the heater at the available pressure, taking into consideration the pressure drop through the supply line. A separate gas-supply line from the meter to the water heater should be employed if the existing line from the meter is too small to supply the combined requirements of the water

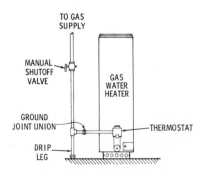

Fig. 8. A typical gas piping installation for an automatic water heater.

heater and such other equipment or appliances as may be connected to it.

Lighting and Operating Instructions

These instructions are supplied by the manufacturer. They should be read and thoroughly understood before attempting to adjust the gas-input rate, the thermostat, and/or any other automatic safety controls, especially if the serviceman is not completely familiar with the particular design and construction of the controls. Before leaving the premises, it is imperative that the serviceman hang the operating-instructions card at a point convenient to the heater where it will be available for ready reference whenever needed by the owner or serviceman.

SERVICING AND REPAIRS

An automatic hot water heater, whether electric or gas, is one of the most trouble-free home appliances when properly installed. A heater upon arriving from the manufacturer may be perfect, and yet if certain initial installation procedures are not followed, or if other minor functions are not performed, the heater will not operate satisfactorily. The most recurrent service complaints are treated in the following sections. When a serviceman is called on to remedy such complaints, he should first check for electrical or mechanical defects in the heater or its controls.

Not Enough or No Hot Water

The first solution to the problem of insufficient hot water appears to be increasing the temperature setting(s) of the thermostat(s). Frequently a slight increase in temperature will correct this condition; however, it is not a good practice to raise the tank temperature beyond 160°F. If the tank temperature is satisfactory, but the water arriving at the faucet has lost too much heat, the heat deficiency may be caused by a too extensive run of pipe. Often this condition can be corrected by relocating or insulating the exposed hot water pipes. This condition may also exist because of an undersized installation.

In an automatic gas water heater, the foregoing condition may be caused by an extinguished pilot flame. If the pilot flame is extinguished, the heater will shut itself off, due to the function of the safety pilot, and additional water will not be heated until the pilot is relighted. Check and clean the pilot filter, the pilot orifice, and the gas tube to insure a clear gas passage free from all foreign material. Check the pilot flame for correct adjustment, as shown on the instruction plate. Overfiring the heater (too much gas) will cause the pilot flame to be smothered and eventually extinguished. Determine the actual gas flow, or input, entering the heater, and compare it with the input specifications for the particular model on the rating plate. An excessive down draft from the chimney can also extinguish the pilot if the down-draft diverter (supplied with the heater) is not installed; this diverter is extremely necessary for a good installation.

Water Too Hot

When this complaint is received, the condition is usually due to an improperly adjusted thermostat, or it may also be caused by the double-throw thermostat sticking in the closed position. If the double-throw thermostat is stuck in the closed position, the upper heating unit will operate all the time, and the lower unit will be completely cut off. The top quarter of the tank will then become dangerously hot, while the bottom three quarters will remain relatively cool. Since only one-fourth of the tank is hot, the amount of hot water available at any one time will be strictly limited, which will account for the excessive water temperature.

Excessive Cost

High bill complaints are perhaps one of the most common and also the most difficult of all complaints to satisfy. Many of the causes for high bill complaints can be traced to an improper selection of a heater for the hot water requirements of the house. Also, when a family has an automatic hot water heater installed and hot water available all the time, they almost invariably use from two to four times as much hot water as they formerly did. This, in itself, constitutes a major reason for such a complaint. Sometimes a user will demand higher water temperatures than are actually necessary. Temperatures above 140°F. are rarely necessary, and temperatures as low as 120°F. are normally highly satisfactory for most families. The water temperature may easily be checked by holding a thermometer under an open hot water faucet. If the temperature exceeds those mentioned, the thermostat setting(s) can be lowered, and the operating costs can be reduced in proportion.

Another common cause for a high bill complaint is leaking hot water faucets. A small leak can pass a lot of water in 24 hours. Since these losses are continuous, it can easily be seen that one or more leaking hot water faucets can cause a considerable loss in hot water over a period of time.

In larger homes, one common cause for a high bill complaint is the existence of a circulating line, which continuously cools some of the stored water in order to provide instant hot water service at every faucet. In this case, individual runs of copper tubing to each faucet are a better way of providing the same service. If a circulating line is suspected as the cause of a high bill complaint, the line should be closed entirely or at least throttled to a minimum.

The following points may also be helpful in seeking a remedy for high bill complaints:

1. Can spring-type faucets be employed in order to make it necessary to hold the hot water faucet open whenever water is desired, rather than letting the water run? Boarding houses, clubs, restaurants, etc., can often use this type of faucet to reduce bills. These faucets cannot, however, be used for showers, bath tubs, or kitchen faucets.
2. Are tanks and water pipe lines properly insulated?
3. Is the thermostat set at the temperature that will produce the lowest bills?
4. Are appliances that use large quantities of water, such as laundry tubs and washing machines, being used wastefully?
5. Is one pipe riser too large? In some houses, especially the larger ones and two-family houses, there may be an excessively large internal riser, which might permit too much circulation.
6. Are there long runs of piping exposed to cold outside building walls?
7. Will a heat trap on the riser help reduce internal circulation?

CHAPTER 15

Electric Ranges

Great strides have been made over the past few years in electric ranges. Such changes include more efficient heating elements, which gives better cooking control, and the development of such features as self-cleaning ovens, fingertip controls and designs to fit almost every possible kitchen need.

One such modern range is shown in Fig. 1. This is the "Counter That Cooks," developed and manufactured by *Corning Glass Works,* along with the double built-in self-cleaning electric oven. The "Counterange," shown in Fig. 2, is the slip-in style which is designed to match your kitchen decor.

Various colors and styles are available in these ranges which features the beautiful glass-ceramic cooktop. You cook on a smooth tough sheet of glass-ceramic material with no coils contacting the pots and pans. A sunburst design on the glass top indicates the cooking area. The heating elements are hidden underneath, yet they are easily accessible from below if any element needs servicing. Spills and splatters on non-heated areas wipe up quickly. For cooked-on spills, simply wait until the cooking area cools, apply a silicon-base cleaner and rub the spot away with a damp paper towel.

There are two ways to tell when any cooking area is operating. First, each cooking area has its own signal light. Secondly, if you are working at a moderate to high heat, you will notice the sunburst area turn yellow and returns to its original color when cooled. Since almost all of the heat is confined to the sunburst area above the heating element, any part of the cooking surface not in use becomes extra counter space.

Fig. 1. A modern kitchen with a "Counter That Cooks" electric range and a double built-in oven.

HEATING ELEMENTS

Surface elements are made up of a resistive ribbon winding on a mica card. The center of each card has a long narrow opening for the sensor bulb to fit. Brass eyelets are used to support the hold down assembly. See Figs. 3 and 4.

The surface controls are hydraulic type that controls the surface temperatures within $\pm 05°F$. of a given reference setting. The surface control operation is as follows:

Bulb, capillary and reservoir contain silicon oil which expands when heated, pushing the plunger up and breaking the contact on the points. As the oil cools, the plunger retracts and the contact points close to complete the circuit.

The control shaft first releases a mechanical cam which allows the hot side of the contacts to make, then allows the cycling contacts to make and adjust the distance the plunger must travel to open the circuit on the control contacts. When

Courtesy Corning Glass Works

Fig. 2. A "Counterange" built-in electric range.

the shaft is turned "On", the cam lifts up and the contacts close.

PERFORMANCE CHECKS

The following checks must be completed following the installation of new elements, controls and ceramic-heater panels and are used in troubleshooting to determine various problems and complaints.

Boil Time

The boil test is run from a hot start. Turn element to be tested "On" to full high. Also, turn the other element on the same side to full high. Turn large element on the opposite side to full

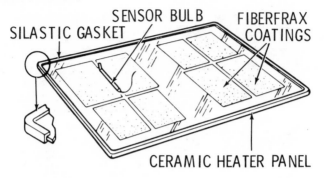

Courtesy Corning Glass Works

Fig. 3. Sensor bulb assembly.

high. (This procedure will help determine if a loose neutral is present as well as checking the performance).

Use a *COOKMATE* 4 quart dutch oven for the large element and fill with 3 quarts of room temperature water (73°F. to 77°F.) Cover and place the vessel on the element and time it until the water

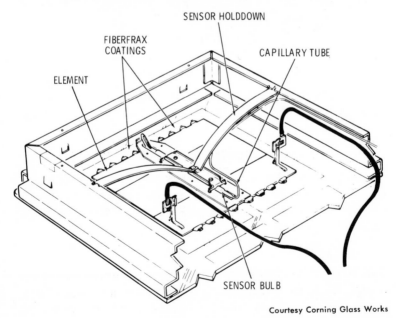

Courtesy Corning Glass Works

Fig. 4. Surface element assembly.

starts to boil. Note the boil time and compare it with the boil time limit chart. For small elements, use the *COOKMATE* 2 quart with 3 pints of room temperature water (73°F. to 77°F.).

Cover the place the *COOKMATE* on the preheated element and time the water until it starts to boil. Note boil time and compare with boil time limit chart.

BOIL TIME LIMITS FOR 120 LINE VOLTAGE

Small Element	14 min., 00 sec.
Large Element	13 min., 00 sec.

In the event the boil time exceeds the above limits, check the line voltage and compare it with the chart below. Boil time depends on line voltage. If the line voltage is found to be correct, check the voltage across each element with any three elements turned on. The voltage should be 120V to each. If one element has a low voltage, a loose neutral is present (see Troubleshooting Guide).

LINE VOLTAGE	SMALL ELEMENT	LARGE ELEMENT
105	16 min., 30 sec.	15 min., 30 sec.
110	15 min., 30 sec.	14 min., 30 sec.
115	14 min., 30 sec.	13 min., 30 sec.
125	13 min., 30 sec.	12 min., 30 sec.

Units which fail the boil time chart will require re-assembly of the element card and sensor location. If assembly is correct and element failed boil time replace the control.

Calibration Response Test

When the boil test has been satisfactorily completed, turn the dial back to the "LO" setting. Wait 6 minutes, boil should cease. This test verifies the integrity of the control system and will identify any runaway controls. In the event the water continues to boil with control turned to the "LO" setting, the control is defective and must be replaced.

Caution—Interior parts will be very hot. Open and allow parts to cool. Check placement of mica element card and sensor bulb. Ensure that the capillary tube is dressed correctly and springs on hold-down are correctly placed. Retest.

COUNTER TOP SAFETY TEST

To guarantee that each appliance is well insulated, every unit should be given a high-voltage test. This is to guard against element ribbons running too close to thermostat bulbs as assembled.

A safety test is given to make certain that the frame of the unit is well grounded after repairs have been made. With proper grounding, the user is completely protected from electrical shock even if insulation is inadequate due to improper reassembly.

In other model electric ranges (also including the earlier models) heating is accomplished by connecting a suitable resistance across an electric potential, thus causing a current to flow. Heat variations are usually obtained by connecting two or more elements in series or parallel and/or by varying the voltage supplied to the elements.

Resistors, or heating elements, as shown in Fig. 5, can be made from various materials, either metallic or nonmetallic, and are selected because of their advantages under certain conditions. Alloys, however, are almost always used in heating elements. For heating elements that are surrounded by air and where temperatures in excess of 1900°F. are not required, a composition of chromium and nickel has been found to be the most durable resistor. Nickel-chromium resistors, when operated within their designed temperature limits, will last for years without any noticeable deterioration.

CURRENT AND VOLTAGE

Electric heating devices may be operated on either direct or alternating current. The heating-element design, however, is not the determining factor as to what type of current the range can be operated on; this is regulated by the type of controls, namely, the switches, thermostats, and timing devices. Generally, electric ranges cannot be operated on direct current, because the switches and thermostats are designed with silver contacts, which only separate by a small gap to open or close the circuit; these contacts would melt together if they were used with direct current. In addition, the timing device (usually an electric clock) is designed for operation on alternating current.

With alternating current, the problem is somewhat different. The electric range can be operated on any frequency that has the same voltage as that for which the unit is built. For frequencies other than 25, 50, and 60 hertz, however, it woud be impossible to

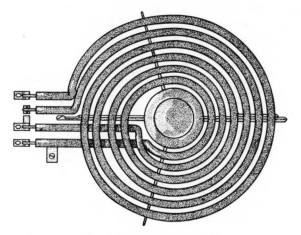

Fig. 5. A typical surface-unit heating element.

install an electric timing device, because they are generally not available for other than these three frequencies.

Power Company Supply Circuits

Lighting power supplies for use in homes are practically all three-wire, single-phase circuits, as shown in Figs. 6 and 7. The ordinary two-wire house lighting circuit is derived from this by using one hot lead and the neutral, or grounded, wire.

A load of 1750 watts is not ordinarily permitted on a circuit of this type, and consequently electric ranges are not available for a two-wire circuit. The reasons for this are:

1. The voltage drop in the lines would be great enough to cause improper operation of the range.
2. The general lighting would be affected.
3. The wiring would not be heavy enough in most cases to carry the load.

There is, however, a distinct value to the use of a three-wire, single-phase circuit; namely, an advantage can be taken of the fact that there are two separate voltages available. Thus, it is possible, by proper switching arrangements, to obtain two different voltages (heating values) from the same heating element. Because most electric ranges take advantage of two different voltages, they cannot be used on either a 230-volt, two-wire, single-phase system or on a 230-volt, three-wire, three-phase system such as is commonly used for power. A circuit known as the 120/208-volt, four-

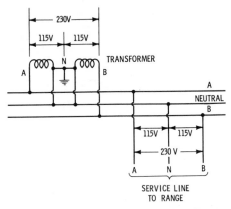

Fig. 6. A three-wire, single-phase, AC power-distribution system with a grounded neutral that is obtained from the transformer secondary. This system is commonly used to provide 115/230 volts on modern electric ranges, thus giving two voltages with one-half the maximum voltage to ground, or neutral.

wire, three-phase, or network, system is being used in some cities to particularly replace the older direct-current systems. The load is connected to this system as follows:

1. General lighting at 120 volts is connected between one phase wire and neutral.
2. Single-phase motor loads may be connected to the circuit by using any two of the phase lines.
3. Three-phase motor loads are connected to the circuit by using the three phase lines.
4. Electric ranges are connected to this system by using the neutral and any two of the phase wires.

The serviceman should obtain an explanation of the types of power-supply systems used in his area from his local electrical

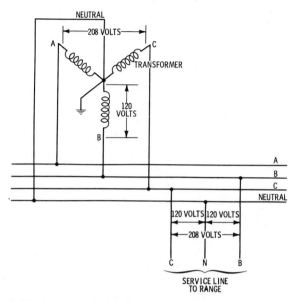

Fig. 7 A Wye-connected, four-wire, three-phase, AC power-distribution system with a grounded neutral. This type of system provides 120/208 volts for electric range service.

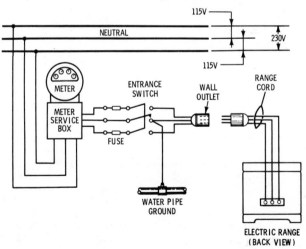

Fig. 8. The wiring method for a single range installation when connected to a 115/230-volt AC distribution system.

utility or power company, so that he might better understand the particular network system with which he will be working.

Power Line to Range Wiring

All wiring to an electric range must be in accordance with the requirements of the National Electrical Code, in addition to any local requirements in effect in the locality of the installation. In the case of a single range installation, as shown in Fig. 8, a 115/230-volt range as a rule requires two No. 6 wires, with a No. 8 neutral or ground wire. In the case of multiple range installation, the wiring should be carefully calculated to avoid excessive voltage drops, since the proper operation of each range depends on the assumption that the voltage supplied to each range under all load conditions corresponds to that given on the range nameplate.

ARRANGEMENT OF UNITS

Depending on the size of the electric range, the heating elements may conveniently be divided into two parts or units, namely, surface units and oven units.

Surface Units

Surface units, as shown in Fig. 9, are located on top of the range, and each is controlled by an individual switch located conveniently on the range. The number of surface units may differ, depending on the size of the range. Some standard ranges, however, are equipped with four surface units of which one is (in some cases) a deep-well-type cooker.

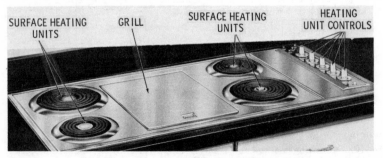

SURFACE HEATING UNITS   GRILL   SURFACE HEATING UNITS   HEATING UNIT CONTROLS

Fig. 9. A typical surface-unit arrangement.

A few years ago the deep-well cooker, shown in Fig. 10, was actually used as the fourth surface unit. In reality, the deep-well cooker is a small insulated oven into which a cooking pail is fitted. Smaller pots or pans are usually designed to be placed into the deep-well cooker, thus making it possible to prepare an entire meal simultaneously.

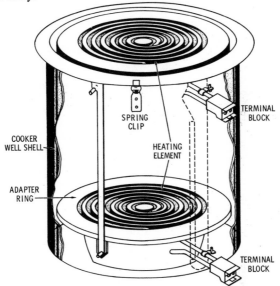

Fig. 10. A deep-well cooker unit shown as the fourth surface unit.

A surface unit, to be desirable, must have several features incorporated in its design. It should be insulated and protected from spillage and mechanical injury; it should also permit the free flow of heat from the resistance wire and should be enclosed in a material that will withstand the operating temperature without deteriorating.

SURFACE HEATING UNIT WIRING

Although the modern electric range with its numerous control switches, timers, pilot lamps, and thermostats seems to be a rather complicated electrical appliance, its operation and control, as will be observed from the following circuit analysis, is quite simple.

In Fig. 11, it can be seen that each surface unit is controlled by an individual switch. The various switch positions are shown in Fig. 12. When the heat-regulation switch is turned to the positions shown, the temperature of the heating unit can be made to vary from "high" to "medium" to "low," assuming that element A is of a higher ohmic resistance value than element B.

Modern electric ranges, however, are usually equipped with surface unit switches that have four to seven heat positions, as illustrated in Figs. 13, 14, and 15. In the four-heat system of Fig. 13, a variety of different arrangements have been designed to give the desired number of heat positions. For "high" heat, each element has a voltage of 115 volts applied across it; for "medium" heat, only one element is connected across the 115-volt circuit; and for "low" heat, the two heating elements are connected in series at 115 volts. On some ranges, a "simmer" position is provided that utilizes a third unit coil, which is connected in series with the other two heating elements in switch position No. 4. This

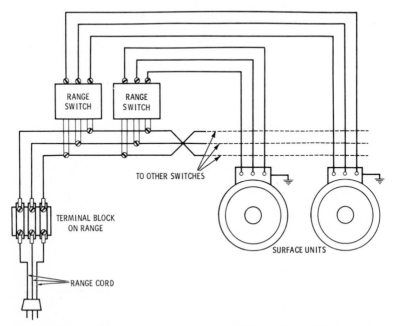

Fig. 11. Switch connections for a typical electric range.

last position of the heating switch connects the elements in series across the 115-volt circuit.

Fig. 14 illustrates diagrammatically the switch positions necessary to obtain seven different heat values on a typical surface unit. In the "simmer" position of the switch, both elements are connected in series across the 115-volt circuit. For "very low" heat, the outer element is connected in series across 115 volts. "Low" heat is obtained by connecting the inner coil, or element, across 115 volts. For "medium low" heat, both elements are connected

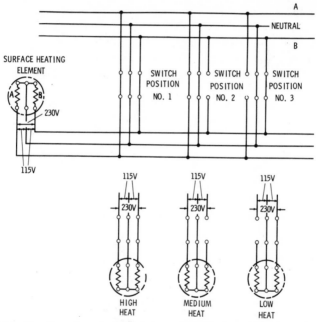

Fig. 12. The three heat positions of a surface heating element switch.

across the 115-volt circuit. Tracing the circuit further, it will be noted that for "medium" heat only, the outer element is connected across 230 volts; for "medium high" heat, the inner element is connected across 230 volts; and for "high" heat, both heating elements are connected in parallel across the 230-volt power-supply line.

The wiring diagrams of Fig. 15 illustrate a surface unit switching arrangement providing five heat positions. Each surface unit consists of two heating elements with a common connection, so that three leads are brought out to the heat-selector switch. For "high" heat, the two heating elements are connected in parallel across 230 volts. In the "medium high" heat position, only the inner element is connected across 230 volts. For "low" heat, the two heating elements are connected in series across 230 volts. "Medium low" heat is obtained by connecting the inner element across 115 volts. The "simmer," or very low heat, position of the surface unit is obtained by turning the heat-selector switch in such a position that the two elements are connected in series across the 115-volt power-supply line.

Numerous ranges of modern design may have one or more of their surface units wired for connection to the oven timer, thus permitting the unit to operate automatically with respect to its

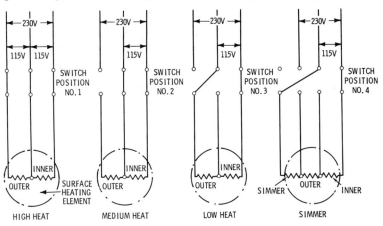

Fig. 13. *A four-heat switching system for a surface heating element.*

cooking period. A variation of this control scheme utilizes a timer connected to one particular surface unit and switch.

Load Balancing—Three-Heat Switch

The purpose of the three-wire load-balancing switch for surface cooking units is to obtain the advantages of the three-wire 115/230-volt distribution system. Thus, by means of the load-balancing switch, the range load is distributed between both legs of the three-

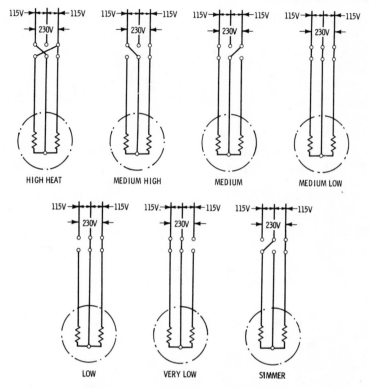

Fig. 14. A seven-heat switching system for a surface heating element.

wire system. As shown in Fig. 16, it is possible to trace the circuit on either the cooking or deep-well unit in any switch position.

On "high" and "medium" heat positions, the green wire is the neutral. With the switch in the "high" heat position, there will be a potential of 230 volts applied between the black and red wires and a potential of 115 volts between the black and green wires or the red and green wires. In the "medium" heat position, there will be a potential of 115 volts between the red and green wires. In the "low" heat position, the black wire becomes the neutral because of the jumper (A) between the two contacts; there will then be a potential of only 115 volts applied between the black and red wires.

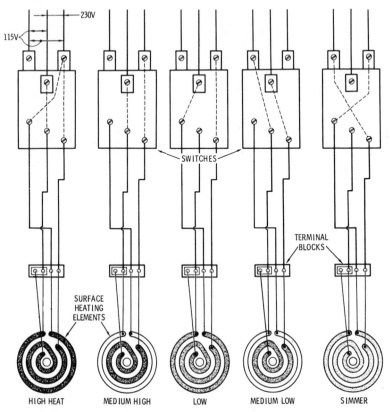

Fig. 15. A five-heat switching system for a surface heating element.

Load Balancing—Five-Heat Switch

The five-heat load-balancing switch, shown in Fig. 17, is similar to the previously described three-heat switch; however, it can supply a greater flexibility of heating temperatures because of the additional heat-control steps.

With the aid of Fig. 17, it is possible to trace the various circuits with the heating-unit switch in any position. The green wire, as indicated, is the neutral in all positions except "low" and "medium high" heat. In the "low" heat switch position, the black wire becomes the neutral because of the jumper (A) between the two contacts on the black wire switch terminal. In the "medium high" heat

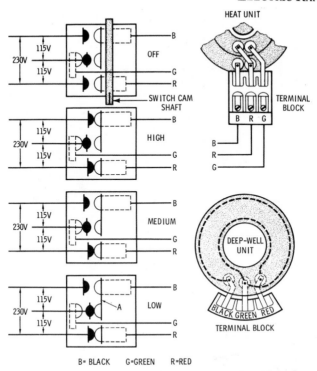

B= BLACK   G=GREEN   R=RED

Fig. 16. The switch positions of a typical surface unit that provides three different heating temperatures.

position, both the black and the green wire terminals are connected to the neutral, thereby producing two 115-volt circuits.

The switch positions and the wattages consumed in the cooking units of a typical electric range are as follows:

Switch Position	6⅝ in. Dia. Unit	8 in. Dia. Unit
High	1200 watts	2000 watts
Medium-high	600 watts	950 watts
Medium-low	400 watts	600 watts
Low	200 watts	350 watts
Simmer	135 watts	220 watts

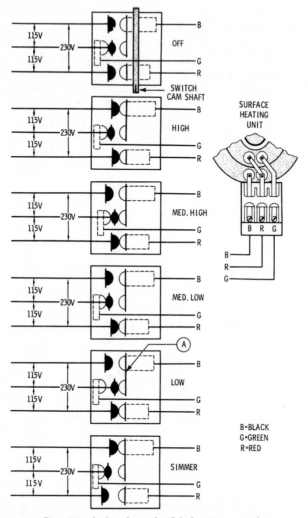

Fig. 17. A five-heat load-balancing switch.

Other electric range units may differ in wattage as well as in wiring methods, although in principle their operation will be similar.

OVEN UNIT WIRING

The oven unit normally consists of two heating elements, an upper element and a lower element. The oven heat is normally controlled by a centrally located thermostat and timing device. The design of an oven unit is different than that of a surface unit primarily because it has a different function to perform. The primary purpose of an oven unit is to heat the oven to a definite temperature, which is controlled by the thermostat, and to provide sufficient heat to produce browning. In addition, the oven unit should be protected from spill-overs and should operate under the maximum temperature possible.

A typical oven circuit is shown in Fig. 18. The broil unit is constructed by stringing the element through the frame in two separate coils, whereas the bake unit is strung with only one coil.

The oven temperature is controlled in much the same manner as that previously described for surface units. In a two-unit oven, the "preheat" switch position provides the highest temperature,

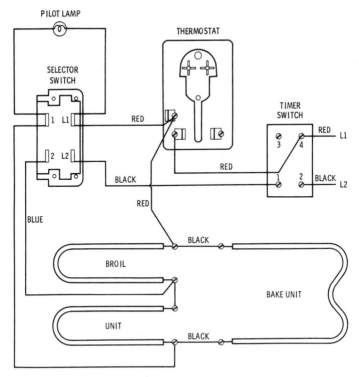

Fig. 18. A typical oven heating circuit with two heating elements.

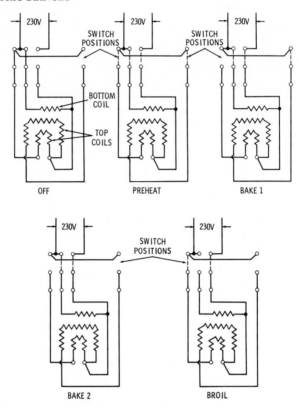

Fig. 19. A switching arrangement that is used to obtain four different heat values in the oven of a typical electric range.

since the inner coil of the top unit and the entire bottom unit are both connected in parallel across 230 volts, as shown in Fig. 19. At "bake 1," the bottom unit and the outer coil of the top unit are both connected in parallel across 230 volts. At "bake 2," a still lower temperature is obtained by a switch arrangement that connects only the bottom element across 230 volts. Finally, in the "broil" position of the switch, only the inner element of the top unit is connected across 230 volts. Although oven heating elements may differ in arrangements, depending on the manufacturer, the heat-switching scheme used will, in almost all ranges, be similar to the circuits shown.

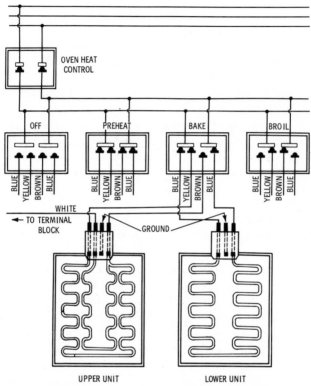

Fig. 20. *The switch positions for three different heat values in the oven of an electric range.*

A wiring diagram of another typical oven unit is shown in Fig. 20. At "preheat," the outer coil of the upper unit and the lower coil unit are both connected across the voltage supply, thus providing the maximum temperature. At "bake," the inner coil of the upper unit and the lower coil unit are connected in parallel across the source, thereby providing a lower oven temperature. A still lower temperature will be reached with the oven switch set at "broil"; in this position, only the outer coil of the upper unit is connected across the circuit. After selecting the desired oven temperature, the preset oven thermostat will automatically disconnect the oven heating units when the temperature exceeds that of the thermostat setting and will close the circuit when the temperature falls below the preset value.

OVEN UNIT CONTROL

By incorporating automatic control devices in electric ranges, the task of cooking has become greatly simplified. In modern ranges, both the temperature to be maintained and the time of starting and terminating the heat are automatically controlled. This control is usually affected by means of a thermostat, which automatically controls the heat, and an electric clock mechanism, which may be set to switch the oven on and off. This type of control makes it possible to prepare a roast, for example, put it into the oven, preset the time and heat controls, and leave the entire cooking process to the thermostat and timer.

Thermostat Operation

It is a well-known fact that practically all automatic heating appliances depend for their functioning on a simple device known as the *thermostat*. This is actually a temperature switch that is actuated by the expansion and contraction of certain liquids or metals with a change in temperature, thereby controlling the heat. For example, if two metal strips of different temperature coefficients are joined together, as shown in Fig. 21, a change in temperature will cause a movement of the unsupported end of the assembly. The direction of travel during an increase in temperature will always be in the direction of that strip whose expansion coefficient is the smaller. By providing suitable contacts, as illustrated, an electric circuit will be established that will open and close its contacts at a predetermined temperature change. It is by this simple principle that a thermostatic switch operates to control the temperature of an electric heating appliance.

Thermostatic switches employed in electric range ovens usually are of the capillary-tube-and-diaphragm, or bellows, type. The switch corresponds in its operation to the thermostatic switch control used on household refrigerators. This type of thermostat consists of two distinct but inseparable assemblies, as shown in Fig. 22. One is the sealed liquid-charged bulb with its capillary tube, and the other is the diaphragm assembly. The liquid-charged bulb is usually placed horizontally, midway up the rear interior wall of the oven. The bulb must be kept free from contact with either the oven wall, the shelves, or pans on the shelves in order to receive

266

its heat from the air circulating in the oven rather than by conduction from any metal objects.

The bulb line, or capillary tube, is connected between the bulb and the diaphragm, or bellows, in the thermostat. When heat is applied to the oven bulb, it causes a liquid expansion within the

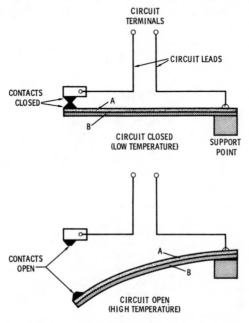

Fig. 21. The operation of a typical thermostatic switch. When heated by the current flow, metal A expands more than metal B, thereby causing the contact points to separate and open the electric circuit. With the current cut off, the metal strips cool and are restored to their original position, thus closing the circuit once more.

bulb, which, in turn, increases the pressure. Since the bulb and line are solidly filled with liquid, the resultant movement on the diaphragm from the application of heat is steady and uniform. The hydraulic pressure is transmitted to the hollow diaphragm, which expands and contracts with varying oven temperatures. This movement is multiplied through a lever action and is used to open and close the contacts of the thermostatic switch.

The temperature setting of the switch can be changed by turning the threaded thermostat shaft, thereby controlling the point at which the expanding diaphragm actuates the switch contacts. Turning the thermostat shaft to the right (clockwise) necessitates more movement of the diaphragm before the switch will cut off. Turning the shaft to the left (counterclockwise) permits the diaphragm to move the switch lever and break the switch contacts with less movement. Large silver contacts that make and break

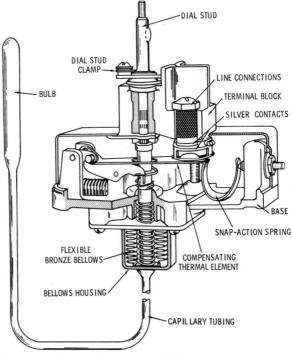

Fig. 22. A typical capillary-tube-and-diaphragm thermostatic switch.

with a magnetically controlled snap action assure long life and a minimum of switch trouble.

Timing Devices

Proper timing of cooking is accomplished in one of two ways, namely, by the use of electric or spring-wound clocks. Spring-wound timing devices are used on electric circuits other than the 60-Hertz AC type. Electric timers are used on all 60-Hertz AC circuits having standard range voltages. See Fig. 23.

Electric timers are simple electric clocks that are fitted with a built-in switch. These are usually built into the range, although some early ranges have their timing devices installed separately. On modern electric ranges, timing devices are always incorporated in the oven circuits, but they may also be employed on surface

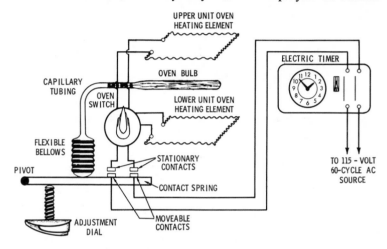

Fig. 23. The schematic diagram of a typical oven-control arrangement.

units and appliance outlets. In addition, some ranges have a separate timing device for surface units.

The oven timer serves three functions; it automatically controls the oven and appliance outlets; it serves as an interval timer for periods of 1 to 60 minutes; and it serves as a conventional clock. For fully automatic operation, set the timer, such as the one shown in Fig. 24, on automatic oven control, and turn knob A to the right (clockwise) until the pointer indicates the desired time at which cooking is to begin. Then, turn knob C in either direction to the number of hours required for the cooking time, and set the oven thermostat to the desired cooking temperature. When knob A is properly set, window B will show red to indicate that the timer is ready to operate on an automatic cycle. The red indicator will disappear when the timing operation begins. Knob C will snap off at the completion of the timed heating cycle. For manual oven operation, knob C must be turned to *MAN* (manual), and window

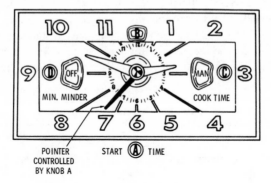

Fig. 24. An oven timer provides fully automatic or manual oven-operation control.

B should not indicate red. Knob D controls the interval timer, which may be set for any period from 1 to 60 minutes; this timer does not control the heating cycle but merely informs the user of the elapsed heating time.

Special Timers

In certain cooking processes that require a comparatively short period of time for completion, a special timer is used. These usually have a maximum timing cycle of 60 minutes, have a spring-wound clock movement, and operate on the principle of release by friction. Thus, by turning the indicator to "20," for example, the frictional resistance to the spring tension of the clock mechanism will result in a period of 20 minutes being required for the handle to return to zero. The dial is graduated to represent 60 minutes. Each time the handle returns to zero, the circuit of a bell is energized, thereby causing the bell to ring and notifying the user that the specific time of the setting has elapsed.

Mechanical Oven Conditions That Affect Baking

Mechanical conditions in the range oven that affect baking are as follows:

1. Bent thermostat bulb,
2. Differential in the thermostat,
3. Improper oven door clearance,

4. Oven units improperly installed,
5. Levelness of the oven.

Thermostat Bulb—A thermostat bulb that is bent so that it touches the oven liner will cause some variation in the oven temperature as compared to the thermostat setting; that is, the temperature will be slightly higher in the oven than that for which the thermostat is set. This condition is caused by the fact that the oven liner conducts the heat away from the thermostat bulb, thereby requiring more heat in the oven to bring the thermostat up to the temperature at which it is set or at which it will cut off. Therefore, if this condition exists, it may result in complaints of drying or burning, caused by the excess temperature.

Differential—Too wide a differential in the thermostat will have its effect on oven results. While this will not be a common complaint, it is not, however, entirely impossible for the differential to change in a thermostat. The differential of the thermostat should be checked as explained later in this chapter. Due to the wide variations in temperature, favorable results in all probability will not be obtained. These unfavorable results may show up either in the length of time to perform the cooking operation or in the appearance and texture of the food being cooked.

Oven Door Clearance—The door clearance at the bottom of the oven door plays an important part in the cooking results obtained in the oven. The purpose of the door clearance is to take care of the thermal expansion of the oven liner when heated. If this expansion is not allowed for by the proper spacing at the bottom of the oven door, and the oven is heated, the expansion of the oven liner will force the door open at the top, thereby allowing heated air to escape and cool air to enter. Here again it can be seen that improper results might occur if the condition were severe enough. The proper amount of gap at the bottom of the oven door is usually between 0.40 and 0.90 inch. Two flexible gauges should be made, one for the high limit and one for the low limit, to determine whether the clearance of the oven door is correct or not.

Oven Units—There is a possibility of interchanging the upper and lower oven units in an oven after removal for cleaning, etc. This will have an effect on the baking operations in that browning may not be attained. Also, when removing oven units for cleaning, there is a possibility of not reinstalling them so that all three prongs on the unit make contact. If this happens, then the unit will not heat at all. Oven unit voltages other than that on which the range is operating will either raise or lower the wattage of the elements. This will also affect the oven operation but will in all probability be more pronounced when operating in the "bake" position.

Levelness—The levelness of the oven is extremely important; it can determine either success or failure of most baking operations in the oven. To illustrate this point, when a cake is being baked and the oven is not level, the cake batter will run to the low side, thus causing different depths of batter and consequently varied baking times. The results, when a condition of this sort exists, are obvious. To level the oven, place a carpenter's level on the oven shelves and shim the range as much as is necessary to bring the oven back to a level position.

Adjusting the Oven Heat Control

Measuring the high temperature in an oven is difficult. It requires several on and off cycles of the oven before a dependable reading can be made. Time-check methods for recording the length of time required for an oven to reach a given temperature are not always consistent. Thermocouple instruments, which indicate oven temperatures in degrees Fahrenheit, are reliable, but due to their cost and the likelihood of damage, they are usually retained in the shop and used only on especially difficult situations. Many servicemen get into the habit of making an oven-temperature-control adjustment every time the customer has an idea the temperature is not right. Oven complaints are frequently not a result of improper heat-control adjustment. Adjusting the heat controls without first analyzing the source of the trouble may result in improper adjustment of controls, which had satisfactory settings originally; these adjustments should be avoided whenever possible.

The following methods of checking oven temperature are recommended when a questionable control is encountered:

Oven Thermometer Method—Set the heat control at 400°. Put the oven thermometer on a shelf in the middle of the oven. Allow the oven to build up to temperature and cut off. Open the door, so that the heat will escape, until the signal lamp comes on again, thus indicating that the unit is heating a second time. Close the oven door; when the signal lamp goes off for the second time, wait approximately two minutes. Then open the oven door, and read the temperature indicated on the thermometer. A reading between 385° and 415° is satisfactory. If the reading is high, for example 450°, pull the thermostat handle off, and loosen the screws that hold the circular calibrated plate to the assembly. Turn the plate to the left a distance of two marks (each mark represents 25°). Do not turn the shaft when you turn the plate. Tighten the screws, replace the handle, and recheck. If the original reading is low, adjust in the same manner by turning the plate to the right for the required number of marks.

The serviceman should be familiar with one other characteristic of the thermostat operation. After an oven reaches a predetermined temperature and cuts off, it should cut on again when a drop of at least 15° but not more than 40° takes place. When checking for this differential with an oven thermometer, proceed as previously described to check the oven temperature. After a reading has been made of the cutoff temperature, close the oven door, and

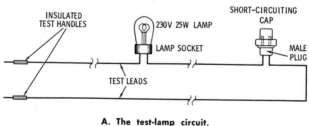

A. The test-lamp circuit.

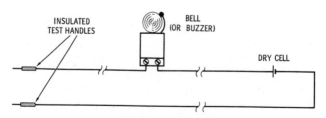

B. The bell (or buzzer) test circuit.

Fig. 25. Testers used for point-to-point continuity tests on all types of electric range circuits.

wait until the signal lamp comes on, thereby indicating the start of another heating cycle. Open the door immediately, and record the temperature. Add 15° to the reading to compensate for the loss of heat during the reading. The differential then will be the difference between the compensated cut-in temperature and the cutoff temperature previously obtained. No adjustment for an improper differential is possible, and the oven thermostat must, therefore, be replaced.

Thermocouple Method—Another type of instrument that may be used to check oven temperatures is the thermocouple. A thermocouple permits temperatures to be checked while the oven door is closed. This instrument is accurate and fast on temperature changes, but care must be exercised to secure the proper readings. When using the thermocouple-type oven tester, do not permit the thermocouple bulb to come in direct contact with the oven shelves. It is recommended that the tube be suspended from a shelf by means of a piece of wire.

Range Wiring Diagrams

In order to make an electric range circuit test, it is essential to employ a light-bulb or battery-buzzer test kit, such as shown in Fig. 25. By using a 230-volt, 25-watt clear glass lamp, as shown in Fig. 25A, burn-out due to high voltage conditions will be avoided, and a visible indication of low voltage will be obtainable. This test-lamp set may be used as a point-to-point tester by connecting a male plug in series with the tubular tips and lamp, as illustrated. When using the device for point-to-point testing, the current to the range should be off, and the ground should be disconnected to avoid the possibility of the lamp lighting when one end of the circuit being tested is open.

When the male plug is not in use for point-to-point testing, a short-circuiting receptacle should be employed to permit testing for "hot" circuits. Fig. 26 shows the test-lamp circuit used for point-to-point circuit checking. Before proceeding with a point-to-point check to locate circuit faults, it may be well to study the circuit diagrams of the range in question. This will provide the serviceman with the color code of the terminal-block location and also the connection between switches, heating units, and automatic control devices.

A typical wiring diagram of an electric range is shown in Fig. 27. This range contains three surface heating units and two oven units

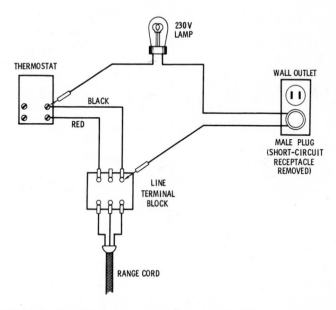

Fig. 26. The method of using the test-lamp assembly for point-to-point circuit tests on an electric range.

with their switches. In addition, the range unit incorporates a thermostat, an oven pilot lamp, and an appliance receptacle with a fuse. The diagram is divided into two parts, one for the surface units and switches and one for the oven heating elements. Since each wire is clearly identified by color, it is a comparatively easy matter to trace each wire from the line terminal block through the various control points and heating elements.

Although wiring diagrams may differ, depending on the particular range manufacturer and the types of control devices employed, all ranges operate on the same general principles and usually have the same number of control elements.

Fig. 28 illustrates an electric range wiring diagram that differs from the one shown in Fig. 27, mainly in that it has four surface heating units instead of three and also has an oven lamp, a timing device, and a fluorescent lamp.

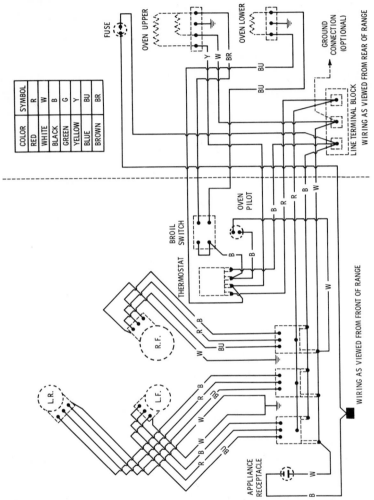

Fig. 27. *The wiring diagram of a typical electric range with three surface heating units.*

SELF-CLEANING OVEN

Special coordinated single- and double-wall ovens are available for today's modern kitchens. Generally, with the double oven, the top oven is self-cleaning and the bottom is a conventional baking oven. See Figs. 1 and 29.

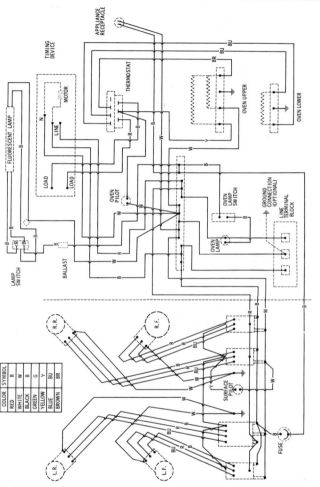

Fig. 28. The wiring diagram of an electric range with four surface heating units.

Various models include clock timers which turns the oven on, cooks the whole meal and keeps it at serving temperature or turns itself off, even if you're not home. Special meat probes with thermostatic roasting control cooks the roast exactly the way you like it. A safety door lock is also featured on some models which locks the oven door and cannot be opened once the temperature goes over 550°F.

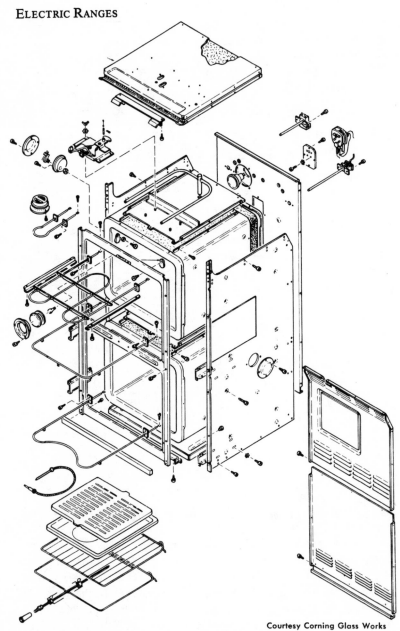

Courtesy Corning Glass Works

Fig. 29. Wall-mounted double-oven assembly.

When troubleshooting an electric range, or oven, it is helpful to follow a sequence of operations in performing various checks. Many possible causes or troubles will require that a specific test be performed.

Range Troubleshooting Guide

COMPLAINT	PROBLEM	DIAGNOSIS
1. Both elements out on one side.	1. One hot line dead due to improper hookup or blown fuse, faulty circuit breaker.	With power on—using a neon indicator-probe, check terminal block hot line to neutral (both red to white and black to white).
2. Element on for a few seconds then fails. Indicator Lights may work.	1. Improper hookup. 240V applied to one side of unit. Hot line tied to neutral (white) wire at junction box.	If above check indicates power is on with hot line to neutral—caused by improper wiring at junction box or main entrance, check resistors on indicator lights for ½ of elements out. If improperly hooked up, the plastic sheath over the resistor will be melted.
		If only one element failed but 240V has been hooked to one side, change both elements and indicators on that side.
3. Indicator light "ON" when control at "OFF".	1. Bad cam on control.	Remove control cover and observe motion of power contacts while turning control to "OFF" position. Ohmmeter should show no power with switch OFF. Arm of power contact can be bent slightly with needle nosed pliers to adjust.

Range Troubleshooting Guide (Cont'd)

COMPLAINT	PROBLEM	DIAGNOSIS
	2. Broken or missing ceramic pin.	Check placement of pin. Missing pin can be replaced by removing phenolic body from control and inserting new ceramic pin. Important— test for calibration and boil time.
4. One element out/no indicator light.	1. Oxidized contacts.	Resistance shows poor contact. With power "ON" use a neon probe— light indicates a poor contact.
	2. Poor terminal contacts.	Neon probe light ON indicates bad connection. Clean contacts if necessary.
5. Element "ON" unable to turn "OFF"—Indicator light on.	1. Two possible failures: Ceramic pin or cam on power contacts and diaphragm failure on thermostat side or points sticking.	Replace Control.
		NOTE: If power contacts fuse and stick together, check contacts on thermostat side of control for failure.
6. Element fails to heat/indicator light works.	1. Element burn out.	Ohmmeter probe to screw hold down on neutral connection. More than 10 ohms resistance indicates failure.
		Failure—replace with new element.
	2. Thermostat side contacts badly burned.	Replace control.

Range Troubleshooting Guide (Cont'd)

COMPLAINT	PROBLEM	DIAGNOSIS
7. Element "hum".	1. Sixty cycle hum caused by iron content of element winding.	May be damped slightly by adding 1" of extra fiber glass in heater box next to cover. The hum is an inherent part of the assembly.
8. Element runs too hot—No temperature control--Always full "ON".	1. Possible failure of control by leakage.	Oil on capillary tube and control. Replace control.
Unit not hot enough	1. Downdrift from burned points. 2. Improper assembly of components.	Will occur only when unit has been in service for a number of years. Perform boil test. If element fails boil test, replace control.
9. Control knob binds, is hard to turn.	1. "O" ring on nylon bushing is out of position	Replace nylon bushing.
10. Broken heater panel.	1. Thermal breakage.	Panel is cracked from one edge of panel to another. Replace. NOTE: A crack will always run from one side of the panel to another—it will never start and stop on surface.
	2. Impact breakage.	Caused by dropping heavy object on unit. Splintered effect. Replace.
11. Scratch on ceramic heater panel.	3. Scratch starts and stops on the same surface.	Caused by diamond ring, abrasive pad, or dirt or grit between vessel and panel.
12. Indicator Light glows very faintly when unit is turned off.	1. Wire from indicator is passing over a hot wire.	Remove the control box cover and visually check the wires leading to the indicator light. Reroute if necessary.

Range Troubleshooting Guide (Cont'd)

COMPLAINT	PROBLEM	DIAGNOSIS
13. Glass ceramic panel doesn't turn a bright yellow.	1. No problem, just a physical property of panel. Does not affect performance.	Turn element on to HIGH and observe.
14. Wavy glass.	1. Result of manufacturing process—will not affect performance.	
15. Red spots or lines showing through panel.	1. Not a defect. New glass is less opaque and element ribbon may show through the glass ceramic panel. Does not affect performance. No danger to homeowners.	Turn unit to HIGH. Turn lights out in room and observe. When power is on, element may show through the glass ceramic panel.
16. Rim of countertop gets too hot.	1. Rim of countertop will get 200°F. The temperature will vary with the size of vessel and type of load on the element.	Place a vessel full of water on the element pattern nearest the size of the vessel. Turn the unit on and allow about 15 to 20 minutes for warmup. Place a surface pyrometer on the rim and observe. Temperature should be 200°F. By moving the load away from the rim, observe the rise in temperature.
17. Erratic operation of surface elements.	1. Loose or burned neutral connections.	Check the neutral connections at the terminal block. Be sure they are secure. Check the junction box for loose neutral connections.

Oven Troubleshooting Guide

1. Oven light fails to operate.	1. Bulb burned out. 2. Inoperative switch.	Replace. Check for continuity across oven light switch terminals. If no continuity, replace switch.

Oven Troubleshooting Guide (Cont'd)

COMPLAINT	PROBLEM	DIAGNOSIS
	3. Solenoid switch in-operative or out of adjustment.	Adjust or replace. Check with lock lever in cook position.
2. Fluorescent light doesn't light at all.	1. Burned out bulb.	Replace.
	2. Open ballast.	Check continuity across ballast. If no continuity, replace ballast.
	3. Lamp not making contact with lamp holder.	Replace lamp holder or adjust holders to make contact.
	4. Defective switch.	Replace.
3. Rotisserie motor doesn't operate.	1. Motor or switch on motor defective.	Replace rotisserie motor.
4. Door locks at proper temperature but lock light does not light.	1. Defective lock light.	Replace.
	2. Defective high limit thermostat.	Replace.
5. Clean indicator light comes on but door does not lock at proper temperature.	1. Misadjusted or mis-aligned mechanical latch mechanism.	Properly adjust or align.
	2. Defective high limit thermostat.	Replace.
6. Oven will not heat with selector set on "BAKE".	1. If indicator light lights, bake element or selector switch is defective.	Check for voltage across bake element. If voltage is present, replace bake element.
	2. If indicator light does not light, relay may be defective.	Should be continuity between terminals 1&2 on relay. If not, replace relay.
	3. If indicator light does not light, selector switch may be defective.	Check continuity between appropriate terminals on selector switch. If no continuity across terminals, replace selector switch.
	4. If indicator light does not light, oven thermo-stat may be defective.	If all above items check good replace oven thermostat.
	5. If indicator light does not light and oven operates properly.	Replace indicator light.

Oven Troubleshooting Guide (Cont'd)

COMPLAINT	PROBLEM	DIAGNOSIS
7. Oven will not bake with selector switch set on "TIMED BAKE."	1. Clock set for "Hold" with hold warm switch off.	Set controls properly
	2. Defective clock.	Check for continuity between appropriate terminals on clock-timer switch.
	3. If oven comes on when selector switch is set on "BAKE", there is a possibility of defective selector switch.	Check for continuity across appropriate terminals. If no continuity, replace selector switch.
8. Oven will not delay in "TIMED BAKE" selection.	1. Defective clock.	Replace clock.
9. Oven will not go into "SERVE" temperature.	1. Hold warm switch off.	Set controls properly.
	2. Defective clock.	Check continuity between appropriate terminals. If no continuity, replace clock.
	3. Defective High-Limit thermostat.	Replace.
	4. Defective Relay.	Replace.
10. Oven does not BROIL.	1. Same causes as in "Oven will not BAKE" and "Oven does not BROIL." Items 6 and 10.	Check continuity of broil element. If no continuity, replace.
11. Oven does not PREHEAT.	1. Same causes as in "Oven will not BAKE," Item 6 except defective BROIL element.	Make same checks as in items 6 and 10 above.
12. Oven will not CLEAN.	1. High limit thermostat defective.	Check out oven as indicated in items 6 and 10 above.
	2. Solenoid switch defective.	Check continuity across switch. If none, replace switch.

Oven Troubleshooting Guide (Cont'd)

COMPLAINT	PROBLEM	DIAGNOSIS
13. Oven cleans, but at temperature too low for acceptable cleaning.	1. If all checks in items 6, 10 and 12 show satisfactory operation, mullion element or smoke eliminator element may be defective.	Check continuity of element. If none replace element.
14. Oven doesn't clean sufficiently.	1. Use and care problem; oven was extremely soiled and clean time wasn't sufficient.	Instruct customer to repeat clean cycle with a minimum of 3 hours. If the oven is not clean after the 3 hour clean cycle, check the following:
	2. High limit thermostat.	Check the oven temperature during the clean cycle. Should be 875°F $\pm$10°F. If temperature is below 865°F, replace the high limit thermostat.
	3. Selector switch.	Check continuity of selector switch. Replace (if found defective) bake element resistance, broil element resistance, mullion element resistance.
	4. Elements burned out.	Note: If the mullion element is burned out, the entire oven liner assembly must be replaced.
15. Oven door locked after clean cycle and will not clean.	1. Burned out solenoid.	To open door, remove lock lever knob. Disassemble control panel and remove end cap and front vent assembly. Lift latchpin up out of locking assembly. Slide lock arm to the cook position and open door. Remove solenoid and replace.

285

RANGE LIGHTING

Signal Lamps

Signal lamps are provided to furnish visual indication to the user of certain oven heat conditions. For example, with the single oven control turned to the desired temperature, the oven signal lamp will light until the oven has attained this value. When this point is reached, the oven signal lamp will go out and will remain out until the temperature of the oven has dropped to the thermostat cut-in point. Numerous electric ranges also incorporate signal lamps for surface units.

Exterior and Interior Lighting

In addition to the various signal lamps, modern electric ranges are usually provided with an interior lamp for the oven and an exterior lamp for the top or surface units. The oven lamp functions in the same manner as the type used in household electric refrigerators to light the interior when the door is opened. A positioning switch is mounted in the oven door mechanism so that the outward movement of the oven door will energize the lamp, which will remain lit until the oven door is closed. Exterior lamps may be of the incandescent or fluorescent type. These are mounted to provide the maximum amount of light where needed.

SERVICING AND REPAIRS

Since an electric range consists of few moving parts, each of which is relatively simple in design, the average range will generally operate satisfactorily for many years without the need of repairs, once it has been properly installed. The manufacturer's instructions accompanying each range give installation methods and details of operation that should be thoroughly read and comprehended before the range is put into service.

Installation Procedure

Since the location and arrangement of all kitchen appliances are usually determined by the architect or builder of the house or apartment, the serviceman's job generally consists of connecting the electric range to the wall outlet provided for it. This should be done only after checking all range circuits for proper operation.

Checking Heating Units

After all the range parts have been properly unpacked and assembled, connect the range to the electric service provided; this service should be close to the location of the range. After connecting the range to the electrical service, it is important that all circuits be checked as follows:

1. Check all surface units for proper heating. This can be accomplished by checking each surface unit to see that the lead

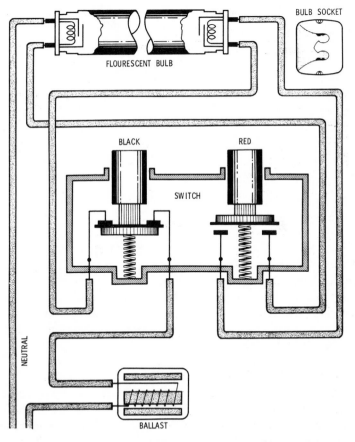

Fig. 30. The schematic representation of a typical range-mounted fluorescent lamp.

wires match the terminal block and that the terminal screws are tight. Next, turn the surface unit switch of each unit to the "low" position and check to see that each outer ring heats; then turn each switch to the "medium low" position to see that each inner ring heats.

2. Check the oven units by turning the oven switch to approximately 350 degrees; the top unit should get warm and the lower unit should get red. At the same time, check the "bake" position signal lamp. Then turn the switch to the "broil" position, and check the upper unit to see that it becomes red and that the "broil" signal lamp is on. Finally, turn the switch to the "preheat" position to see that both units become red and that the "broil" and "bake" signal lamps are on.

3. Check the timing devices through an entire cycle for each proper setting; turn the clock hand manually.

4. Check the convenience outlet by means of a test lamp.

5. Check the warmer unit (if there is one on the range) for heating by turning on the warmer switch, removing the warmer drawer, and feeling the unit.

Checking Fluorescent Lamps

When checking difficulties on inoperative fluorescent lamps, the circuit of which is shown in Fig. 30, first inspect the terminals on the line-terminal block to be sure that the correct voltage and current are available to the lamp. If the proper current is available, then the simplest method of checking from this point is to use a new bulb and starter. First, replace the starter; if this does not correct the difficulty, leave the new starter in the circuit, and replace the lamp. If this replacement corrects the difficulty, remove the new starter and replace it with the old one to ascertain whether or not it is still in good operating condition. If it is, then it can remain in the circuit. If, however, after replacing both the bulb and starter, the lamp still fails to operate, the difficulty is in either the ballast or the wiring. The wiring may be checked by point-to-point testing in the usual manner. If the difficulty is in the ballast, then this part must be replaced. It must also be remembered that the lamp is normally in a fused line, and the possibility of a burned-out fuse must also be checked.

CHAPTER 16

Microwave Ovens

As you may or may not know, any adjustments, substitutions to or modifications of the microwave generating components incorporated in the microwave oven, by other than qualified personnel, may lead to violations of the federal regulations established by the *Bureau of Radiological Health of the Food and Drug Administration (BRH/OBD 73-5)* and the *Federal Communications Commission* which may result in an unsafe condition. Therefore, any defective component parts which must be replaced *must* be replaced with an exact duplicate (not a substitute) and must be replaced by an authorized service representative and should not be serviced by a do-it-yourself home owner. A copy of these regulations can be obtained by writing the *Department of Health, Education, and Welfare, Rockville, Maryland 20852*.

Microwave, or electronic, ovens (Fig. 1) use microwave energy to produce the heat used to cook food. Unlike the conventional ovens, microwave energy cooks food without applying external heat.

The microwaves are short electromagnetic waves of radio frequency (RF) energy that pass through materials such as paper, china, glass, and most plastics. Aluminum and metal materials tend to reflect the microwaves and should not be used. This would generally cover packaged frozen products which you would purchase from grocery stores or supermarkets.

Food with a high moisture content will absorb microwave energy. As the microwave energy (frequency of approximately 2450 MHz) enters the food, the molecules align themselves with the energy. Since the microwaves are changing polarity (or direction) every half cycle, the food molecules are changing direction every half cycle or oscillating back and forth 4,900,000,000 times

Courtesy Sharp Electronics Corp.

Fig. 1. Typical microwave oven assembly.

per second. This high speed oscillation causes friction buildup between the molecules, thereby converting the microwave energy to heat.

OPERATION SEQUENCE

Note: Never operate oven with outside cabinet removed.

The following is a complete description of component function during the three stages of oven operation. The three stages are as follows:

1. Off condition,
2. Idle condition,
3. Cooking condition.

Off Condition—When the pointer on the timer knob is at the 0 position, no component in the oven will operate. See Fig. 2.

Idle Condition—When the door is closed, both door interlock switches are activated; the front latch switch and the rear safety switch. When a cooking time is selected, the timer switch contacts leading to the cook relay coil and timer motor closes. See circuit shown in Fig. 3.

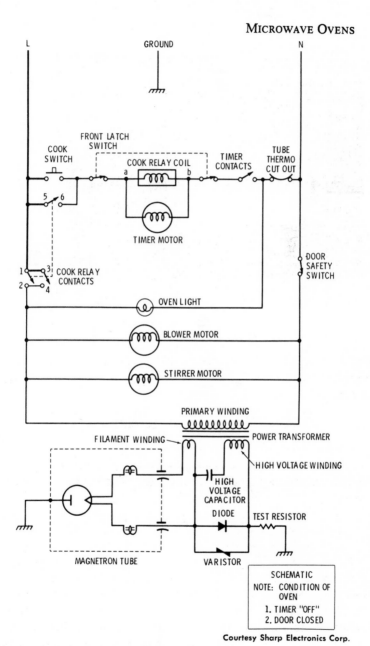

Fig. 2. Oven circuit—off condition.

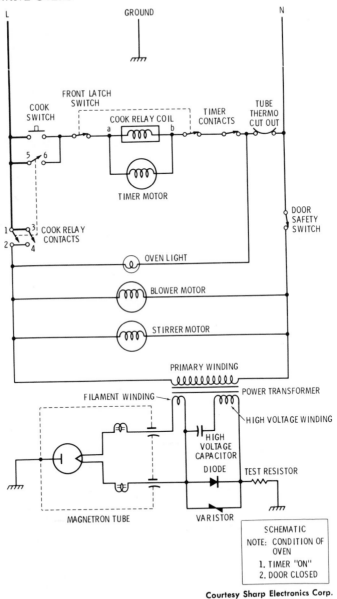

Fig. 3. Oven circuit—idle condition.

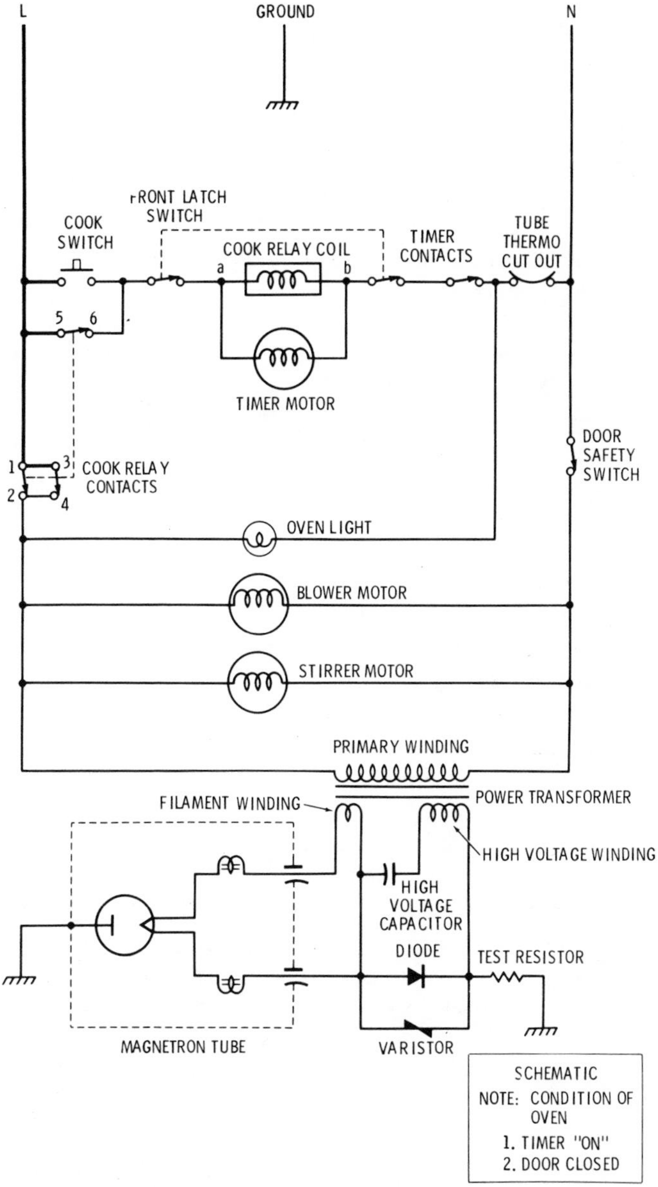

Fig. 4. Oven circuit—cooking condition.

Courtesy Sharp Electronics Corp.

Cooking Condition—When the cook switch is depressed, the following operations occur. See Fig. 4.

1. The coil of the cook relay is energized. Terminals 5 and 6 of the relay contacts close to provide a current path to the timer motor. This also provides the holding circuit which energizes the coil of the cook relay when the start switch is released. Terminals 1,3 and 2,4 close to provide a current path to the oven light, blower motor, stirrer motor, and power transformer.

2. The filament winding of the power transformer (3.2VAC) heats the magnetron filament and the output from the high-voltage winding (1900VAC) is sent to a voltage-doubler circuit consisting of a capacitor and diodes. This 1900 volts is converted to approximately 3800 volts DC (negative) (peak-to-peak) by the voltage doubler circuit and sent to the magnetron tube assembly.

3. The negative 3800 volts DC applied to the cathode of the magnetron tube causes it to oscillate and produce the 2450 MHz cooking frequency.

4. The RF energy produced by the magnetron tube is channeled through a waveguide into the cavity feedbox, past the stirrer blade, and into the cavity where the food is placed to be heated.

When the cooking times expires, the timer switch opens and the coil by the cook relay is de-energized. The contacts of the cook relay open with the following results:

1. The oven light goes out,

2. High voltage is cut off from the magnetron so no **RF** energy is produced.

3. The timer bell rings to indicate the end of the cooking cycle. The oven has reverted to the *off* position.

DESCRIPTION AND FUNCTION OF COMPONENTS

Oven Light—The oven cavity light illuminates the interior of the oven so that the food being heated can be examined visually through the door window without having to open the door. The oven cavity light also serves as the cook indicator.

Blower Motor—The blower motor drives an impeller blade which draws cooling air through the oven base. This cooling air is directed through the air vanes surrounding the tube and cools the magnetron assembly. Most of the air is then exhausted directly through the back vents.

However, a portion of this air is channeled through the cavity to remove steam and vapors given off from the heating foods. It is then exhausted at the top of the oven cavity into a condensation compartment.

Stirrer Motor—The stirrer motor turns the stirrer blade assembly at the top of the oven cavity. The stirrer blade assembly revolves slowly and reflects the electromagnetic energy produced by the magnetron tube. This reflected energy bounces back and forth between the walls, top, and floor of the metal cooking cavity. This allows the RF energy to reach the food from all angles giving a uniform heating pattern.

Dual Latch Switch—Both sections of the front latch switch are activated by the latch on the door handle. When the door is open, one switch opens the line circuit to the cook relay and the timer motor. The other half of the switch interrupts the neutral line circuit to the cook relay and the timer motor. A cook cycle cannot take place until the door is closed and the thumb latch on the door handle closes the front latch switch.

Timer Assembly—(Timer switch contacts) The timer switch contacts are mechanically opened or closed by turning the dial knob located on the timer motor shaft. These contacts control the current path to the timer motor and the cook relay.

(Timer bell) The bell striker is mechanically driven by the timer motor and rings at the end of the cook cycle.

(Timer motor) Cooking time from 0 to 25 minutes may be selected with the timer. The timer motor is energized through the cook relay contacts. When the timer reaches the 0 point on the scale, the timer switch opens the circuit to the cook relay, timer motor and stirrer motor and the cook cycle stops.

Cook Switch—When depressed, the cook switch completes the circuit to the coil of the cook relay through the front latch switch.

Cook Relay—The coil of the cook relay is initially energized by the closing of the cook switch. The cook relay then becomes a holding circuit for the relay coil through its closed contacts (5) and (6).

Energizing the coil of the cook relay also closes contacts (1)(3) and (2)(4) which provide a current path to the stirrer motor, the power transformer and cook light, provided the rear door safety switch is activated (closed by the door cam arm).

Thermo Cut-Out—The thermo cut-out, located on the magnetron assembly, is designed to prevent damage to the magnetron if an overheated condition develops in the tube due to blower failure, obstructed air ducts, dirty or blocked filter, etc.

Under normal operating conditions, the thermo cut-out remains closed. However, when abnormally high temperatures within the magnetron approach a critical level, the thermo cut-out will interrupt the circuit to the cook relay coil and the other control circuits and stop the cycle. When the magnetron has cooled to a safe operating temperature, the thermo cut-out closes, and a cook cycle can be resumed.

Door Safety Switch—The door safety switch is activated (closed) by the cam arm of the oven door and provides a circuit through contacts (1)(3) and (2)(4) of the cook relay to the primary of the power transformer.

When the oven door is opened, the door safety switch provides a secondary function of opening the power transformer circuit in the event the front latch switch failed to open the circuit and de-energize the cook relay coil.

Power Transformer—The power transformer consists of three windings:

1. Primary,
2. Filament,
3. High voltage.

During a cook cycle, the 120 volts AC applied to the primary winding of the transformer is converted to 3.2 volts AC on the filament winding and approximately 1900 volts AC on the high-voltage winding. The 3.2 volts AC heats the magnetron filaments. This causes the tube cathode to readily emit the electrons (negative 3800 volts DC) applied to the cathode. The 1900 volts AC is fed to the voltage-doubler circuit.

Voltage-Doubler Circuit—The voltage-doubler circuit consists of a diode assembly and a capacitor. The 1900 volts AC from the high-voltage winding of the power transformer is applied to the voltage-doubler circuit, where it is rectified and converted to approximately 3800 volts DC (negative) needed for the magnetron operation. (Diode). The diode is a solid-state device that allows current flow in one direction, but prevents current flow in the opposite direction. The diode acts as a rectifier changing alternating current into pulsating direct current.

(High-voltage capacitor) The capacitor stores energy on one half of the power cycle and releases it along with the transformer

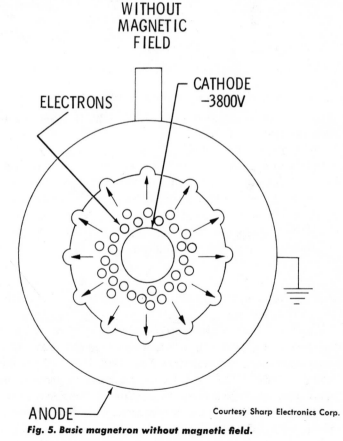

WITHOUT
MAGNETIC
FIELD

ELECTRONS

CATHODE
−3800V

ANODE

Courtesy Sharp Electronics Corp.

Fig. 5. Basic magnetron without magnetic field.

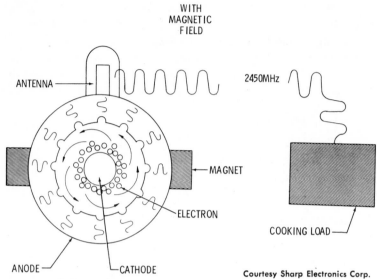

Fig. 6. Basic magnetron with magnetic field.

Courtesy Sharp Electronics Corp.

output to produce approximately 3800 volts DC negative to the magnetron.

Magnetron Tube—The basic magnetron tube is a cylindrical cathode within a cylindrical anode surrounded by a magnetic field. When the cathode is heated by the filament winding of the transformer, electrons are given off by the cathode. These negatively charged electrons are attracted to the more positive anode of the tube when the negative 3800 volts DC is applied to the cathode.

Ordinarily, the electrons would travel in a straight line from the cathode to the anode as shown in Fig. 5. The addition of a magnetic field, provided by permanent magnets surrounding the anode, causes the electrons to take an orbital path between the cathode and anode (Fig. 6).

As the electrons approach the anode, they travel past the small resonant cavities that are part of the anode. Interaction occurs, causing the resonant cavities to oscillate at a very high frequency of 2450 MHz. This RF energy is radiated from the magnetron antenna into the waveguide, into the cooking cavity feedbox, past the stirrer blade assembly, and finally into the cooking cavity where food is placed to be heated.

SERVICING

When troubleshooting the microwave oven, it is helpful to follow the sequence of operation in performing the checks. Many of the possible causes of trouble will require that a specific test be performed. *Note: Never operate oven with outside cabinet removed.*

Troubleshooting Guide

PROBLEM	POSSIBLE CAUSE
	OFF CONDITION
Line fuse blows when power cord is plugged into wall receptacle.	Shorted wire in power cord or wire harness. Replace cord or check wiring.
	COOKING CONDITION
Blower motor inoperative	Defective blower motor—replace. Open wiring in circuit to blower. Check wiring.
Heat produced in oven load, but oven cavity light does not illuminate.	Burned out bulb—replace. Open wiring in circuit to light. Check wiring.
Oven cavity light does not illuminate and no heating.	Defective cook relay. Check and replace if necessary. Defective thermo cut-out. Replace. Open wiring in circuit to thermo cut-out. Check wiring.
Oven does not go into cook cycle when cook switch is activated.	Defective contacts on timer switch. Front latch switch defective or out of adjustment. Defective cook switch. Replace. **Defective cook relay.** **Open wiring between the above components. Check wiring.**
Oven goes into cook cycle but timer does not time out.	Defective timer motor. Replace. Open wiring in circuit to timer motor. Check wiring.

Troubleshooting Guide (Cont'd)

PROBLEM	POSSIBLE CAUSE
	COOKING CONDITION
Oven goes into cook cycle but stirrer motor inoperative.	Defective stirrer motor. Replace. Open wiring in circuit to stirrer motor. Check wiring.
Oven cooking light indicates cycle, but little or no heat is produced in oven load.	Shorted high-voltage circuit between voltage-doubler circuit and magnetron. Check wiring. Defective power transformer. Replace. Defective diode. Replace. Defective high-voltage capacitor. Replace. Defective magnetron. Replace. Rear door safety switch defective or out of adjustment. Check and replace if necessary.
Oven goes into cook cycle but shuts down before end of cycle.	Thermo cut-out opened. Check circuit.
Power source fuse blows when the cook switch is depressed.	Defective power transformer. replace if necessary. Secondary circuit of power transformer is shorted. Replace.

DISASSEMBLY AND REPLACEMENT OF PARTS

There are many models of microwave ovens on the market but their operations are basically the same, disassembly and replacement parts are also similar.

WARNING: TO AVOID POSSIBLE EXPOSURE TO MICROWAVE RADIATION, before operating the unit:

1. Make sure that both dual latch switch and the door safety switch are activated with a click when unlatching and opening the door slowly.

2. Visually check the door and seal for arcing and damage.

Do not operate the unit until after repair if any of the following conditions exist:

1. Door does not close firmly against the appliance front (more than 1/16 inch of movement in latched position).
2. There is a broken door hinge or broken support.
3. The door gasket or the door seal is damaged.
4. The door is bent or warped.
5. If there are any defective parts in the interlock oven door, or microwave generating and transmission assembly.
6. If there is any other visible damage to the oven.

Do not operate the unit without:

1. RF gasket,
2. Waveguide and oven cavity intact,
3. The door closed.

Outer Case Removal

To remove outer case, refer to Fig. 7 and proceed as follows:

1. Disconnect unit from power supply.
2. Remove the screws from the rear and along side edges of the case.

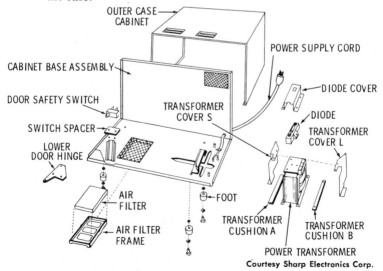

Courtesy Sharp Electronics Corp.

Fig. 7. Outer case and chassis components.

3. Slide entire case back about 1 inch to free it from retaining clips on cavity face plate.
4. Lift the entire case from the unit.

CAUTION: DISCHARGE HIGH-VOLTAGE CAPACITOR BEFORE REMOVING ANY PARTS.

Power Transformer Assembly

The power transformer assembly consists of a power transformer, diode, diode cover, and transformer cover which is attached to the base by nuts (Fig. 7).

1. Disconnect unit from power supply and remove outer case.
2. Discharge high-voltage capacitor.
3. Disconnect primary and secondary leads from the power transformer.
4. After removing terminal cover from secondary terminal of power transformer, disconnect the wire leads from diode.

Diode Removal

To remove the diode refer to Fig. 7 and proceed as follows:

1. Disconnect unit from power supply and remove outer case.
2. Discharge high-voltage capacitor.
3. Disconnect wire leads from diode.
4. Remove two (2) screws on terminal side holding diode cover to power transformer.
5. Remove two (2) screws holding the relay board to cavity for easy access to diode.
6. Remove two (2) screws holding diode to power transformer while lifting up diode cover.

Stirrer Cover Removal

The stirrer cover must be removed to service the stirrer motor and stirrer blade, Fig. 8.

1. Disconnect unit from power supply.
2. Remove three (3) screws holding the cover to top of oven cavity.
3. Remove the stirrer cover by sliding unit toward front of cavity.

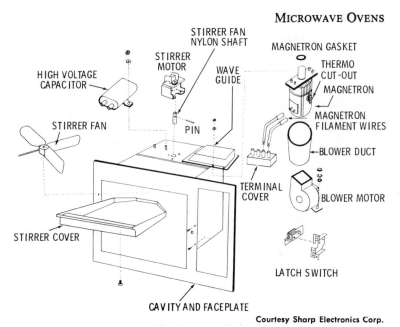

Courtesy Sharp Electronics Corp.

Fig. 8. Stirrer motor and cavity components.

Stirrer Motor Removal

To remove the stirrer motor refer to Fig. 8 and proceed as follows:

1. Disconnect oven from power supply and remove outer case.
2. Remove stirrer cover.
3. Remove stirrer blade assembly by loosening the two (2) screws on the blade hub.
4. Disconnect wire leads from quick connectors.
5. From the inside of oven, remove the three screws holding the stirrer motor bracket to the oven cavity.
6. Separate the stirrer motor from the stirrer fan (nylon shaft) by removing the pin from the nylon shaft.
7. Replace blade assembly so hub is flush with the bottom edge of the nylon shaft and secure mounting screws.

Magnetron Assembly Removal

To remove the magnetron assembly (Fig. 8), first disconnect

303

from power supply, remove outer case, discharge the capacitor.

NOTE: The magnetron can be removed easier if the blower assembly is first loosened.

1. Remove rubber blower duct and exhaust duct.
2. Remove wire leads from thermo cut-out and high-voltage leads from terminals on magnetron assembly.
3. Carefully loosen the four (4) mounting nuts holding the magnetron to the wave guide while supporting the magnetron from below.
4. Lower magnetron assembly until the tube is clear of the waveguide.

CAUTION: WHEN REPLACING THE MAGNETRON, BE SURE THE RF GASKET IS IN PLACE AND MOUNTING NUTS ARE TIGHTENED SECURELY.

Control Panel and Component Removal

The complete control panel should be removed for replacement of the component. To remove the control panel, refer to Fig. 9 and proceed as follows:

1. Disconnect from power supply and remove outer case.
2. Discharge high-voltage capacitor.
3. Disconnect and identify wire leads to on-off switch, cook switch, timer motor and switch and cook light.
4. Remove the two (2) screws holding bottom edge of control panel to oven body and pull panel forward and downward. Panel is now free from face plate.
5. Remove the timer knob and place the panel on the smooth surface and remove the four (4) screws holding back plate. Separate back plate with timer and control switches attached.

Replacement of individual components is as follows:

TIMER:

1. Remove the timer pointer.
2. Remove mounting screws to free timer assembly.
3. After timer is replaced, position mounting screws at original position. Remount all other components and reconnect wiring. Check timer operation by turning timer knob

to 5 minutes mark and returning toward "0" position until bell rings. If timer pointer does not correctly come to "0" position, pour a small quantity of thinner into two holes under the boss of pointer and adjust the pointer.

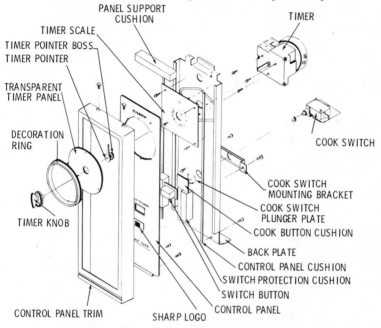

Fig. 9. Control panel and component removal.

COOK SWITCH:

1. Remove screw and cook switch plunger plate from cook switch.
2. Remove two (2) screws holding cook switch mounting bracket to back plate.
3. Loosen the nut to free cook switch.
4. Securely tighten nut on replacement switch and replace plunger plate and mounting bracket.

Cook Relay Removal

To remove cook relay refer to Fig. 10 and proceed as follows:
1. Disconnect from power supply and remove outer case.

2. Discharge the high-voltage capacitor.
3. Disconnect all wires to the cook relay.
4. Remove two (2) screws holding cook relay to relay board.

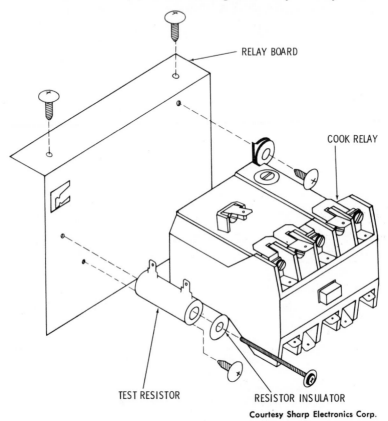

RELAY BOARD

COOK RELAY

TEST RESISTOR

RESISTOR INSULATOR

Courtesy Sharp Electronics Corp.

Fig. 10. Cook relay board assembly.

Test Resistor Removal

To remove test resistor refer to Fig. 10 and proceed as follows:
1. Disconnect from power supply and remove outer case.
2. Discharge the high-voltage capacitor.
3. Remove the screw holding the test resistor to the relay board.
4. Remove test resistor by unsoldering or cutting wire leads.

Dual Latch Switch Removal

1. Disconnect from power supply and remove outer case.
2. Discharge the high-voltage capacitor.
3. Disconnect lead wires from the switch.
4. After control panel assembly is free, remove two nuts from steel subpanel to free the latch switch, Fig. 8.
5. Because the latch switch is in two sections, make sure both switches operate when door is closed.

Oven Light Removal

1. Disconnect from power supply.
2. Remove outer case.
3. Remove bulb.

NOTE: Screws and nuts hold the light bracket to the oven cavity and screws hold the individual light socket.

Door Safety Switch Removal

1. Disconnect from power supply and remove outer case.
2. Discharge high-voltage capacitor.
3. Disconnect lead wires from the switch.
4. Remove the two (2) screws holding switch to insulating plate to free the switch Fig. 7.

High Voltage Capacitor Removal

1. Disconnect from power supply and remove outer case.
2. Discharge the capacitor.
3. Disconnect lead wires.
4. Remove two nuts holding high voltage capacitor, Fig. 8.

Door Replacement and Adjustment

1. Disconnect oven from power supply.
2. Remove door cam trim and two screws holding door cam holder to door. Note position of cam holder.
3. Place the oven on side to gain access to lower door hinge screws.
4. Remove three screws holding lower door hinge to the chassis, while holding the door in place.

5. Carefully move door assembly down off of upper hinge pin being careful not to lose the spacer.

6. On reinstalling new door assembly, insert hinge pins with spacer into the upper door hinge and latch the door. Remove lower door hinge with three mounting screws making sure the spacer between hinge plate and door is in position. Make sure the door is parallel with the decoration sash and latch head passes through the latch hole correctly.

7. Replace the door cam arm with the mounting screws and trim strip.

DOOR HANDLE:

To replace the door handle, remove the two screws holding the latch trim. Remove two screws and washers holding door handle to door casting and remove handle being careful not to drop and damage the latch plunger.

DOOR LATCH SPRING:

To remove the door latch spring, remove the two screws holding the latch trim. Unhook spring from latch and remove screw and washer holding spring to latch bracket. When replacing spring, first hook it to the latch. Carefully stretch spring until the lower hook and screw line up with hole in latch bracket and replace screw. *NOTE:* After any service to the door, an approved RF measuring device should be used to assure compliance with proper RF emission standards.

CHAPTER 17

Gas Ranges

Proper maintenance of the modern automatic gas range, such as the one shown in Fig. 1, requires a thorough and fundamental knowledge of gas heat controls. Prior to the introduction of the automatic gas range, servicing of the old style range was quite simple; it consisted of elementary gas-orifice and air-shutter adjustments or cleaning the top burners. Simple adjustments of this sort naturally did not require extensive training, particularly since the burner designs were rarely changed.

AUTOMATIC GAS RANGE FEATURES

Modern automatic gas ranges include several features that have been perfected for practically effortless methods of cookery; they are:

1. Automatic lighting of top and oven burners with precision heat control,
2. Improved oven insulation that assures better heat distribution and prevents heat from entering the kitchen immediately,
3. New low temperature oven burners that assure the maintenance of a low oven temperature,
4. New round, horizontal-flame top burners with individual simmers that enable customers to use the waterless cooking method.

SHAPE AND LOCATION OF BURNERS

The burner should be located as closely as possible to the surface that is to be heated; however, this minimum distance is fixed by the proper combustion of the flames, which depends, in turn, on the circulation of the secondary air. For top burner work and

Fig. 1. An automatic gas range.

the heating of small utensils, the ring and star-shaped burners are preferred, because they conform more nearly to the shape of the vessels than do other forms of burners, as shown in Fig. 2.

Of these two burners, experiments seem to show that when given equally good design in the form and size of the air mixers, the drilled-ring burners possess a slightly greater efficiency; that is, they enable a larger proportion of the total heat given off by the consumed gas to be utilized in useful work. The star-shaped burner offers the advantage over the ring burner in that it is possible to drill more ports on the star-shaped surface, thus obtaining a more even distribution and greater capacity.

The advantage of the ring burner is that it provides a better supply of secondary air to all parts of the flame, since the air is

Fig. 2. A typical surface burner arrangement on a modern gas range.

free to rise on the inside of the circle of ports as well as on the outside. In the star-shaped burner, however, unless it is properly designed, the access of the secondary air to the gas issuing from the central portions of the burner is greatly obstructed. For spherical surfaces, the ring burner should be used, because the flames burning from the circle of ports will more nearly conform to the surface of the vessel than either a star-shaped or pipe burner. The ring burner can, therefore, be brought closer to the vessel than either of the other two. In certain forms of water heaters and steam boilers, where a large quantity of gas is used and a relatively small space is provided, the pedestal-type burner is used. The arrangement of these burners easily conforms to the surfaces to be heated, as shown in Fig. 3, and, at the same time, operates over a wider range of gas consumption than the other forms of burners.

On large surfaces of almost any shape such as may be found on tanks, trays, ovens, etc., the drilled-pipe burner is used. With this type of burner, an even distribution of heat over the entire surface may be obtained. If either the star-shaped or ring burner were used for this type of work, it would be almost impossible to heat the vessel evenly, and overheating directly above the burner would result.

GAS RANGE CONTROLS AND ADJUSTMENTS

The various controls associated with the modern gas range consist generally of the following:

1. Burner valves,
2. Lighters (manual and automatic),
3. Oven heat-control thermostats,
4. Automatic pan control,
5. Time controls (automatic time clock, etc.).

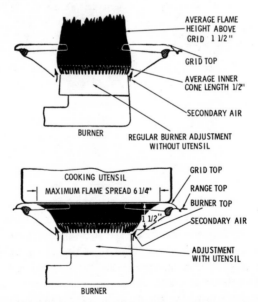

AVERAGE FLAME HEIGHT ABOVE GRID 1 1/2"

GRID TOP

AVERAGE INNER CONE LENGTH 1/2"

SECONDARY AIR

BURNER

REGULAR BURNER ADJUSTMENT WITHOUT UTENSIL

COOKING UTENSIL

MAXIMUM FLAME SPREAD 6 1/4"

GRID TOP

RANGE TOP

BURNER TOP

1 1/2"

SECONDARY AIR

ADJUSTMENT WITH UTENSIL

BURNER

Fig. 3. With a cooking utensil on the surface burner, the flame spreads out to conform to the bottom surface of the vessel.

Burner Valves

These valves control the gas supply to the burners; they are of two general types, the rotor type and the plug-and-barrel type.

These valves may be of the single- or double-duty type, depending on the particular gas volume control involved. Specifically, a double-duty gas valve is so constructed that it offers gas volume control of the center simmer, the main burner parts, and the combination of the simmer and main burner section.

A typical gas valve of the rotor type is shown in Fig. 4. These valves are right and left handed, depending on their location on the manifold assembly. Their direction of rotation is clockwise.

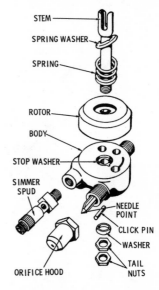

Fig. 4. The principal parts of a typical rotor-type gas valve.

Fig. 5 illustrates a typical single-duty top burner gas valve; the principal parts of the burner are a tapered plug; a barrel, or body; a stop pin; a spring; a tail nut, or cap; and an adjustable or fixed orifice hood.

The principal servicing of burner valves involves regreasing or tightening loose nuts so as to make the plug tight in the barrel. Designs and lubricants for the newer types of valves have been so improved that they usually require no regreasing. The grease used will not break down, even when subjected to temperatures of as high as 300°F. In common with all orifices, the gas passages occasionally become clogged. When this situation is found, it will be necessary to remove the valve and clean it. More commonly, however, faulty lubrication causes the valve plug to become so badly worn that the spring is no longer effective in holding the plug tight. In this case, the adjustable distance allowed for wear is completely used, and the bottom of the plug rests on the shoulder of the body. Gas valves in this condition must be replaced. Caps and spuds for gas range valves are seldom interchangeable, because special threads are cut by each manufacturer. Consequently, when

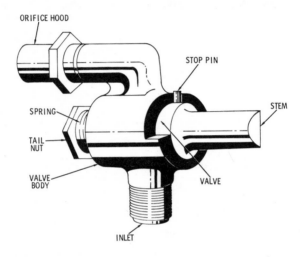

Fig. 5. A typical single-duty plug-and-barrel gas valve.

replacing gas valves, a new orifice must be supplied with the new valve. If the valve is satisfactory but is merely sticking, it can be lubricated with a suitable high-temperature grease. Both the body and the plug should be wiped to remove dirt; the gas passages are cleaned with a wire or special brush to remove any accumulation. Only the plug should be lubricated, and that sparingly. Just enough grease should be used to form a thin film of lubricant between the plug and the body wall. Excess grease tends to accumulate in the gas passages and interferes with the proper flow of gas.

Top Burner Adjustments—The correct and incorrect top burner adjustments are illustrated in Fig. 6. When correctly adjusted, the total flame should extend a short distance above the burner grates. With the change in burner adjustments comes the

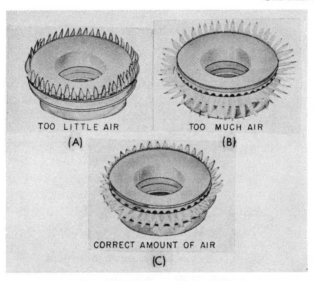

Fig. 6. Proper and improper burner flame adjustments. Adjust the orifice hood on the valve for proper flame size, and adjust the air shutter on the manifold for correct flame character.

problem of establishing a simple method of adjusting these burners to the correct input. It has been found impractical to determine the correct adjustment by measurement of the inner cone flames or the total flame, since these measurements vary on different types of burners. In few cases has the distance from the top burner ports to the top of the burner grates or the number and size of the burner ports been found to be the same. The flame spread for a correctly adjusted top burner is shown in Fig. 6C. This flame spread should be determined with a cooking vessel placed on the burner, as shown in the illustration.

Simmer Burner Adjustments—The simmer burner, shown in Fig. 7, is provided for the purpose of maintaining boiling temperature; therefore, it is important that as much attention be paid to the simmer adjustment as to the regular, or giant, burner adjustment. If this burner is adjusted properly, it will maintain any amount of water up to six quarts at boiling temperature, provided the vessel is covered and there is just enough heat applied to the cooking vessel. The simmer flame adjustment should not exceed

GAS RANGES

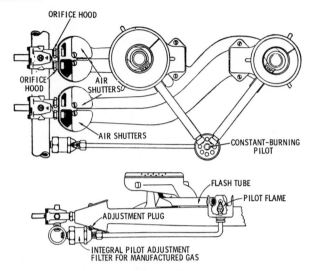

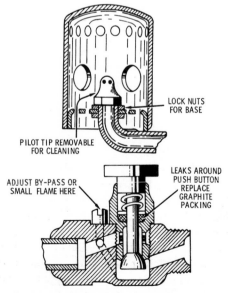

Fig. 7. A typical simmer burner; this burner is controlled by a special double-outlet gas valve that supplies gas to a small independent inner cooking burner and a large outer starting burner.

Fig. 8. A typical push-button-type gas range lighter.

the top of the burner grid. However, this adjustment is also governed by the ignition of the flames on the simmer burner by the flame lighter; the simmer section of the burner must provide immediate ignition from the pilot.

Lighters

Lighters may be divided into two groups or classes, depending on the method used to ignite the gas. These groups are the top lighters (push-button or swivel-joint-operated) and the automatic lighters. The older ranges were usually equipped with top lighters, as shown in Fig. 8, while the modern gas ranges generally employ automatic lighters.

With top lighters, pilot flames generally should be adjusted so that the top of the flame does not come in contact with any part of the lighter. An impingement of the flame on cold metal will usually produce either an undesirable odor from the incomplete combustion of the gas or an undesirable carbon deposit. Pilot outage and odor are sometimes experienced with lighters that have a solid top lighter hood; these conditions are caused by an accumulation of carbon and dirt in the dome and around the air ports of the hood and can usually be corrected by drilling a ½-inch-diameter hole in the top of the hood. If the pilot goes out and the tip is clean, the push-button assembly will have to be removed to clean the gas passages. Pilots are particularly sensitive to pressure variation or internal stoppage, which may be the cause of outage troubles. If a leak exists around the lighter valve stem, it will be necessary to remove the nut and replace the graphite packing. The base for the lighter hood can be adjusted and located by means of the jam, or lock nut, so that the gas will burn properly through the center of the openings in the hood.

Automatic lighters, such as the one shown in Fig. 9, are constructed with flash tubes that extend from the central protecting chamber, which contains the permanent pilot, to each of the burners. The speed and reliability of the flash ignition are dependent on top burner adjustment and correct positioning of all lighter parts. When the burner gas valve is opened, gas flows through the burner orifice into the burner mixer throat, and in so doing injects a certain amount of primary air. This gas-air mixture arrives in the burner head, and a portion of it flows out through a drilled

317

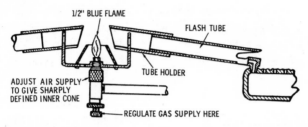

Fig. 9. The construction details of a typical automatic gas range lighter.

port, which is usually designated as the lighter port. This lighter port is generally located at the end of one of the fingers of a star-type burner and below the level of the main burner ports. On circular-type burners, it is located on the side of the burner head and below the main ports. The lighter port is usually supplied with gas before the main burner ports. The stream of gas-air mixture leaving the lighter port is directed through an air gap; it then enters the flash tube and finally emerges at the far end of the tube next to the ignition pilot flame. If the burner gas and primary air supplies have been properly adjusted, the gas-air mixture will light from the pilot, and the flame will immediately travel back through the flash tube, thereby igniting the gas mixture. The main burner ports are then ignited, either by direct communication with the lighter port or by the instantaneous flash created by the interruption of the backward travel of the flame. The gas at the lighter port remains lighted as long as the burner is operated. In general, the flame length should be such that the flame extends to the center of the tubes, which converge in the permanent pilot. Delayed ignition or complete failure may be caused by pilot flames that are either too low or too high.

Oven Heat-Control Thermostats

An oven heat control, as shown in Fig. 10, consists essentially of a device actuated by temperature changes that is designed to control the gas supply to the oven burner, thus maintaining a constant oven temperature. This device is known as a *thermostat*; its thermal element is sensitive to temperature changes and, through physical changes thus produced, originates the motion directly and indirectly to control the movement of the thermostatic valve. While gas range thermostats operate on the same principles, they may be divided into two classes—graduating and snap-action. A graduating-type thermostat is one in which the motion of the thermostatic valve is directly proportional to the effective motion of the

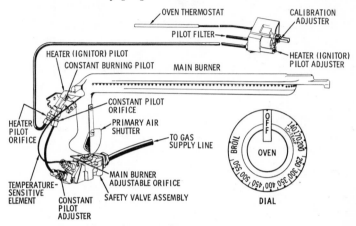

Fig. 10. A typical thermostat arrangement in an oven heat-control system. The gas supply to the heater pilot comes from, and is controlled by, the oven thermostat; the thermostat gas supply is connected to the pilot filter on the supply line.

thermal element induced by the temperature change; in the snap-action-type thermostat, the thermostatic valve travels instantly from the closed to the open position, or vice versa. The temperature dial is that part of the thermostat by means of which the position of the thermostatic valve in reference to the valve seat may be normally adjusted. The bypass is a passage that permits a flow of gas from the inlet to the outlet connection, entirely independent of the thermostatic valve.

Assist Pilot Adjustment—All oven thermostats that regulate the flame of a manually lighted oven burner are equipped with an assist, or precautionary, pilot. This pilot is supplied with gas from the inlet side of the oven heat control and burns only when the oven valve is turned on. Its purpose is to reignite the bypass flame of the main burner if the flame should become extinguished. The light-weight construction of the assist pilot makes it possible to move the tip of the pilot. The position of the tip should be checked and should be set so that the tip does not project into the oven burner flame; if it does overlap into the flame, the pilot tip might burn off. The proper length of the pilot flame after the tip has been accurately located depends on its position with respect to the oven burner flame. A flame length of between ¼ and 1 inch is usually maintained. This length can be adjusted by loosening the lock nut or protecting cap and then turning the pilot adjusting screw in the body of the oven heat control. The pilot flame should easily ignite the oven burner but should not impinge on any part of the burner itself. Ranges equipped with automatic oven ignition do not require an assist pilot due to the presence of the constant burning pilot.

Bypass Flame Adjustment—Oven thermostats, regardless of type, are equipped with an adjustable bypass valve. This valve is used to control the minimum amount of gas to the oven burner, even though the main thermostat valve is closed. Although there are two methods of adjusting the bypass flame (oven cold and oven hot), it is first necessary to light the oven burner with the thermostat in a hot position (approximately 400°F.). The thermostatic valve must then be completely closed. Some controls are designed to permit manual closing of the main valve by moving the temperature wheel, or dial, to a stop pin that corresponds to a temperature of approximately 70°F. The main valve will then be closed tight, provided the temperature scale adjustment is correct.

Bypass Adjustment With Cold Oven—Oven heat controls that have a room-temperature or 70° marking are so designed that the bypass flame can be adjusted when the oven is cold. Light the oven burner in the usual manner; then turn the wheel or dial back to a position below the 70° mark, so that the oven will not heat up. Quick and convenient adjustment can be made, if the temperature-scale adjustment is correct, by simply turning the bypass screw to the right or left until the smallest possible stable flame is obtained on the oven burner. This flame must not go out when the oven and broiler doors are quickly opened or closed. The bypass screw should then be locked in position, or the protecting cap should be replaced.

Bypass Adjustment With Hot Oven—Assuming that the burner has been lighted, set the control at a temperature well above the minimum marking on the scale, and wait until the gas is automatically adjusted to the minimum flame size. Then, change the temperature setting to a point well below that just used to be certain that the main valve is completely closed. By removing or releasing the protecting cap or lock nut, the bypass screw can be adjusted until the smallest possible stable flame is obtained on the oven burner; this, then, is the proper adjustment. Be sure that the flame does not go out when the oven and broiler doors are quickly opened or closed.

Automatic Pan Control

The gas range pan control consists essentially of a sending element in contact with the cooking utensil. As the pan temperature approaches the preset dial temperature, the sensor reduces the gas volume to the burner. At the lower dial settings, the gas volume is slowly reduced to zero, finally extinguishing the burner flame. Maximum flame size can be preset with the riser by slowly rotating the dial "flame-set" position and then turning the dial to the desired temperature. The pan-control burner is initially ignited by its pilot. The orifice hood and air shutter adjustments are the same as on nonautomatic burners. The adjustments should be made rapidly, because the pan-control sensor will be influenced by the burner flame and will automatically reduce the gas flow.

Automatic Time Controls

Automatic electric clock controls, as shown in Fig. 11, should not be confused with the spring-operated timers that are sometimes used on gas ranges. These electrically operated timing devices act like an ordinary alarm clock and issue a signal at a predetermined time. The clock control used with automatic oven ignition systems directly controls the operation of a shutoff valve, which is placed either in the main oven gas supply or in the lighter-arm supply line. The safety valve, or automatic shutoff, is used to terminate the gas supply to the oven burner if the constant-burning pilot light goes out. The methods of connecting the controls vary with different range models, but the principles of operation are identical in each case.

When the range is installed, the correct adjustments of the sizes of the constant-burning pilot flame, lighter-arm flame, oven-bypass flame, and main-burner flame must be made. The constant-burning pilot flame is supplied from a separate gas line that comes directly from the manifold, usually ahead of all other oven controls. The lighter-arm flame is supplied through either a fixed orifice or a needle valve, which should be adjusted to get the proper flame size. The oven-bypass-flame adjustment is located on the body of the oven heat control. The constant-burning pilot flame should be adjusted so that it is capable of effectively lighting both the main burner and the flame on the lighter arm. The timing adjustment should not be made unless it is absolutely necessary. The timing mechanism has been adjusted at the factory by the manufacturer to provide proper operation when the lighter arm is correctly adjusted. In any case, the adjustment of the pilot and lighter-arm flame should be checked carefully before going ahead with the timing adjustment.

Field adjustment of the timing device should only be made by experienced persons, since a full understanding of the cycle and the mechanism is necessary. In the instructions given here, the

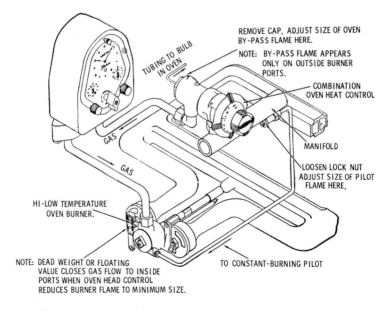

Fig. 11. A typical automatic electric clock control for the oven burner. In this system, the electric clock automatically causes the oven burner to be ignited and shut off at specific preset times.

time required for the valve to open is understood to be the period of time when the thermal element is cold. It is obvious that if the oven has been in operation for some time, and the parts of the control are heated up, the time required for the valve to open will be shorter than this prescribed time, and a proper adjustment, therefore, cannot be made.

Range Regulation

In locations where gas pressure is not regulated, it may be necessary to install a pressure regulator at the range. The function of the range regulator is to maintain a constant gas pressure at the burner orifices in order to assure uniform heat on each burner, thus improving the operation of the automatic heat controls and reducing pilot outages. The pressure regulator consists essentially of a diaphragm in the gas supply line that is governed by the pressure on the gas outlet. This pressure is opposed by the force exerted on the top of the diaphragm by weights or a spring that may be adjusted to vary the force on the diaphragm, thereby changing the gas outlet pressure. The pressure regulator, when used, should be installed as close to the range as possible, to eliminate the effect of a variable pressure drop in the house piping, and in a cool place, since excessive heat will dry out the leather diaphragm. In addition, the regulator should be so placed that the essential ports are readily accessible for cleaning and adjustment.

GAS RANGE INSTALLATION

Before starting the gas range installation, turn off all other gas appliances, such as the gas furnace, gas water heater, gas space heater, etc. Next, shut off the gas supply at the main inlet, which is usually located at the gas meter.

Gas Piping

The gas supply to the range must be of adequate size to allow a full supply of gas to the appliance. A ¾- or ⅝-inch-diameter pipe or flexible tubing will usually give satisfactory service, provided the gas supply line ahead of the reduction point is of adequate size. Do not use soldered connecting pieces. To prevent leakage around threaded joints, use a joint compound and gas line gaskets

323

that are resistant to the actions of liquified petroleum gases. Where regulations permit, use a suitable flexible connector to attach the range to the gas piping; this type of connector will avoid the necessity of aligning the gas supply to the gas inlet on the range and will generally make installation simpler. If the range is equipped with an electric clock, signal timer, and/or an oven lamp, the power-supply cord should be plugged into the wall terminal but not connected at the disconnect device until all joints are ascertained to be leak-proof.

Checking for Gas Leaks

With the gas range in place, and the gas supply line connected to it, the complete system must now be checked for gas leaks. Under no circumstances should a range be used until all leaks have been eliminated. **Do not attempt to test for leaks by using a match or other type of open flame.** Proceed as follows:

1. Turn off all gas cocks on the range.
2. Turn on the gas supply at the main inlet or meter outlet.
3. Apply a thick solution of soap and water to all connections in the supply line. If bubbles appear at any connection, a leak is indicated and should be repaired.
4. All leaks should be eliminated before the range is lighted. Do not allow any exposed flame when repairing leaks.

Another method of testing for leaks is to shut off all gas cocks and pilots on the range as well as on all gas equipment in the house and then check the position of the test dials on the gas meter. If the dial positions have not moved in one hour, it is safe to assume that there is no leak in the gas supply system. If, however, the dials have moved, the first method (outlined above) should be carried out, since this will locate the exact position of the leak.

Clearance From Combustible Material

Domestic kitchen ranges, when installed on combustible floors, should be set on their own bases or legs and should be installed with clearances of not less than those shown on the marking plate and in the manufacturer's instructions. In no case should the clearance be small enough to interfere with requirements for the combustion of air and accessibility for operation and servicing. Domestic kitchen ranges should have a vertical clearance above the cooking top of not less than 30 inches to combustible materials or metal cabinets. When the underside of such combustible materials or metal cabinets is protected with asbestos millboard that is at least ¼-inch thick and covered with sheet metal of not less than No. 28 U.S. gauge, the clearance should not be less than 24 inches. This protection should extend at least 9 inches beyond the sides of the range.

CHAPTER 18

Mixers, Blenders and Can Openers

The electrically operated food mixer is a most useful kitchen appliance that can quickly and efficiently perform numerous mixing and food-preparation tasks. Fig. 1 shows a modern electric food mixer; it consists essentially of a stand or pedestal to which a speed-regulated electric motor is fitted, a pair of food beaters, and a large bowl in which the food to be processed is placed. In operation, the food is beaten or mixed by the revolving beaters, which are attached to the motor shaft and gear assembly.

OPERATION PRINCIPLES

Food mixers, like any other household appliance, are simple to operate once the fundamentals have been thoroughly understood. As shown in Fig. 1, the bowl is held in position by the turntable assembly. The complete motor and the transmission case are attached to the pedestal and base by a hinge pin. This hinge allows the head of the mixer to be swung up and back, so that the beater and bowl can easily be removed and replaced. The mixer head is locked into position prior to operation.

When it is desired to operate the mixer, the cord plug is inserted into the electric socket in the same manner as that of a toaster or any other electrical appliance. The speed control energizes the motor and regulates the speed of the beater. Be sure that the control is in the "off" position prior to inserting the cord plug. When the speed control is turned to its first speed step (usually the "low" speed position), the beaters begin to revolve slowly; when turned to the next position, the speed increases noticeably, until the last

Fig. 1. A typical electric mixer with a meat-grinder attachment.

speed step (or "high" speed) on the mixer has been reached. The control snaps into place at each speed step. After a certain period of continuous usage, the motor head may become warm, but this is only normal; no harm will result to the mixer because of the increased heat. It is quite normal for the temperature of the motor to rise if it is used steadily for a long time or when mixing heavy batters.

In mixers equipped with bowl turntables, or revolving discs, the position of the beaters is adjusted by means of a lever arrangement, which positions and also gives the proper bowl speed according to the size of the bowl used. Beaters are removed or ejected from their sockets either by a slight pull on their stems, or, as in some mixers, by turning the handle down to the side of the motor. They are reinserted by merely pushing them back up until they click in place.

The proper beater speed to be selected depends on the type of food to be processed; thus, food containing heavy ingredients should be stirred slowly, while light ingredients, such as eggs, should be beaten or stirred rapidly. Most recipes call for a number of different speeds and also recommend the required time in minutes for each speed as additional ingredients are added; the manufacturer's instructions accompanying each machine should be followed in each instance.

Portable Mixers

Most models of stationary food mixers may be detached from their base or pedestal, thus enabling the user to beat or whip the food directly in the cooking vessel by using the handle support attached to the machine. This procedure will, in many instances, simplify as well as facilitate the amount of work involved, particularly since there will be fewer dishes to wash and dry after the cooking or food-preparation operation.

Portable mixers, as shown in Fig. 2, are food mixers without stands or pedestals. They are available in most well-equipped appliance stores. These are employed in the same manner as stationary mixers, the only difference being that the mixer must be supported by hand or on an improvised stand during operation.

Motors

The universal-type motor is most commonly used for food mixers. It may be used on either alternating or direct current. The voltage specified for the motor is usually 115 volts, and the mixer should not be used on a power supply whose voltage varies more than 10% of that figure. Universal motors can be distinguished from induction motors by the fact that universal motors have a commutator and carbon brushes like an ordinary DC motor. Most universal motors used in food mixers are of the concentrated-

329

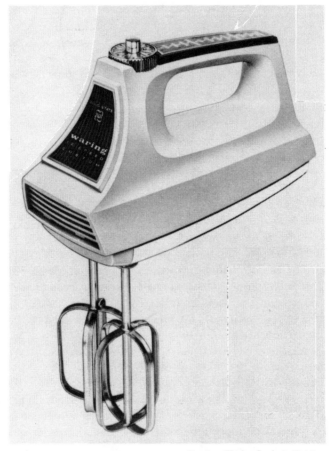

Courtesy Waring Products Division

Fig. 2. A typical portable electric food mixer.

pole, noncompensated type, while those with a higher horsepower rating are of the distributed-field, compensated type. In a universal motor, all the iron in the magnetic circuit must be laminated; if this were not done and the motor was operated on AC, the eddy currents would quickly cause excessive overheating. Since the armature and the field are series-connected, this type of motor has certain inherent characteristics of the DC series motor, such as high speed without load, good torque, etc.

Motor Speed Control

Because of the difference in fluidity (thickness) of the various food mixtures and in the amount of stirring action required, modern food mixers are closely regulated with respect to motor speeds. The various speed ranges are usually tabulated in a number of increasing speeds that vary from approximately 60 revolutions per minute to several hundred. Since all universal motors are series-wound, their performance characteristics are much like those of the usual DC series motor. Without speed control, the no-load speed is quite high, but seldom high enough to damage the motor, as is the case with larger DC series motors. When a load is placed on the motor, the speed decreases and continues to decrease as the load increases. Although universal motors of several construction types are manufactured, they all have these varying speed characteristics.

There are three general methods of speed control; they are:

1. Governor type,
2. Tapped-field type,
3. Adjustable-brush type.

Governor Control—Governor speed control of the motor in a typical food mixer is attained through the use of an electrical governor assembly that is normally mounted at the rear of the armature shaft against the control-plate assembly. The electrical circuit, as shown in Fig. 3, is made and broken by the action of the governor as it revolves against the control plate. When the switch lever is moved to the "on" position, the position of the control plate with respect to the governor is changed by the action of the switch-control mechanism, which is built into the bottom of the gear case and motor housing, thereby changing the speed of the motor. Thus, when the control plate is set close to the governor, a relatively low motor speed causes the governor to make or break the electrical circuit through the control-plate contact points. When the control plate is set farther apart, a greater motor speed is required before the governor can break the circuit.

The action of the governor is such that the speed of the motor will remain constant for a given setting of the control plate regardless of the load imposed on the mixer. The speed-control mechanism contains a resistor that is connected in parallel with the control-plate contact points. By this means, the electrical circuit is not completely broken when the contact points are opened through the action of the governor; the circuit is then shunted through the

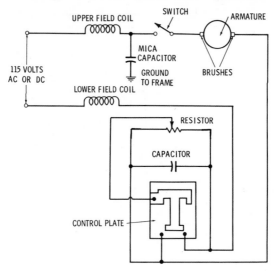

Courtesy Hobart Manufacturing Company

Fig. 3. The electrical wiring connections between the motor and the speed-control components.

resistor. A capacitor is connected across the control-plate breaker points to suppress sparking.

Tapped-Field Control—The tapped-field method of speed control is illustrated in the circuit arrangement of Fig. 4. With the speed-control switch in the "low" position, the field windings are in series, thus providing the lowest possible speed obtainable. With the speed-control switch in the "medium" position, part of the field winding is disconnected from the circuit, and an intermediate speed will be furnished. Finally, a shift of the speed-control switch to the "high" position cuts out an additional portion of the field winding with a resultant increase in the current and motor speed. In this manner, any desirable number of speeds may be obtained from the universal-type motor by merely adding the desired number of field-winding taps.

332

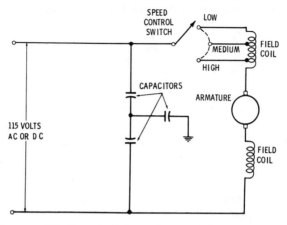

Fig. 4. The tapped-field method of speed control as employed in universal motors.

Adjustable-Brush Control—The adjustable-brush speed-control method is shown in Fig. 5 and is widely employed in food mixers. This method consists of a simultaneous brush displacement as the brushes ride on the motor commutator. This operation is performed by means of an externally located brush shift lever. Since there is only one location of the brushes, with respect to the motor field, at which the motor will develop its maximum speed and power, any movement away from this position will result in a continued decrease in both speed and torque; it is by this means that speed control is obtained in adjustable-brush-type universal motors.

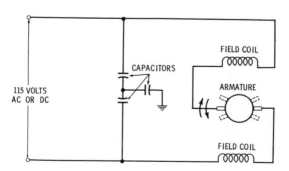

Fig. 5. The adjustable-brush method of motor speed control.

SERVICING AND REPAIRS

In order for the serviceman to intelligently diagnose a defective food mixer, he must be familiar with the functional parts and operation of a normal mixer.

Trouble Chart

The charts on the next three pages lists the more common troubles encountered in the repair of typical food mixers. The trouble and possible causes are given with the method used for remedying the defect.

Lubrication

Every food mixer requires oiling and greasing at certain time intervals for the best performance. Certain types of mixers are equipped with oil holes located on top of the motor casing; these mixers require from two to three drops of an approved oil once

Food Mixer Trouble Chart

Trouble	Possible Cause	Remedy
Motor will not run (assume correct voltage available at wall outlet).	Open electrical circuit.	Progressively disassemble speed-control mechanism and motor, and perform following checks until open circuit is found: Defective plug on cord set. Loose connection in speed-control switch or mechanism. Wire-lead clip detached from brush holder. Bad connection between field and cord. Open-circuited field.
Planetary turns, but beater does not revolve.	Pinion gear drive pin broken.	Remove planetary, and take off pinion gear. Replace drive pin.
Mixer runs with raspy, bumpy noise at planetary.	Gear case cover or internal gear teeth worn or broken.	Remove planetary and gear case cover. Complete gear case cover assembly must be replaced.
Mixer runs with bad vibrations and rumbling noise.	Defective governor.	Replace governor.

Food Mixer Trouble Chart (Continued)

Trouble	Possible Cause	Remedy
Mixer runs on "low" speed but has no power.	Dirty contacts. Bad electrical connections. If mixer still has no power after cleaning contacts and checking connections, contacts are defective.	Clean or replace. Check connections, tighten or repair. Replace control-plate assembly.
Mixer runs only on high speed.	Control-plate spring unhooked.	Remove trim and cover, and check control-plate spring. If unhooked, attach to top of control plate and squeeze spring end.
	Welded contacts.	Observe operation of control-plate contacts when switch is turned from "high" to "low" speed. If contacts do not separate, they are welded together. Replace control-plate assembly.
	Short-circuited capacitor.	Disconnect one capacitor lead, and turn switch to "low" speed position. If mixer goes into "high" speed when the loose lead is touched to other leads, capacitor is shorted and should be replaced.
Mixer runs with loud screeching noise.	Ball in end of armature shaft worn or flat.	Disassemble motor, and replace ball.
	Thrust plug receiving thrust from worn ball.	Disassemble mixer, and replace thrust plug.
Mixer has no power on "low" speed, but has normal power on "high" speed.	"Low" speed adjustment improperly set.	Remove end cover, and plug in cord. Planetary should revolve at approximately 60 rpm at "low" speed position. To adjust, turn in control-plate adjusting screws by an equal amount.
	Defective governor.	Turn switch to highest speed position, and hold out control plate as far as possible. Turn speed selector to "off" position, and observe governor as it recedes when armature slows down. Outer plate must move in smoothly without sticking, until

Food Mixer Trouble Chart (Continued)

Trouble	Possible Cause	Remedy
		it almost touches middle plate. If governor is defective, replace with a new one.
Mixer will not shut off.	Switch not correctly adjusted with control link and cam assembly, or defective switch.	Remove trim and adjust. Replace defective switch.
Operator receives shock when mixer is touched.	Bare wire in contact with mixer housing.	Pull out plug, and, with switch "on," check for ground with series test lamp. Touch one prong of test lamp with prong of plug, and touch other prong to an unpainted spot on housing. If lamp lights, mixer is grounded. Examine all wiring in order of its accessibility until grounded wire is found and repaired.
Mixer will not run, although switch clicks and motor hums.	Frozen bearing.	Examine all bearings in order of their accessibility until frozen bearing is found and repaired or replaced.

every month. Occasional oiling is also required for the mixing-bowl bearing on mixers furnished with turntables for bowl rotation. Under normal service conditions, most modern food mixers do not need disassembly for the purpose of greasing and oiling for many years. When a mixer is subjected to abnormally severe usage, it is advisable to check the gear case and the planetary and internal gears for the proper amount of lubrication. Also see that the planetary beater shaft is free-running; place a few drops of light- or medium-weight oil on the beater shaft, and wipe off the excess oil.

In food mixers that are equipped with juice extractors, the gear case lubricant should be checked for thinness due to over-oiling or carelessness in extracting fruit juices. If water or other liquids have been allowed to overflow the juice-extractor bowl and have penetrated the gear compartment, all of the old lubricant should be cleaned out, and the gear compartment should be packed with an approved medium-weight grease.

Disassembly of Food Mixers

It will rarely be necessary to completely disassemble and reassemble a mixer, since most repair operations can be confined to the immediate area affected. If it is necessary to disassemble a mixer, there is no particular sequence to follow, since a close inspection of the mixer in question will immediately reveal the method of sequence to be used in the disassembly procedure. When assembling the mixer, however, a definite procedure should be followed so that the work can be checked at the various stages of the assembly. In the case of a complete disassembly, it is well to check and reassemble the gear case first, then the motor, the speed-control mechanism, and finally the trim strip and handle assembly. When checking or repairing the electrical system, particular attention should be given to the wiring diagram, so that all connections are made as they were prior to disassembly; this will insure the proper working condition of the mixer.

Disassembly of a General Electric Portable Mixer— These are light-weight appliances designed to meet the demand for a mixer of maximum portability and service. They are equipped with a powerful two-speed motor, which is connected as shown in Fig. 6. With the beaters removed, the motor unit may be set down on any convenient counter; with the beaters in place, the mixer may be rested on its heel stand in the same manner as an electric iron. The beaters are of a special tear-drop design, which is suitable for use in almost any type of bowl. The requirement for a special bowl has thus been eliminated.

The disassembled view of a typical mixer is shown in Fig. 7. To completely disassemble this appliance, proceed as follows:

1. Raise the escutcheon plate (1), and remove.
2. Remove two screws (4), and separate the case and handle assembly from the motor and base assembly.
3. Remove two screws (9), and lift up the switch (8). Loosen the soldered connections to replace.
4. Remove two screws (13), and lift off the thrust plate (14).

337

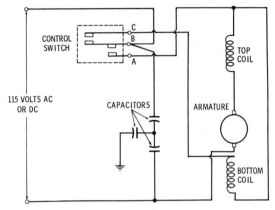

Courtesy General Electric Company

Fig. 6. The electrical connections of a General Electric two-speed food mixer. With the control switch in the OFF position, all switch contacts are open. With switch contacts B and C closed, the motor will operate at low speed. When contacts A and B are closed, the motor will operate at high speed.

Push up from the bottom, and lift out the spindle assemblies (18 and 19). Loosen the setscrew (23) to adjust or replace the thrust bearings (30).

5. Remove two screws (33) to release the brush-holder assemblies (26 and 34), brushes (28), and springs (27).

6. Remove one screw (24) holding the cord clip (22) to release the cord set (25). Remove three additional screws (24) and two screws (11) that hold the bearing cap (12) in place. Lift out the bearing cap and then, in a group, the armature (17), the field (16), and the front bearing assembly (15). The oil wick (29) can now be removed for replacement.

7. Pull the gasket (38) from the motor base (32), and remove two screws (37) and the base cover plate (36). The spindle washers (31) may now be removed.

8. Remove screw (7) and screw (10) to separate the handle (3) from the case (6).

If it is necessary to loosen the switch but not replace it, the leads must be handled with extreme care, since they are quite fine and brittle at the soldered connection.

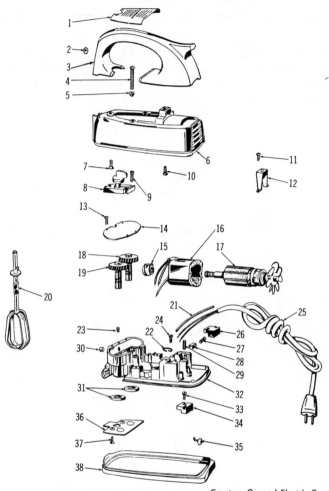

Courtesy General Electric Company

Fig. 7. The disassembled view of a General Electric portable electric food mixer.

The reassembly of this General Electric portable mixer is simply a reversal of the foregoing disassembly procedure. When reassembling the mixer, it should be noted that the left- and right-hand spindles (18 and 19) are physically interchangeable but not functionally interchangeable, and extreme care must be exercised to make sure that each is installed in its correct location. Each spindle has a spiral groove on the shaft that acts as an oil pump. If installed in the wrong location, the oil groove will tend to pump oil down into the beater instead of up into the gear case and will cause an oil leak. When spindles are removed and replaced, make sure that the beater slots are at an angle of 45° relative to each other in order to avoid a clash between the beater blades.

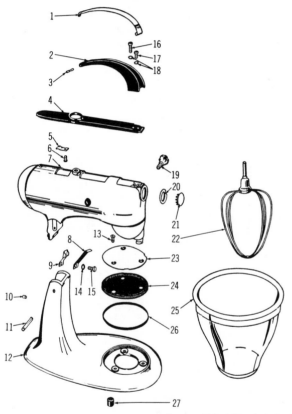

Fig. 8. The disassembled view of a Kitchen Aid food mixer. In the illustration, 1 represents the handle cover; 2, the gear case handle; 3, a pin; 4, the trim strip; 5, the trim strip retainer; 6, a screw; 7, the gear case and motor; 8, the spring clip; 9, the gear case stop; 10, the setscrew; 11, the hinge pin; 12, the pedestal; 13, a screw; 14, a washer; 15, a screw; 16, a screw; 17, a screw; 18, a washer; 19, a thumb screw; 20, the attachment hub ring; 21, the attachment hub cap; 22, the wire whip assembly; 23, the clamp disc; 24, the screw cap; 25, the bowl; 26, the rubber seal; and 27, the base foot.

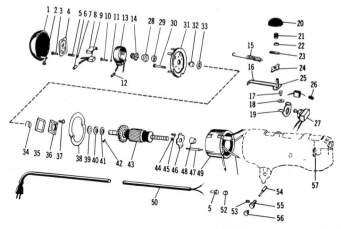

Fig. 9. The disassembled view of a Kitchen Aid mixer motor and control unit. In the illustration, 1 represents the end cover; 2, a screw; 3, an adjusting screw; 4, the control-plate assembly; 5, the adjusting spring; 6, a wire connector; 7, a lock nut; 8, the capacitor insulator; 9, a capacitor; 10, a screw; 11, a washer; 12, a wire connector; 13, a resistor; 14, the governor assembly; 15, the control-plate spring; 16, the control link and cam assembly; 17, a screw; 18, a washer; 19, the switch bracket; 20, the switch knob; 21, the knob spring; 22, the spacing washer; 23, a pin; 24, the compression spring; 25, the radio-interference capacitor; 26, a wire connector; 27, the toggle switch; 28, the motor bearing cap; 29, a felt washer; 30, a screw; 31, the bearing bracket; 32, the bearing; 33, the bearing retainer; 34, a spring; 35, the oil shield gasket; 36, the oil shield; 37, a screw; 38, the seal ring; 39, the spacing washer; 40, the keyed washer; 41, a felt washer; 42, the governor drive stud; 43, the armature; 44, the motor shaft ball; 45, a screw; 46, the bearing retainer; 47, a screw; 48, the bearing; 49, the motor field; 50, the cord set and plug; 51, the bushing; 52, the strain relief clip; 53, a setscrew; 54, a brush; 55, a brush holder; 56, a screw cap; and 57, the thrust plug.

Disassembly of a Kitchen Aid Mixer—In the disassembly of this mixer, reference is made to Figs. 8, 9, and 10, which illustrate the parts relationship and how they are assembled. The disassembly of the handle, trim strip, and switch knob may best be accomplished by referring to Fig. 8. The handle (2) must be removed from the mixer before the switch knob can be removed. To disassemble the handle (2), drive out the pin (3) in its back tip. Next, lift up the handle cover (1), and remove the two screws (16 and 17) and lock washers (18) that attach the handle to the gear case and motor frame (7). Pull the switch knob from its shaft, and remove the trim strip (4).

341

Now, with reference to Fig. 9, remove two screws (17) and lock washers (18) that hold the toggle switch (27) in place. Lift out the switch, and break the electrical connection. One toggle-switch lead

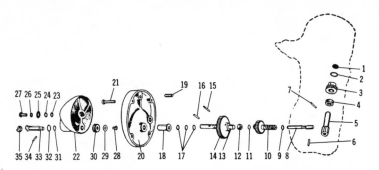

Courtesy Hobart Manufacturing Company

Fig. 10. The disassembled view of a Kitchen Aid mixer gear and planetary unit. In the illustration, 1 represents the expansion plug; 2, a washer; 3, the attachment hub gear; 4, the attachment hub bearing; 5, the attachment hub shaft; 6, a pin; 7, a pin; 8, the worm gear shaft; 9, a fiber washer; 10, worm and worm gear assembly; 11, a steel washer; 12, a ball bearing; 13, the vertical center shaft; 14, the helical gear; 15, a pin; 16, a pin; 17, a fiber washer; 18, the center shaft bearing; 19, a dowel; 20, the gear case cover; 21, a screw; 22, the planetary; 23, a shim washer; 24, a shim washer; 26, a lock washer; 27, a screw; 28, a screw; 29, a steel washer; 30, a pinion; 31, a steel washer; 32, a washer; 33, the beater shaft; 34, a pin; and 35, the beater shaft spring.

is attached to the black lead from the motor field assembly by means of a small Bakelite screw connector (26). The other toggle-switch lead has a U-shaped spring cap attached to it, which, in turn, is fastened to the inside end of the right-hand brush holder (viewed from the front of mixer). A red lead with a U-shaped clip is fastened to the left-hand brush holder; this lead runs to the speed-control mechanism at the rear of the motor. Removal of these clips will be facilitated by the use of a pair of long-nosed pliers.

When reassembling this portion of the mixer, several precautions must be carefully observed. The two lead clips must be pushed into place in the slots at the ends of the brush holders. The leads themselves must be pushed back so that they do not make contact with the commutator. When reconnecting the switch lead to the motor field lead, the connection must be tight, and no bare wire should be exposed which might cause a ground. When using wire connectors, the best results are obtained by laying the bare ends of the wires alongside each other with their insulation even; then place the connector over the bare ends and screw them together firmly.

Assemble the toggle switch (27 in Fig. 9) to the mixer with the two screws (17) and lock washer (18). The lock washer is used tightly, but be careful not to strip the threads in the casting. The toggle switch (27) must be assembled in the proper relation to toggle switch (27) must be assembled in tne proper relation to the cam (16), which is attached to the lower end of the shaft where the switch knob (20) is mounted. When the switch knob turns the cam from the "off" position, the point of the cam must just barely clear the two points of the toggle-switch lever. If this clearance is not held to a minimum, the mixer may not turn "on" or "off" when the switch knob is operated.

Slide the trim strip (4 in Fig. 8) into position. Place the switch knob on its shaft, and check for a tight fit. If the switch knob does not grip the shaft firmly, remove and replace the spring inside the knob. When positioning the switch knob, make certain that the toggle switch is in the "off" position; the word "OFF" on the knob should be toward the front of the mixer and in line with an imaginary line through the center of the gear case and motor frame. Attach the gear-case handle (2) to the mixer with two screws (16 and 17) and lock washers (18). Position the handle cover (1) on the handle (2), and insert the pin (3).

The handle, switch knob, and trim strip should be removed if the motor is to be completely disassembled. If the work, however, is confined to the speed mechanism only, do not remove the trim strip. Access to the speed-control mechanism and motor, Fig. 9, is through the rear of the mixer. To reach the speed control, remove the screw (2 in Fig. 9), which holds the end cover (1) to the gear case and motor frame, and lift off the end cover (1). Break the electrical connections by removing the three-wire connectors (6 and 12). Turn the switch knob (20) to the high speed position, so that the push link of the control-link-and-cam assembly (16) extends as far as possible. Unhook the control-plate spring (15) from the top of the control-plate assembly (4), and hook it into the small hole in the end of the push link; this step will prevent the spring from flipping back into the gear case. Detach the control-plate assembly (4) by loosening the lock nuts (7) and removing the adjusting screws (3) and spring (5). Lift out the capacitor (9) and the insulator (8). To remove the governor (14), insert two screwdrivers under the oblong flange, and pry it from the shaft. Remove the drive stud (42) from the end of the armature shaft. The resistor (13) may be removed by unscrewing two screws.

To disassemble the motor, remove the two brush screw caps (56), and pull the springs and brushes (54) from the brush holders (55). Next, remove the rear bearing bracket (31) to remove the armature. The field assembly (49) and front bearing (48) may then be removed in the usual manner. The reassembly of the motor and speed-control mechanism is, in general, a reversal of the disassembly procedure. Precautions should be observed to see that the leads are not pinched in the reassembly process, and that they are not placed in such a manner as to be damaged by the ventilating fan.

If a new motor-shaft ball (44) is required in the end of the armature shaft, press it into position. If the ball has a tendency to fall out, it can be held in position with a small portion of heavy grease. Once the armature (43) is in position, the ball cannot fall out. If the bearing bracket (31) is disassembled, and a new gasket (35) is required, cement it into position with shellac. Place the bearing bracket (31) in the gear case and motor frame. Check the motor armature (43) for free running and end play. Occasionally the self-aligning bearings are cocked slightly at this stage; to realign the bearings, tap the outside of the casting gently with a mallet. The armature should turn freely and should have a slight end play. To adjust the end play, remove the bearing bracket (31), and add or remove spacing washers (39) on the end of the armature shaft. If excessive end play is encountered, the speed-control mechanism will not function properly.

It will seldom be necessary to disassemble the gear case, and then only to replace damaged or worn parts. The exploded parts view of the gearing and planetary unit is shown in Fig. 10. To disassemble, turn the mixer upside down in a padded cradle or other suitable device padded with soft cloth to protect the mixer finish. Remove screw (27) and washers (23, 24, 25, and 26), which hold the planetary (22) to the vertical center shaft (13).

One screwdriver will not be sufficient, as the planetary will bind on the shaft if pried from one side only. Remove the planetary drive pin (16) from the vertical shaft (13). To disassemble the beater shaft (33), remove screw (28) and washer (29) from the top, and lift off the pinion (30). Take out the pin (34), and lift off the washer (31). Pull the beater shaft (33) from the planetary (22), and remove the beater shaft (35) and washer (32). The gear case cover (20) can easily be removed by unscrewing four screws (21), which hold the cover in place. Insert the juicer reamer and shaft from the juicer attachment into the attachment hub. Grasp the reamer, and turn it to the right (clockwise) until the gear-case cover (20) comes out of its seat. Remove the grease from the housing; if the grease is in good condition, place it aside to use at reassembly.

During the reassembly of the gear case, each part should be lubricated. Lubrication of the gear case consists of approximately five ounces of specially selected grease, which should be approved by the manufacturer. Make certain that the inner space of bearing (12) fits tightly against the steel washer (11). Pin the helical gear (14) to the vertical center shaft (13). Place the same number of fiber washers (17) on the shaft as were removed at disassembly. Position the vertical center shaft (13) in the housing, and assemble the gear-case cover (20) to the housing with the four screws (21). The gear case should be tested at this time to make certain that the bearings are running freely and the gears are properly en-meshed.

Disassembly of a Sunbeam Mixmaster—This food mixer is shown in Fig. 11. The speed control for this machine is obtained by means of a centrifugal-type governor that is mounted on the motor shaft and connected in the motor circuit, which is illustrated in Fig. 12.

To disassemble the Mixmaster, pry out the center disc in the speed-control dial, and remove the lock nut under it; this will allow the dial unit to be removed. Two screws will be found in the rim of the next section, and after their removal, this part can be removed. Next, a rotating switch and the accompanying brushes that bear on a pair of collector rings will be found. Remove the brushes, and loosen the Allen setscrew; the rotating-switch element can then be pulled off the shaft.

345

As previously mentioned, speed control is achieved by means of a governor, and the chief controlling element is the rotating switch. The centrifugal-force action on the pivoted arms causes

Fig. 11. An automatic Sunbeam Mixmaster mixer.

the contacts to open and close. This serves to cut resistance in and out of the circuit as required to maintain the desired speed. The action, therefore, in moving the speed-control dial on the end of the machine moves a conically shaped slotted piece in and out; this affects the contacts in their opening cycles, thus regulating the speed of the machine. With this sliding piece in the fully-in position, the lowest speed is obtained, because the contacts are held open, thus permitting the complete use of the circuit resistance. With the contacts held closed, the highest speed is delivered, since the circuit resistance is shunted out. Further dial action actuates a plunger rod, which is connected to an "on-off" switch.

As in most circuits of this type, a capacitor and a resistor are connected across the switch contacts to eliminate radio interference and reduce contact arcing. Both the resistor and the capacitor may be replaced when found defective by simply spreading the brass supporting strips apart and lifting the raised buttons out of the holes in the strips.

The motor armature can be removed in the customary manner by removing the rear-bracket and motor-bearing assembly. Access

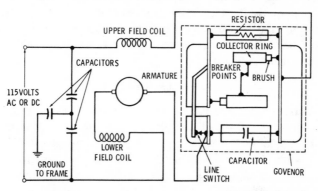

Courtesy Sunbeam Corporation

Fig. 12. The wiring diagram of a Sunbeam Mixmaster mixer. Turning the switch dial closes the line switch, thereby permitting the current to pass through the circuit and allowing the control points to make and break intermittently. As the switch dial is turned to different speed settings, one of the control points is moved in or out between the high and low positions. When the motor increases its speed, the governor causes a sliding pin to move outward; this pin pushes a control spring with its contact outward, thereby breaking the circuit at a speed corresponding to the contact setting.

to the gear box at the front of the machine can be obtained by removing a single screw, which holds the handle in place, and then by removing four machine screws in the cover. After cleaning and inspecting the gears, fresh grease should be added as required. In some food mixers, motor bearings may be worn loose after a long period of use. This is usually indicated by a noisy motor; in some cases, this condition may prevent the motor from attaining normal speed. If bearing wear is suspected, the armature shaft should be tried in the sleeve bearings; if a loose fit is found, new sleeves should be installed. Proper lubrication at certain specified time intervals can eliminate this condition in most instances.

To reassemble the mixer, reverse the disassembly procedure. Be careful when reassembling the governor to replace the parts in the same location as they were originally. A short pin with one square end fits in a hole in the metal cover piece, its square end fitting in the circular groove in the inside of the dial control. The long push rod also fits in a hole in this cover, with its insulating end resting in a recess in the "on-off" switch arm. When replacing this cover on the end of the motor, make certain that the slotted sliding piece fits in place in the grooves in the center of the rotating governor unit. Brushes should be replaced if worn, and new brushes should be shaped to the diameter of the commutator with a piece of fine-grade sandpaper.

BLENDERS

Various types and models of blenders are on the market ranging from the deluxe 14 different speeds with a 60 second timer to the economy 2 speed model. Fig. 13 illustrates a Waring 7 speed with 60 second automatic timer. Attachments are available to automatically peel fruit and vegetables and crush ice.

The method of drive will vary due to design and manufacture. Some manufacturers use direct drive while other manufacturers use a reinforced toothed belt drive. Whichever drive is used, the disassembly and service procedures will follow a general pattern when service is required.

Disassembly of the blender is confined to the base which houses the motor and all component parts. The blending jar contains the cutting knives which is the heart of the blending action. Always check the condition of the blending knives before you disassemble the base—many times the knife and shaft assembly will become loose and slip on the drive coupling. A typical blender jar with knife assembly is shown in Fig. 14. If nut (10) in Fig. 14B becomes loose, the cutter or knife assembly will loosen and liquid will leak at the bottom of the blender jar. Make sure the shaft and seal assembly is tight and not binding before proceeding any farther.

Courtesy Waring Products Division

Fig. 13. A 7-speed blender with 60 second automatic timer.

SERVICING PROCEDURE

Blender noisy. Check knife and seal assembly on blender jar. See Fig. 14. A good check for noise is to remove blender jar from base and then turn on motor. If noise has stopped, the trouble

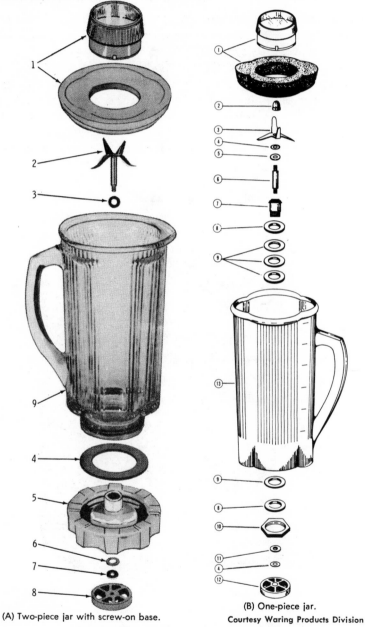

(A) Two-piece jar with screw-on base.

(B) One-piece jar.

Courtesy Waring Products Division

Fig. 14. Typical blender jar with knife assembly.

would naturally be in the knife and seal assembly. Check fan blade clearance at back or bottom cover. Check for loose belt or worn rubber drive coupling.

Motor will not run. Check all wire connections. Check line cord for continuity. Check carbon brushes to make sure they are not hanging up in the brush holder. With line cord removed from socket and power switch in the "off" position, check diode with an ohmmeter. Diode should read 4 to 10 ohms and infinite reading with test loads reversed. Check armature for open or shorted windings. Check field coil.

Courtesy Waring Products Division

Fig. 15. A typical electric can opener.

Motor runs but will not turn. Check drive assembly for broken or worn belt or worn drive coupling.

CAN OPENERS

An electric can opener (Fig. 15) is one of the most widely used appliances. There are many makes and models which include automatic shut-off and knife sharpeners plus easy clean removal cutter wheel and operating lever as added features.

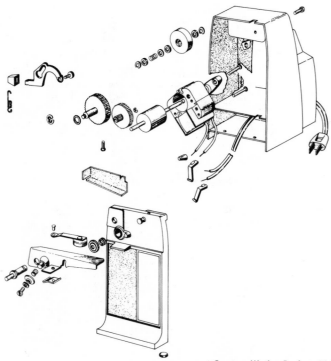

Courtesy Waring Products Division

Fig. 16. A disassembled view of a typical can opener.

OPERATION PRINCIPLES

A can is placed with the lip on the drive wheel and with the piercing lever in the up position. As the lever is lowered, the cutting wheel engages the can and the latch contacts the latch stud forcing the lever to the maximum right position. As the can lid is pierced and the switch button is depressed by the lever, the motor starts and the can is rotated in a counerclockwise direction by the drive wheel.

The pressure of the can lid against the cutting wheel shifts the lever back to the maximum left position and locks the latch into the lock stud until the lid is completely severed. When the lid is completely cut, the cutting force disappears and the pierce lever is allowed to shift to the right. The switch blade spring pressure and the hold down spring pressure then raises the lever enough to stop the motor. The can is then removed by raising the lever.

SERVICING PROCEDURE

Motor runs continuously. Check switch blades for spring return and replace if defective. Also check pierce lever assembly for free movement.

Motor will not run. Check for continuity in electric cord, switch and motor. If motor repeatedly overheats or stalls, check drive wheel for contamination and proper lubrication. Check gears for broken or binding teeth.

Can stalls or does not cut. Check drive wheel for worn teeth. Check cutter wheel for worn or nicked cutting edge and for free movement. Check mounting screw and spring for contamination, rust, food, etc. Check cutter to driver gap—should be between .005″ to .012″.

Can drops. Check for proper latching. Check cutter wheel to driver gap by holding piercing lever assembly in operating position while twisting lever so as to obtain maximum gap which should be between .005″ to .012″. Correct gap by adding or removing spacing washers on drive wheel shaft.

Disassembly

Although these disassembly instructions are for a specific can opener (General Electric Model EIEC32) disassembly methods, in general, will be similar for all models.

Cutter wheel. Withdraw the cutter wheel screw, cutter wheel, and spring. Reinstall so the recess in the cutter wheel is toward the spring; compress the spring so the wheel is on the stud. A small amount of lubricant should be placed behind the cutter wheel to assure free movement.

Drive wheel. Use a pair of "dykes" or side cutters fitted into teeth to remove drive wheel.

Motor and gear train. Open the case by removing one screw in back and prying rear housing off retaining ears on bottom. Remove three screws holding motor. Lift out rotor. Remove "C" clip holding primary gear. With drive wheel previously removed, lift out drive gear. Remove washers.

Switch. The switch blades are press fitted into receptacles molded into the plastic front housing. Cut away the heat staked boss and lift the fish paper shroud. Each blade may be removed by slightly spreading the receptacle. Leads can be soldered to the replacement blades and the spare boss can be heat staked to again retain the shroud.

Lead dress advice. The cord enters the rear housing, is knotted around the post supporting the motor and then splits. One lead goes to the switch, the other lead to the motor. The lead from the motor to the switch completes the circuit. Dress snugly so no leads are near (or could touch) moving parts.

CHAPTER 19

Electric Fans and Blowers

The electric fan is one of the most common household appliances. It is designed to circulate the air within a room, particularly during the hot season of the year. Two typical electric fans are illustrated in Fig. 1. In its simplest form, an electric fan consists essentially of a small electric motor having (usually) four propeller-like blades mounted on its shaft. When the motor is actuated, the rotary motion of the shaft forces the blades to circulate the surrounding air.

The electric blower differs from the fan mainly in that the blower is designed to move the air along a guided path, usually outward, as the motor spins the blades. Because of this requirement, the blower is commonly furnished with a suitable housing, the function of which is to guide the air stream in a given direction. The air inlet and outlet of a blower are termed the air-intake end and the air-exhaust end, respectively.

The electric fan is used in homes primarily to circulate the air, whereas the blower is employed in attic ventilators, air-conditioning systems, oil burners, etc. Electric fans and blowers are sometimes furnished with heating elements to heat the air as it is moved and circulated. A typical example of an air-heating blower is the familiar electric hair dryer, which produces either cool or heated air by means of a convenient heat-control switching device and one or more resistance heating elements.

FAN TYPES

Modern electrically operated household fans may be divided into several classes, depending on their operation and method of mounting; these classes are:

Fig. 1. Two typical modern electric fans.

1. Desk fans (oscillating or nonoscillating),
2. Floor fans,
3. Window fans.

Desk Fans

The familiar oscillating and nonoscillating desk fans are commonly mounted on a heavy base or pedestal and are furnished with a set of blades. The blades are protected by a suitable wire guard. The type of motor used depends on the size of the fan and may, in the case of the smaller fans, be of the shading-pole or universal type; the larger fans are usually furnished with a capacitor- or split-phase-type motor. The only fans that may be operated on either alternating or direct current are those that are driven by a universal-type motor.

Oscillating fans are so termed because they oscillate in a back-and-forth motion as the motor and fan rotate. In this manner they can move a large volume of air in the room or area in which the fan is placed. The oscillating mechanism consists essentially of a worm gear on the motor shaft that engages a gear on a short jack shaft. This shaft has a worm on the other end and is enmeshed with a gear on a vertical shaft. A disc attached to the lower end of the vertical shaft rotates at a very slow speed, and, by means of a short lever attached to the disc at one end and the motor at the other end, the fan is caused to rotate back and forth. This principle is employed in most oscillating fans, although some models employ a vertical shaft with a knob that is built into the gear mechanism with a clutch device. This design permits the fan to be used either as a stationary or oscillating model.

Floor Fans

Floor fans, such as the one shown in Fig. 2, are so termed because their mountings are such that they may be placed on the floor and may be designed for horizontal or vertical operation. The horizontal fan is commonly supported by a heavy base and an adjustable support, which provides suitable height adjustment. As the name implies, the vertical-type floor fan is mounted for vertical operation and moves the air from the floor or lower part of the room outward in a circular motion.

Window Fans

Window-type fans are designed for installation in or above windows, usually in the kitchen, and when so used are often called kitchen-ventilating fans. Whether they are of the built-in or the temporary type (some of these fans can be installed in the top sash of a window), most window fans are available with reversible motors or fan positions, so that the fan can either remove cooking odors and stale air by blowing the inside air outward or bring in clean, fresh air from the outside. A typical window fan is shown in Fig. 3. These fans usually do an excellent job of ventilating any room or area in the home. Window fans are of two varieties—permanent, or stationary, and portable. The permanent-type window fan is usually bolted to an accompanying frame, which is securely fastened to the window frame itself, whereas the portable-type window fan may be transferred from one room to another

Courtesy Westinghouse Electric Corporation

Fig. 2. A typical floor-type utility fan with air deflectors that direct the flow of air throughout the room or area being served.

as conditions or desires dictate. Most window fans in use today are of the portable type because of convenience.

BLOWERS

As previously described, blowers have a somewhat different area of employment than that of fans. Typical blower installations are the forced-air circulation arrangements in warm-air-heating and air-conditioning systems, such as that illustrated in Fig. 4. The blower consists of a multiblade wheel with the blades mounted parallel to the shaft of the blower. The blower is belt-driven by an electric motor, the rotation of which forces the cooled air through the air-discharge grille located in the upper part of the air-conditioning unit.

Courtesy Westinghouse Electric Corporation

Fig. 3. A typical window fan. This type of fan is ideal for ventilating any room or comparable area in the home.

ATTIC VENTILATORS

These may be of either the fan or blower type and are employed to provide air circulation by means of expelling heated stagnant air from the attic. Attic ventilators are effective in cooling a home, particularly when the days are hot and the nights are cool. Ordinarily under these climatic conditions, the heated air in the house has no chance to cool quickly. A fan or blower exhausts the air from the attic, or top floor, and simultaneously draws cool air from the outside into the house. By opening certain doors and

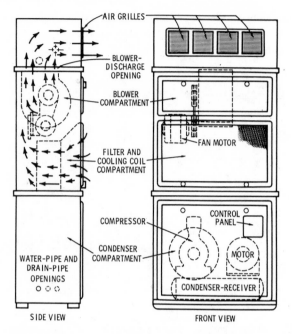

AIR GRILLES

BLOWER-DISCHARGE OPENING

BLOWER COMPARTMENT

FAN MOTOR

FILTER AND COOLING COIL COMPARTMENT

COMPRESSOR

CONTROL PANEL

CONDENSER COMPARTMENT

MOTOR

WATER-PIPE AND DRAIN-PIPE OPENINGS

CONDENSER-RECEIVER

SIDE VIEW

FRONT VIEW

Fig. 4. A typical blower installation in a self-contained air-conditioning system. The fan is a 12-inch diameter, centrifugal, multiblade type, and is driven by a ¾-horsepower motor.

windows in the different rooms, air circulation can thus be greatly facilitated.

Two different installation systems are shown in Figs. 5 and 6. In Fig. 5, the suction side of the blower is connected to exhaust ducts, which are connected to grilles that are placed in the ceilings of the two bedrooms. The air exchange is accomplished by admitting fresh air through open windows; this air is then drawn up through the suction side of the blower and is finally discharged through louvers, as shown by the arrows.

In the installation shown in Fig. 6, the blower is of the centrifugal curved-blade type. It is mounted on a light angle-iron frame; this frame supports the blower, shaft, and bearings with the motor, which supplies the motive power to the blower through a belt-drive system. The air inlet in this installation is placed close to a circular opening cut in an airtight board partition, which divides the attic space into a suction chamber and a discharge chamber. The air is admitted through open windows and doors, drawn up the attic stairway by the blower into the discharge

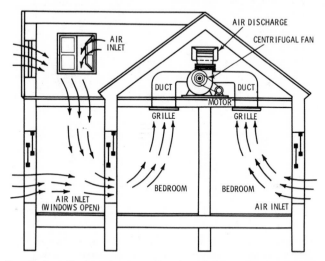

Fig. 5. *The installation of a centrifugal multiblade fan in the attic of a single-family dwelling. This type of system is ideal for ventilating an entire home.*

chamber, and is finally forced to flow through the open attic window to the outside.

INSTALLATION AND SERVICING

Fan Operation (Portable Types)

To obtain the maximum service from the various air-circulating devices, it is necessary to actually move and expel the heated and stagnant air through a window or other external outlet in the home. With every window in a room closed, a fan can only churn the existing hot air around; it cannot cool the air, since the enclosed air is not allowed proper entrance and exit. This fact may conveniently be demonstrated by noting the thermometer readings with the fan in an enclosed room. The proper method of using an air-circulating fan permits the cooler outside air to enter through one window and be expelled through another, with the assistance of

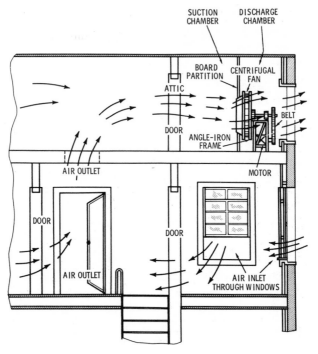

Fig. 6. The installation of a centrifugal curved-blade blower, driven by a V-belt, in a typical attic location.

a properly positioned fan between the two windows whenever possible. This is the only practical method of achieving a cooling effect by means of a fan or blower.

Attic-Fan Installation

Because of the low static pressure involved (usually less than ⅛ inch of water), disc or propeller fans are generally used instead of the blower or housed type. The fans should have quiet operating characteristics, and they should have sufficient capacity to provide at least 30 air changes per hour. The type of fan to use for a particular installation, its size, and location should preferably be determined by a heating and air-conditioning engineer in order to secure uniform air distribution in the individual rooms or areas to be served.

Attic-Fan Operation

To secure the best and most efficient results with an attic fan, the routine of operation is important. A typical operating routine might require that in the late afternoon when the outdoor temperature begins to decrease, the windows in the first floor and the grilles in the ceiling on the attic floor should be opened, and the second floor windows should be closed. This procedure will place the principal cooling effect in the first floor rooms. Shortly before bedtime, the first floor windows can be closed and those on the second floor opened to transfer the cooling effect to the bedrooms. A suitable time clock can be used to shut off the fan motor before the predetermined arising time, or the motor may be stopped manually later.

Attic-Fan Noise Control

To decrease the noise associated with attic-fan installations, the following rules should be observed:

1. The air-exchange equipment should be judiciously located with respect to important rooms in order to be a reasonable distance from them.
2. The fan should be of proper size and capacity to obtain reasonable operating speed.
3. Equipment may be mounted on rubber or other resilient bases. These materials assist in preventing transmission of noise through the building. A typical noise-reduction platform that is commonly used in attic-fan installations is shown in Fig. 7.

If the attic air-exchange equipment must be located above the bedrooms, it is essential that every precaution is taken to reduce the equipment noise to the lowest possible level. Since high-speed AC motors run somewhat quieter than low-speed motors, it is preferable to use a high-speed motor connected to the fan by means of a V-belt, where the available floor space permits such an arrangement. It has been found by experience that the top speed of a fan should not exceed 3300 feet per minute, if quiet operation is to be obtained. For example, if a fan operates at a speed of 570 rpm and has a 22-inch diameter fan, the top speed of the fan will be (22/12) × π × 570, or 3280 feet per minute. It is then evident that great care should be taken when contemplating installations with directly connected fan motors having speeds in excess of 1000 rpm, since in most instances the top speed of the fan will then be too high to insure quiet operation.

Speed Control

Speed control of fans and blowers may be achieved in various ways, the most common of which are the tapped-reactor and the two-winding motor arrangements, shown in Figs. 8 and 9, respec-

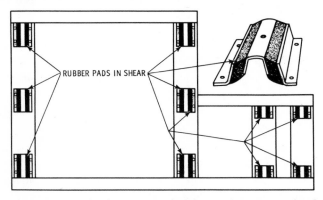

RUBBER PADS IN SHEAR

Fig. 7. A typical sound-reduction fan platform used in attic and industrial installations.

tively. The speed-control method shown in Fig. 8 is commonly used for small portable fans, and it can provide two or three definite speeds as conditions require. Fig. 9 illustrates a speed-control arrangement suitable for belted blowers, attic ventilators, and air-conditioning apparatus. Where ratings above ¼ horsepower are required, these motors are usually of the capacitor-start type. In the speed-control method shown, the motor is always started on the "high-speed" switch position; the transfer to the "low-speed" position is made by the starting switch. Other speed-control methods involve the use of an autotransformer and a two-speed switch; this method is sometimes employed on capacitor-type motors or one-value permanently-split induction motors.

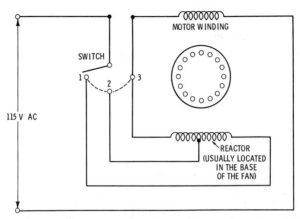

Fig. 8. Fan speed control may be obtained by means of tapped-reactor (induction coil) circuit, which is normally used in the smaller fans.

Fan Disassembly (Desk Types)

Before taking the fan apart, it should be properly checked for possible trouble in the external wiring circuit. Examine the line cord carefully for damaged insulation and shorts; also examine the plug connections to make certain that the operation failure of the fan is not due to any external openings in the circuit. A worn or damaged plug or cord should be replaced immediately, since these conditions not only impair the proper operation of the fan but are also a constant fire hazard. If the splices or terminals where the cord joins the winding can easily be reached, open these joints when necessary so the cord and motor winding can be checked separately. An open motor winding usually involves a rewinding job.

A typical nonoscillating fan can be easily disassembled by first removing the screws or nuts holding the wire guards in place and then removing the fan unit. On most fans, the fan unit is removed by loosening the setscrew on the hub that holds the fan proper to the motor-shaft extension. On certain desk fans, the motor has a fixed hollow shaft on which the rotor turns, and the fan unit screws on an extended hub of the rotor so that it will turn the rotor. Disassembly of such a fan is accomplished as follows: The motor end cap is removed as far as possible to allow a pair of

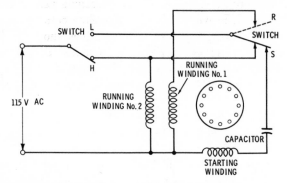

Fig. 9. The schematic diagram of a two-speed, two-winding fan motor. This type of speed-control method is often used where two definite speeds, independent of load, are required. In ratings above ¼ horsepower, the motors are usually of the capacitor-start type.

pliers to grip the hub of the rotor; the fan unit can then be unscrewed; the motor and cap can then be entirely removed after unscrewing a special locking nut on the end of the shaft, after which the rotor can be removed. Carefully note the number of washers and their location on each end of the rotor, so that they may be replaced exactly in the same order when reassembling. These washers serve the double purpose of properly aligning the rotor with the stator and taking up excessive end play.

With other types of desk fans, it may merely be necessary to remove the screws holding the end plates to the motor frame, which, after a light tapping, may be conveniently removed. After removal of the end plates, the rotor can be withdrawn from the motor unit. On some desk-fan motors, only one end cap is removable. When removing the end cap, a plastic or fiber hammer should always be used in order to prevent damage to the fan.

During the disassembly procedure, the various fan parts should be placed in a shallow metal tray in which the parts may conveniently be cleaned by the use of a small amount of carbon tetrachloride or kerosene; apply the cleaning fluid to the parts by means of a small brush suitable for the purpose. After the parts have been thoroughly cleaned, they should be dried with a clean, soft cloth. The cleaned parts are then examined for wear, oil passages to wearing surfaces are cleaned, and the rotor is tested on its shaft for proper clearance. Wear exceeding two or three thousandths of an inch will usually cause noisy operation; in the case of excessive bearing looseness, new bearings should be installed. These new bearings can usually be obtained directly from the manufacturer or made on a shop lathe. After pressing in the new bearings, a hand reamer should be used to obtain a perfect fit.

The stator winding can easily be tested for shorts, grounds, or open circuits by means of a series test lamp in the conventional manner. A grounded motor should be rewound unless the damaged part of the insulation can be found and repaired.

If the motor-winding tests indicate normal operation, the fan may be reassembled after thoroughly cleaning the winding, fan blades, and wire guard. A few drops of oil placed on the shaft before inserting the rotor will assure an effective lubrication. After reassembling the fan, which is a reversal of the disassembly process, oil is provided in the oil cup or cups. Grease-lubricated fans usually have a grease cup at each end that requires cleaning and refilling with fresh grease of the type specified by the fan manufacturer. After starting the fan, make certain that the fan runs smoothly and quietly with no imbalance or undue vibration. Imbalance is usually caused by fan blades being out of line due to mishandling or shock treatment. The track or pitch of the blades may be checked by placing the fan blades face down on a smooth surface and measuring each blade individually to its highest point.

Oscillating Fan Repairs

Failure to oscillate is usually caused by stripped worm or gear teeth; frequently this condition occurs in the short shaft piece. Replacement with a new part is the only remedy in such a case. Oscillation failure might also be caused by wear in the clutch end of the control shaft, where this type of mechanism is employed.

When disassembling a fan of this type, the oscillating gear box should be removed first. Inspect the parts carefully, and note their condition. Remove the guard and fan unit, and disconnect the lever from the driving disc; then loosen the motor-clamping screws in the end cap. Before the rotor can be withdrawn, it is usually necessary to drop the short shaft in the gear box down so that its gear is out of mesh with the worm in the rotor shaft; by loosening a setscrew and pulling out a sleeve, this step can be accomplished. The end cap can then be tapped off, and the rotor can be removed. In other respects, the disassembly process is similar to that already explained for nonoscillating fans.

Numerous fans of this type are equipped with a speed-regulation switch located on the base. Switches of this type occasionally present trouble due to insufficient pressure, burned or broken spring contacts, etc. In case of a loose contact, bend the contacts together with a pair of long-nosed pliers or other appropriate tool until a sufficient spring pressure has been established. If the switch springs are broken, a new switch must be installed. One frequent trouble with oscillating fans is that the short wire that connects the motor to its base receives excessive bending during operation. This continuous bending may eventually cause the conductor to break, which will result in an open circuit. This condition can be checked with a test lamp; if an open-circuit condition does exist, the broken wire must be replaced.

After all tests and inspections have been made, assembly of the fan is quite simple. Fill the gear box with medium-weight grease, and refill the grease cup at the other end of the motor, if such a cup is provided. Apply a drop of oil to each of the pivot screws in the main outer ring, and adjust these screws so that there is no lost motion, while the motor is allowed to swing freely. Also apply a drop of oil to each end of the operating lever at its screws.

If a malfunction occurs in the speed-control reactor located in the base of the fan, it is usually caused by an open circuit in the reactor proper or in any one of its connecting leads or by a burned-out reactor winding. In any case, a test lamp may conveniently be employed to check any of these conditions. A burned-out reactor may be rewound, using the same size wire and number of winding turns as that of the original reactor.

CHAPTER 20

Electric Vacuum Cleaners and Floor Polishers

All vacuum cleaners, irrespective of type and working principles employed, contain a line cord to supply electric power, a power-control switch, a motor-operated fan to provide the necessary air suction, a nozzle for the collection of dirt and dust, and a container where the dust and dirt are allowed to accumulate. In operation, the air enters through the nozzle and is carried into the dust container or filtering device. When the nozzle is pressed against the rug or other surface to be cleaned, the suction created by the fan pulls the dirt through the fan and into the bag, allowing the air to escape through the bag's cloth mesh. Some cleaners employ a water filter instead of using the bag as a filtering device. The air stream passes through the water and deposits the dirt or dust on the water surface.

TYPES OF VACUUM CLEANERS

Although there are many vacuum cleaners in use, they all fall into two general classifications with respect to their operation; they are the upright type and the cylinder and pot types. The upright vacuum cleaner, as shown in Fig. 1, sometimes termed the rotating-brush type, depends for its cleaning action on air suction in addition to the whirling-dirt-dislodging function of the revolving brush, whereas the cylinder and pot types operate on suction alone.

The Upright Vacuum Cleaner

The principal parts of an upright cleaner, shown schematically in Fig. 2, are a motor, a suction fan, a filter bag, and a nozzle with or without stationary and revolving brushes. Suction created by the fan draws the carpet to the nozzle. The dislodged dirt is picked up by the air suction and carried to the bag where it is retained, while clean air is returned to the room.

The motor, usually of the universal type, is directly connected to the suction-fan assembly, as shown in Fig. 3, and is connected by means of a belt-and-pulley arrangement to the revolving rotary-type brush. The fork handle is attached to the cleaner and serves to guide the vacuum cleaner over the carpet surface. Wheels attached to the casting incorporate an adjustment screw by means of which the wheels may be raised or lowered for cleaning rugs of different thicknesses.

Fig. 1. A typical upright vacuum cleaner.

Courtesy Eureka Williams Corporation

Other features incorporated in the upright-type vacuum cleaner are a headlight lamp to illuminate the area to be cleaned and a set of accessories, or cleaning attachments, some of which are shown in Fig. 4. Depending on the particular cleaner under consideration, the accessories consist generally of a flexible hose and extension wands to which a nozzle, dusting brush, upholstering brush, and several other devices may be attached for removing dust and embedded grit from upholstery, draperies, mattresses, stair carpets, automobile interiors, etc. A special attachment for spraying insecticides, liquid wax, or paint is also included as an accessory in most upright vacuum cleaners.

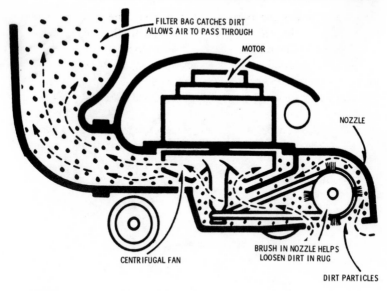

Fig. 2. The essential parts and working principles of a modern upright vacuum cleaner.

The Cylinder-Type Vacuum Cleaner

A cylinder, or tank, vacuum cleaner, shown schematically in Fig. 5, consists essentially of a cylinder; in one end of the cylinder is a motor that drives a set of powerful suction fans, while the other end of the cylinder contains a fine mesh dust-collector bag. In operation, suction created by the fan draws the carpet or rug to the nozzle. The dislodged dirt is picked up by air suction and carried to the bag through a flexible hose. The dirt or dust is retained in the bag, while the clean air is returned to the room.

Fig. 6 illustrates one popular type of cylinder, or tank, vacuum cleaner. The usual attachments accompanying a cylinder-type cleaner provide additional cleaning tools used for dusty surfaces, such as upholstery, draperies, walls, table tops, lamp shades, book shelves, etc. Numerous vacuum cleaners provide an attachment for spraying paint, liquid insecticides, and floor wax. The special spray gun furnished for this purpose consists of a liquid container with a detachable spray-gun top to which the air-pressure cleaner

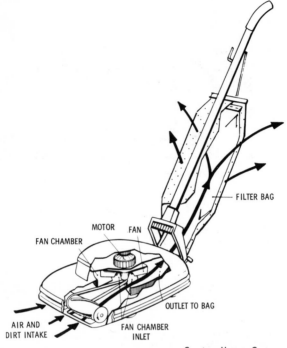

FILTER BAG

MOTOR FAN

FAN CHAMBER

OUTLET TO BAG

AIR AND
DIRT INTAKE

FAN CHAMBER
INLET

Courtesy Hoover Company

Fig. 3. The construction of a typical upright vacuum cleaner.

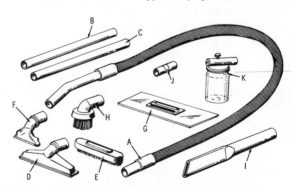

Fig. 4. Typical vacuum-cleaner attachments that are used for a wide variety of cleaning tasks. In the illustration, A represents the hose assembly; B and C, the extension wands; D, the floor nozzle; E, the wall brush; F, the upholstery nozzle and swivel assembly; G, the floor polisher; H, the dusting brush; I, the crevice tool and demother; J, the moth repellant; and K, the sprayer.

hose may be attached, much in the same manner as on the upright cleaner. Fig. 7 illustrates the disassembled view of a typical cylinder-type vacuum cleaner.

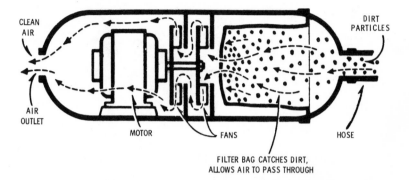

Fig. 5. The schematic representation of the cylinder-type vacuum cleaner, showing the essential parts and working principles.

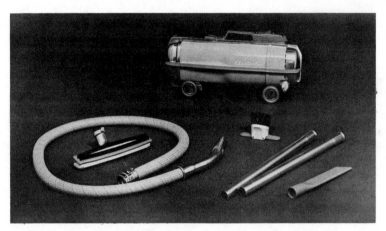

Courtesy Electrolux Corporation

Fig. 6. A typical cylinder-type vacuum cleaner with its attachment cleaning tools.

The Pot-Type Vacuum Cleaner

A pot-type, or vertical-tank, cleaner is schematically illustrated in Fig. 8. This type of cleaner is similar, with respect to its operation, to the cylinder-type vacuum cleaner. As previously noted,

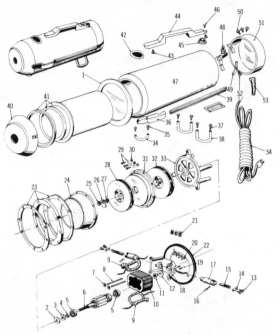

Fig. 7. The disassembled view of a typical cylinder-type vacuum cleaner. In the illustration, 1 represents the filter assembly; 2, the bearing cap; 3, a felt washer; 4, a spring collar; 5, a ball bearing; 6, the motor armature; 7, a nut; 8, a field stud; 9, the field coil; 10, the pole piece; 11, a thrust washer; 12, a bearing cap; 13, screws; 14, a brush terminal; 15, a carbon brush and spring; 16, a cotter pin; 17, a brush holder; 18, the motor frame; 19, setscrews; 20, screws; 21, a connector; 22, the motor mounting plate and gasket assembly; 23, the support assembly; 24, the fan casing; 25, a nut; 26, the counter balance; 27, a nut; 28, the revolving fan and hub assembly; 29 and 30, screws, nuts, and washers; 31, the stationary fan assembly; 32, the revolving fan and hub; 33, the motor frame; 34, the front skid; 35, screws for skids; 36, the intake fan clamp; 37 and 38, the rear skid and attachment screws; 39, moulding; 40, the intake flange; 41, the dust bag; 42, the nameplate; 43, screws; 44, the handle; 45, the switch knob; 46, a screw; 47, the tank; 48, the switch; 49, the motor mounting segment; 50, screws; 51, the exhaust flange; 52, a screw; 53, the strain relief; 54, the cord and plug assembly; and 55, the nomenclature plate.

the principal parts of the pot-type cleaner are a motor, a suction fan, a filter, and a nozzle. Suction created by the fan draws the carpet to the nozzle. The dislodged dirt is then picked up by this air suction and carried through the filter in the container, where it is retained, while the clean air is returned to the room. Fig. 9

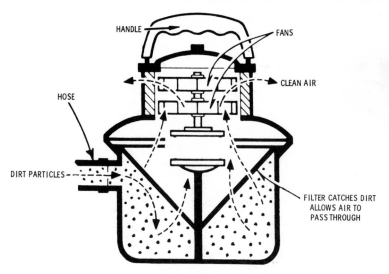

Fig. 8. *The essential parts and working principles of a pot-type vacuum cleaner.*

represents the disassembled view of a typical pot-type vacuum cleaner.

The Portable Cleaner

A typical portable cleaner is illustrated in Fig. 10. This is a simplified version of the upright cleaner and operates in a similar manner. Most cleaners of this type are provided with a set of cleaning attachments to increase their usefulness in various cleaning tasks. This type of cleaner is often preferred over the conventional upright cleaner for above-the-floor cleaning and dusting because of its light-weight portability.

SERVICING AND REPAIRS

Almost all vacuum-cleaner complaints, whether the cleaner is of the upright, cylinder, or pot type, fall into one or more of the following categories:

1. Motor will not run.
2. Motor overheats.

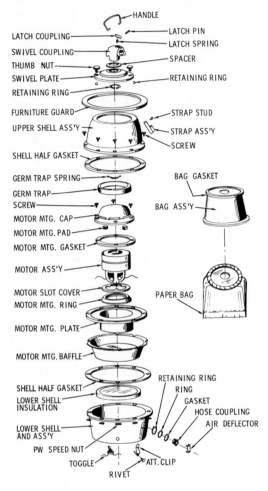

Fig. 9. The disassembled view of a pot-type vacuum cleaner.

3. Motor is noisy.
4. Cleaner does not pick up properly.
5. Operator receives a shock.

Motor Will Not Run

The failure of the cleaner motor to operate does not necessarily indicate motor trouble; this failure may be caused by a circuit

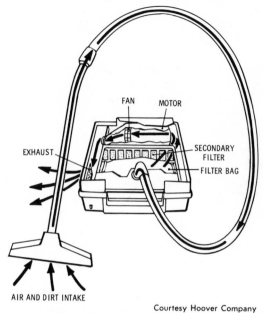

FAN   MOTOR

EXHAUST

SECONDARY
FILTER

FILTER BAG

AIR AND DIRT INTAKE

Courtesy Hoover Company

*Fig. 10. The essential parts and working principles of a typical portable
vacuum cleaner.*

interruption between the power outlet and the motor terminals.
The procedure to follow in checking such a circuit for continuity
is as follows:

1. If the cleaner is of the upright type, check the handle for
circuit continuity by first plugging a special test lamp into the
plug in the bottom of the handle and then plugging the cord
into the power outlet. As an alternative method, the receptacle terminals in the bottom of the handle may be shorted
together, and a series test lamp may be used across the terminals on the connecting cord plug. Turn the switch to the
"on" position. If the test lamp lights, check the cord and
the connections by twisting and flexing the cord to see that
broken wires do not exist. Check particularly at the point
where the cord enters the handle and switch. Check the entire
length of the power-supply cord for defects in insulation and
wire breakage.

2. Check the switch for proper operation; be sure to check both positions, if it is a two-speed switch.

3. Inspect the terminal connections at the male plug, and also note the condition of the terminals.

4. If the test lamp does not light (in section 1 above), the handle group must be disassembled, and the defective parts must be located and replaced. If the circuit through the handle extension, including the switch and switch connection, is found to be in good order, but the motor still does not operate, connect the power supply to the motor leads, and check the motor operation in the following manner:

   a. Check the power cord for breakage where it enters the motor housing.

   b. Check the carbon brushes to make sure they are making contact with the commutator. The brushes may be worn, may be sticking in brush holders, may have weak springs, or may be dirty.

   c. Check to see if the armature shaft is frozen in the bearings. Also check the front bearing for wear by working the end of the shaft sideways.

Motor Overheats

Motor overheating may be due to several causes; proceed with a thorough check as follows:

1. Check to see that the ventilation openings are not filled or clogged up with dirt.

2. Check the motor field coils for continuity. A shorted coil may give an excessively high wattage reading at no-load conditions. Also, one coil may become unusually hot if it is shorted.

3. Check to see that the windings are not grounded.

4. Check the armature for contact with the field, in which case worn bearings may be the cause. Thoroughly examine the surface of the armature and the bore of the field for rubbing.

Motor Is Noisy

If the motor is noisy, check for the following:

1. Check for a bent or broken suction fan. This will throw the

378

entire assembly out of balance. Fans with bent blades can often be straightened and reused. If this is impossible, the fan will have to be replaced.

2. Check for a bent armature shaft. This can usually be corrected by means of a special armature-straightening tool.

3. Check the motor bearings. Generally, the front bearing receives the greatest wear due to the load of the belt. Sometimes defective bearings can be detected by operating the motor at low speed. To test a single-speed motor of 190 to 350 watts at reduced speed, use a 100-watt lamp in series with one of the line leads. When testing larger motors, use lamps of higher wattage to produce greater resistance. For 600-watt motors, a 150-watt and a 200-watt lamp should be used in series with each other.

4. The bearing cup on the fan end may be too large, thereby allowing the bearing race to rotate. This must be a close slip fit.

5. The bearing cup on the commutator end may be too small. If so, the bearing will not slide freely, and the bearing compression spring will be ineffective.

6. The bearing compression spring may be weak, thus allowing the armature shaft to weave back and forth.

7. The brushes must ride freely and with sufficient pressure on the commutator. The brush holders must be tight. If adjustable types are sprung so that they do not hold the proper adjustment, bend the moveable arms so that they will hold the brushes firmly. Where bearings stick or drag, it will be necessary to clean and relubricate or replace them entirely.

Cleaner Does Not Pick Up Properly

This common complaint is generally due to several minor causes, such as failure to empty the dust bag, which results in a clogged cleaner condition, a misadjusted or worn rotary brush, etc. If the cleaner does not pick up properly, check the following:

1. Check the nozzle-adjustment mechanism and handle-tension spring. There are three types of nozzle adjustments—thumb screw, lever, and automatic types. Make sure these parts are

not worn, since they could cause improper nozzle adjustment. The tension spring must be replaced if broken.

2. Check the floor brush; see that the bristles are flush with the bottom casting lip. If of the adjustable type, the brush may be set to the next lower position; otherwise, the brush should be replaced. Brushes must, of course, be clean to operate at proper efficiency.

3. Check the belt. Make sure the proper belt is used for the cleaner. A belt with too much tension exerts additional load on the front bearing, whereas one with too little tension will slip when the brush contacts the carpet or rug. Belt tension may be checked by setting the cleaner on the edge of the rug and placing the hand underneath the rug. If vibration is felt, the brush is turning.

4. Check the rear caster mechanism; see that it swings freely and that the wheel revolves freely. The height of the caster governs the position of the nozzle above the rug; therefore, if the wheels are worn, they should be replaced.

5. Check the dust bag. If it is clogged, its efficiency is greatly reduced, since the air flow through the bag is then reduced.

If all the foregoing items are in satisfactory condition, and the motor is not operating at its rated speed, the vacuum obtained will not be up to standard. The wattage of the motor is a fair test of vacuum efficiency and speed. Pick-up failure in a cylinder- or pot-type vacuum cleaner, assuming the motor functions normally, is usually due to loss of vacuum; to a clogged hose, bag, or filter; and sometimes to the unfamiliarity of the user with the cleaning tool.

In the absence of special laboratory equipment, a vacuum check may be easily improvised by employing a suitable glass bottle, a glass tube, a rubber tube, and a ruler, assembled in the manner illustrated in Fig. 11. With the hose attached to the cleaner suction outlet, make an air-tight connection between the free, or cleaning-tool-attachment end of the hose and the rubber tube fastened to the upper end of the glass tube. When the cleaner is turned on, the height of the water lifted from the bottle into the glass tube is an accurate measurement of the vacuum condition of the cleaner, because the stronger the vacuum, the higher the water will rise in the glass tube. The water lift in inches may be obtained from the vacuum-cleaner manufacturer for a particular type of cleaner, or the vacuum strength of the cleaner under test may be compared to that of a new cleaner of an identical model or type. If the cleaner shows a low vacuum, which is not due to a clogged condition, the motor speed should be checked; if the speed is too low, the cause should be ascertained and corrected.

Operator Receives a Shock

If the operator receives a shock when handling the cleaner, the insulation in the electrical system is faulty, and the cleaner is

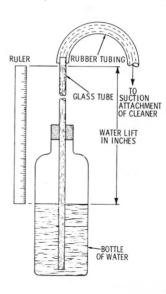

Fig. 11. A simple testing arrangement for measuring the amount of vacuum developed in a vacuum cleaner. The water lift, in inches, depends on the vacuum being pulled; this lift usually ranges from 15 inches for small portable models to 45 inches for large domestic cleaners.

dangerous to use. To test for grounded parts on the vacuum cleaner, proceed as follows:

1. Check the handle assembly and motor unit separately by means of a high-voltage transformer. Apply 1000 volts across the handle assembly and 500 volts across the motor.
2. Check by first connecting one lead and then the other with one insulated test prod and metal parts with the other. Make

contact for *one second only*. If a ground is indicated, the cleaner must be disassembled, and the parts must be checked individually.

3. Parts most likely to cause such troubles are the carbon brush holders, the motor housing, the cord, handles, the switch and handle assembly, the motor field, and the armature.

FLOOR POLISHER

Electric floor polishers and scrubbers, such as the one shown in Fig. 1, consist essentially of a high-speed motor that drives a set of buffing wheels or brushes through a spur gear train. Floor polishing motors are usually of the 60-cycle, 115-volt AC type and are rated at from 200 to 300 watts. In a typical 16,500-rpm high-speed motor, the twin brushes are driven at 500 rpm through a spur gear train with a speed reduction of 33 to 1. The base of the polisher is usually made of die-cast aluminum with a hood of molded thermoplastic. When the motor is energized, the brushes or wheels rotate, and the polisher can be moved over the floor area to be treated with little effort exerted by the operator. Normally the polisher is equipped with a liquid cleaner tank or dispenser by means of which the carpet can be shampooed during the cleaning process.

Attachments And Their Uses

Typical attachments for scrubbing, waxing, polishing, and buffing are shown in Fig. 2. A special combination of bristles, mounted in thermoplastic supports, provides thorough floor scrubbing action. Special plastic mesh pads can be used to apply both paste and liquid polishing waxes. Felt buffing pads are used to provide extra shine or to bring back the luster between waxings.

Maintenance

Because of the simplicity in construction, the maintenance of modern floor polishers consists simply in keeping the unit and accompanying attachments clean at all times; this is particularly important with respect to the brushes after shampooing rugs. Thus, if the brushes are found to have an excessive accumulation of lint,

Fig. 1. A modern electric floor polisher and scrubber that can also be used to shampoo rugs and carpets.

Courtesy Hoover Company

they should be cleaned immediately. Hold the brush with the bristles pointed up; insert the long tapered tail of a curl comb between the bristle tufts, as close to the bottom of the tufts as possible, and push from the outside into the center of the brush. Lift the tail straight up through the bristles. Repeat this process over the entire brush until all the lint at the bottom of the bristle tufts has been

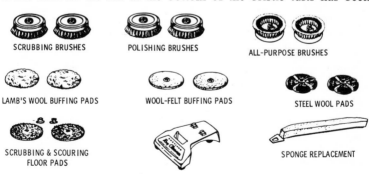

SCRUBBING BRUSHES

POLISHING BRUSHES

ALL-PURPOSE BRUSHES

LAMB'S WOOL BUFFING PADS

WOOL-FELT BUFFING PADS

STEEL WOOL PADS

SCRUBBING & SCOURING FLOOR PADS

SPONGE REPLACEMENT

Courtesy Shetland Company

Fig. 2. Some of the common attachments that are provided for electric floor polishers.

loosened. Then, with the comb end of the curl down into the bristles, comb the loosened lint up and out with a scooping action. Repeat this procedure until all the lint is removed. If the brushes have dried, wet them thoroughly in warm water, and shake the excessive water off before cleaning.

There is little in the way of additional service and maintenance that a floor polisher requires except for regular service of the motor. This consists generally in shaft-bearing lubrication at periodic intervals, depending on the frequency of use. The carbon brushes of the motor may become worn and will have to be replaced. Brush replacement may easily be accomplished by removing the motor brush clamps to check the carbon brushes and springs. If the brushes are broken or worn too short, or if the brush springs have been damaged, replace the brush assembly with a suitable substitute, as specified by the manufacturer.

CHAPTER 21

Electric Washing Machines

The purpose of any washing machine, regardless of its construction, is to force the water in combination with soap or detergents through the clothes in order to thoroughly clean them. In the modern machine, the washing action is accomplished by means of various motions between the water and soap or detergent and the clothes to be washed. See Fig. 1.

CLASSIFICATION

To produce the necessary motion, various principles of operation are employed. The most common types of washing machines may be classified with respect to their operation as:

1. The agitator type, Fig. 2A,
2. The cylinder type, Fig. 2B,
3. The vacuum-cup type, Fig. 3A,
4. The pulsator type, Fig. 3B.

Each of these types of washing machines uses varying methods for excessive moisture removal. Some are equipped with wringers, while others employ a special spinner basket in which the clothes are rotated at high speed after the washing process has been completed. In yet another type, the wash is spun at high speed in the same tub in which the washing takes place.

The heart of the agitator-type washing machine is the agitator, which consists of a vertically fitted impeller that derives its oscillating motion from the rotary movement of the motor shaft. In order to convert the straight rotary motion of the motor to that of the back-and-forth oscillating rotary motion of the agitator, a

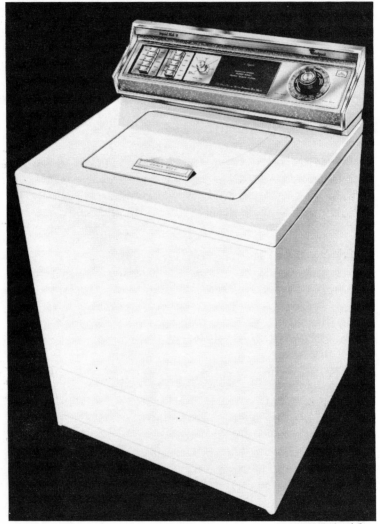

Fig. 1. A typical modern automatic washing machine.

special gear-case mechanism is required. The motor and gear case are usually installed in the lower part of the machine enclosure. Some agitator-type washers employ a gear mechanism that con-

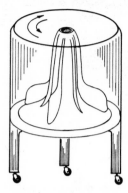

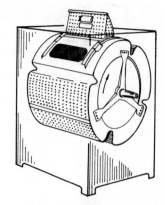

A. The agitator type.   B. The cylinder type.

Fig. 2. The working principles of two common washing machines.

verts the rotary motion of the motor shaft into vertical pulsations instead of a back-and-forth movement during the wash period. This principle affects the rapid up-and-down movement of the pulsator, which causes the water to swirl or circulate from the top to the bottom of the tub, thus creating the washing action.

In wringer-equipped machines, the gear case also contains a wringer roll drive. This mechanism includes an enclosed vertical shaft whose rotary motion is transferred to the wringer rolls on top of the washer. In addition, numerous washers contain a motor-driven pump, whose function it is to pump the water from the machine, thus enabling the machine to pump out the water through a hose and into an adjacent sink or tub, rather than depending on a gravity drain.

From this brief description, the conclusion may be reached that a washing machine is a comparatively simple appliance. This is true only insofar as its operation and reliability are concerned. A further discussion of the various mechanical and electrical mechanisms involved in an automatic electric washing machine will convince the reader of the great amount of engineering research work involved to produce the present.product. Therefore, despite the reliability and ease of servicing of today's modern washing machines, the serviceman will need complete instructions concerning construction, operation, and servicing in order to cope

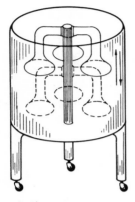 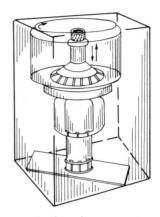

A. The vacuum-cup type. **B. The pulsator type.**

Fig. 3. The operating principles of two common washing machines.

with any service problems that may arise. There are definite code regulations covering the installation of washing machines in all localities, and the serviceman must acquaint himself with the regulations that are in effect for his area.

WASHING MACHINE CONSTRUCTION

Since it is obviously impossible to attempt to describe every type of washing machine on the market, only the principles involved in each of the previous classifications will be discussed. It is self-evident that if the operating mechanisms of one type of machine are clearly understood, another washer will only differ in construction to some slight degree. This difference should not cause any great amount of service difficulties, since the basic principles involved are the same in each instance.

Agitator-Type Washing Machines

The principal assemblies of any agitator-type washer are as follows:

1. Cabinet,
2. Gear case and agitator drive,
3. Agitator (pulsator),
4. Motor,

5. Wringer (if used),
6. Pump,
7. Timer(s).

Washing machines of this type may be either manual or automatic; the difference is that in the manual type, the washing, rinsing, and excess moisture removal are performed by manual control, whereas in an automatic machine, an electric timer provides complete control, starting and ending the various cycles by a predetermined setting as indicated on a timing dial.

Cabinet Arrangement—All cabinets, whether of the rectangular or cylindrical type, are finished in a synthetic baked enamel. The structural details of a typical agitator-type automatic washer are shown in Fig. 4. The front and two sides of the cabinet are of conventional one-piece wrap-around construction, with a back panel attached to the two sides by a roll weld. The corners are reinforced by four gussets welded across the top corners and four gussets welded across the bottom corners. A water-tight bulkhead separates the motor compartment from the tub-assembly compartment. This bulkhead is located on flanges, which are welded to the inside of the cabinet, and is held in place by four draw rods, which run from the underside of the bulkhead to the four bottom corner gussets. These rods can be individually adjusted to square the mechanism mounting flange of the bulkhead with the cabinet; this is necessary to prevent undue tub-assembly oscillation during the spin operations and to insure sufficient clearance between the tub and the cabinet top. On early models, the bulkhead seal was made by applying rubber-like mastic materials in the bulkhead support flanges and over the flange formation crevices and then drawing the bulkhead flanges down into the mastic.

An opening in the back panel, protected by a guard plate, permits an ample flow of air to the motor cooling fan, which is located at the base of the washer-mechanism assembly. The cabinet top is fastened to the cabinet by four oval-head, stainless-steel, sheet-metal screws; the joint is made water-tight by a rubber seal, which is located on the upper flange of the cabinet. The cabinet lid is hinged to the cabinet top and is held in the open position by lid stops, which are located on each side of the lid. Four adjustable feet are located at the bottom corners of the cabinet and are used

389

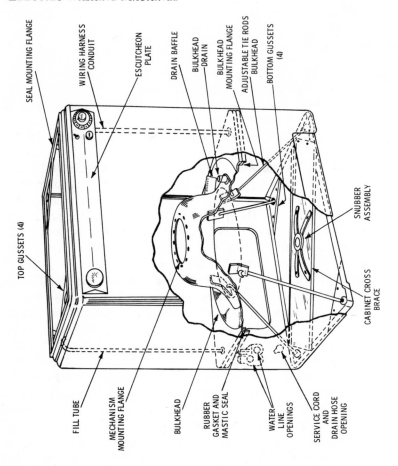

Fig. 4. The cabinet construction of a typical automatic washing machine.

to level the washer. Holes are provided in the back panel for the hot and cold water lines, the drain connection, and the service cord. The cabinet exterior finish is usually porcelain on heavy gauge enameled iron. The cabinet interior finish is a ground coat of porcelain.

Gear-Case Assembly—The main function of the gear assembly is to convert the rotary motion of the motor shaft into the

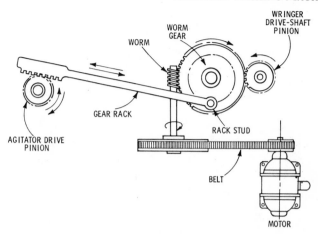

Fig. 5. The schematic representation of the gear-case principles employed in washing machines.

oscillatory motion required by the agitator. This motion change is accomplished in a gear case, which is shown schematically in Fig. 5. In this unit, the belt-driven pulley of the worm imparts its motion to the worm gear, to which is fitted a gear rack; this rack, in turn, drives the agitator, thereby causing the agitator drive pinion to rotate back and forth. This gear assembly also includes the wringer drive shaft and water-pump connection.

Another type of gear-case assembly is shown in Fig. 6. In this system, the rotation of the worm gear by means of an eccentric and connecting rod imparts a back-and-forth movement to a segment gear. The segment gear, in turn, engages the agitator pinion and shaft, thereby producing the desired agitator motion.

A third type of washing machine gear case is shown in Fig. 7. This assembly differs from those previously shown, in that the worm gear engages a gear rack through a crank arrangement and in this manner transmits the rotary motion of the motor to the oscillating back-and-forth motion of the agitator. The bottom view of this gear-case assembly is shown in Fig. 8.

Agitators—These are fitted in a vertical position on the agitator shaft, which extends from the gear-case assembly through the center of the washing machine cabinet, as illustrated in Fig. 9. Principally, the agitator consists of a base and vertical or spiral

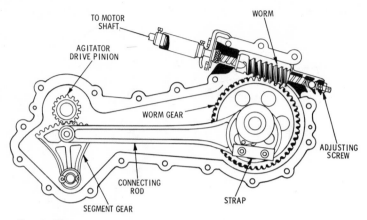

Fig. 6. The principal parts of a typical washing machine gear case.

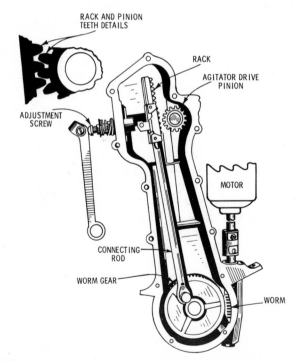

Fig. 7. This gear-case mechanism employs a gear rack and a connecting rod to transfer motion from the motor shaft to the agitator shaft.

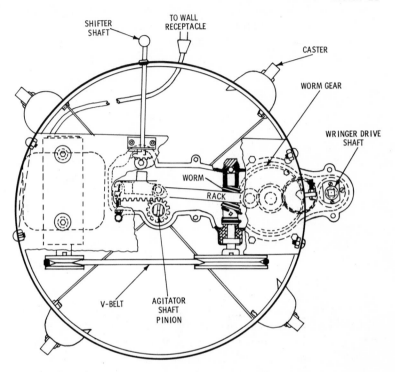

SHIFTER SHAFT

TO WALL RECEPTACLE

CASTER

WORM GEAR

WRINGER DRIVE SHAFT

WORM

RACK

V-BELT

AGITATOR SHAFT PINION

Fig. 8. The bottom view of a typical washing machine, with its gear-case assembly.

vanes, which are usually made of plastic or metal. The oscillatory (back-and-forth) motion derived from the gear-case agitator drive pinion introduces a great amount of turbulence in the water, thereby causing the washing action. The agitator action ceases for the rinsing and drying cycles.

Motors—These are commonly of the capacitor-start type for connection to the conventional 60-cycle, 115-volt AC house circuit. The motor is in operation during all wash, rinse, and spin periods, and it is electrically protected by fuses or overload-relay devices. The motor is connected either directly or by means of belts and pulleys to the drive shaft of the gear-case mechanism, water pump, wringer shaft, and fan assemblies. The motor sizes differ, depending on the size of the washer; the average size is usually ⅓ horsepower. Although the motor controls are quite simple in the manual-type washers, they require additional knowledge in automatic machines, since these are equipped with timers, water-temperature switches, and regulating solenoids.

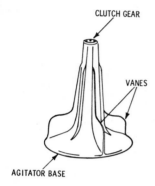

Fig. 9. A typical washing-machine agitator.

Wringers—A wringer, as shown in Fig. 10, consists essentially of two closely fitted rolls that are suspended in pressure-controlled bearings; these bearings are commonly driven by a vertical extension shaft, which is coupled either to a worm-gear shaft or to an extra gear that is meshed with a gear on the worm-gear shaft. An additional matching gear, or gears, on the lower roll shaft can be

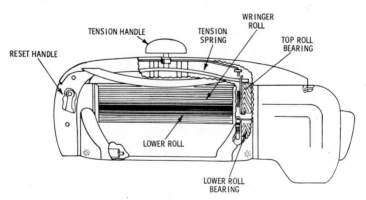

Fig. 10. The principal parts of a typical wringer assembly.

made to mesh one way or another with the extension shaft gear or gears. This meshing action is controlled by a centrally located shift lever or handle and provides alternate roll directions. By turning the shift lever to an intermediate or neutral position, the gears can be unmeshed, and the roll can then be stopped. Although there are numerous wringer-drive head mechanisms, they differ only in the method used to accomplish this operation, irrespective of manufacture.

The rotating wringer rolls are made of soft rubber and may be adjusted for various pressures by means of knobs on top of the wringer. As a safety measure, a special release is provided that functions either by a slight hand pressure or when the rolls become overloaded. This safety device releases one end of the upper roll carrier, thus permitting the rolls to become separated.

Wringers are usually provided to swing from one position to another to suit various laundry arrangements of tubs and baskets and may be locked in any desired position. The operation of the safety attachment on the rolls and wringers should be clearly understood, so that the rolls may be released in the event that too much material is passed through at one time. A typical wringer-type washing machine is illustrated in Fig. 11.

Pumps—The function of the water-pump assembly is to draw off all water entering the washing machine cabinet and to expel it through the drain connection and hose to the house drain. All pumpless washing machines have special provisions for the expulsion of water by hose connections and a gravity drain.

Depending on the particular manufacture of the washer, the water pump may be driven directly from the motor shaft or gearcase assembly, or it may be belt-driven. The water impeller commonly employed is of the turbine type and has two or more blades mounted on it, as illustrated in Fig. 12. When the pump is in operation, the water in the pump housing is thrown outward by centrifugal force and, in this manner, is expelled through the pump housing.

On some washers, the pump is manually operated by a special lever, whereas on fully automatic machines, the pump is operated during the wash, spin, and rinse periods to expel the water during the spin operation or to expel any overflow during the wash and rinse operations. On other machines, the pump runs constantly

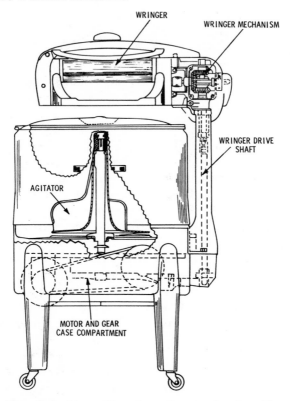

Fig. 11. The principal assemblies of a wringer-equipped washing machine.

but starts the pumping action only when the drain valve is opened by the automatic timer or by the cam action.

Timers—These are incorporated on automatic washing machines to control the washing process through its complete cycle, thereby automatically washing, rinsing, and damp-drying the clothes at certain predetermined times. Timers may be operated either by an electric motor or by a special cam-operated device that is driven by the washer motor to provide complete control of the entire washing cycle.

Pulsator-Type Washing Machines

In an automatic washing machine, the various washing processes are performed automatically, so that no attention need be given

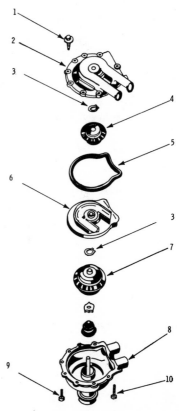

Fig. 12. A typical washing-machine water pump. In the illustration, 1 represents a screw; 2, the upper pump housing; 3, a retainer ring (2); 4, the upper impeller; 5, the gasket; 6, the pump partition; 7, the lower impeller; 8, the lower pump housing; 9, a screw; and 10, a screw.

the machine after the correct amount of laundry has been deposited in the cabinet together with the required amount of soap or detergent. All that is required is to supply the machine with hot and cold water and to set the timer according to the manufacturer's instructions accompanying it.

The operating mechanism of a pulsator-type washing machine is shown in Fig. 13. The vertical agitator, instead of rotating back and forth, moves up and down in a pulsating action. The moving member is therefore termed a *pulsator*. In Fig. 13, the mechanism converts the rotary motion of the motor shaft into the vertical pulsations of the pulsator shaft during the wash periods. This same pulsating action is also produced during the first and second rinse periods. At the end of each pulsation period, the pulsating action is stopped, and the rotary motion of the motor is employed directly to spin the tub assembly. This spinning action is accomplished by a helical clutch spring, which, by its position, determines whether pulsating or spinning action is produced. At the end of each spin period, the tub assembly is brought to a stop by a brake assembly, which is part of the operating mechanism. This stopping is accomplished without the aid of motive power, since the motor is idle during all brake periods.

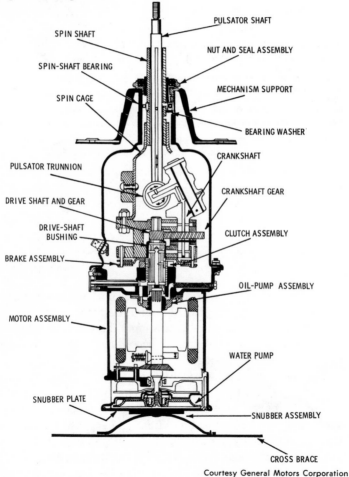

Courtesy General Motors Corporation

Fig. 13. The operating mechanism of a pulsator-type automatic washer.

A water pump, connected directly to the motor shaft, expels all water loads that enter the washer cabinet. The entire mechanism is usually driven by a ⅓-horsepower, single-phase, capacitor-start motor, which connects directly to the other components of the mechanism, thus eliminating the use of belts and pulleys which tend to decrease the overall efficiency of the washer.

Dial Timer—The dial timer is the heart of the automatic mechanism and is responsible for automatically selecting and timing each washer operation. The timer mechanism is powered by a small, self-starting synchronous motor, which works in conjunction with a cam system that contains three double sets of contacts to control the washer throughout its entire cycle. This cycle varies in length, according to the water supply pressure available at the water valve inlet. Installations where flowing pressures over 18 pounds are present use the standard 29½-minute timer, while installations where the flowing water pressure is as low as 10 pounds use a 34-minute timer. A contact arm rides on each cam and, when directed, makes contact with either the upper or lower contact. Fig. 14 illustrates the cam, contact arm, and contact arrangement in the timer and shows the lead from each contact to the part it energizes. As shown in the illustration, the lead from the upper No. 2 contact runs directly to the temperature-selector switch, which, in turn, relays the current to the hot- or cold-water solenoids, depending on the switch setting. The water-temperature-selector switch provides the user with a choice of hot or tempered water for the wash period.

The forward movement of the timer mechanism is not constant; it moves ahead approximately 5° every 30 seconds. This motion is accomplished as the timer motor winds a spring, which is fastened to a trip arm. At the end of every 30 seconds, this arm trips, relieves the pressure on the spring, and moves the timer mechanism forward. This operation eliminates the slow breaking of contacts and the resultant arcing. The last 30-second period breaks the circuit, which energizes the timer motor. The washer mechanism and the washer then remain inactive until the dial timer is manually reset. Between the first and second cams is a line-contact switch, which is controlled by the push-pull dial-and-knob assembly. Pushing in on the dial closes the contact; pulling the dial outward opens the contact. This contact lies between the motor

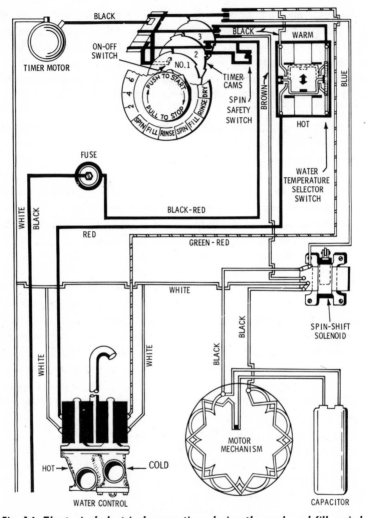

Fig. 14. The typical electrical connections during the wash and fill periods.

protector and the first contact arm, and no electrical energy can be supplied to the washer mechanism when the contact is open. The 15-ampere motor protector is specially constructed to withstand heavy sudden loads but will blow out or trip if the load is sustained because of a malfunction in the motor or washer mechanism.

Although the dial timer is completely automatic in operation, its sequence of operation can be interrupted by the manual control of the timer dial. To make any alterations of the timer sequence, the dial should be pulled out and rotated in a clockwise direction until the desired point on the dial is reached. In some cases, it will be necessary to rotate the dial past the dial stop; added pressure will move the dial through this position. The dial should neither be rotated counterclockwise nor rotated clockwise with the dial pushed in.

If extra washing time is desired to handle badly soiled clothes, the operator can permit the washer to wash for the full 10-minute period by pulling out the dial and rotating it clockwise until it reaches the number on the wash section of the dial that corresponds to the amount of extra wash time desired. Rinse or spin periods can be omitted, repeated, or altered by manual control of the timer.

Cycle of Operation—The dial timer is so constructed that once the timer dial is turned to the start position and pushed in to start the timer motor, the timer motor takes over, drives the motor through all the operations, and shuts itself off at the end of the completed cycle. The complete cycle of operation, shown in Fig. 15, is as follows:

Fill, 2½ minutes; wash, 10 minutes; spin, 1½ minutes; brake, ½ minute; fill, 2½ minutes; rinse, 2½ minutes; spin, 1½ minutes; brake, ½ minute; fill, 2½ minutes; rinse, 2½ minutes; and spin, 3 minutes. The total washing operation takes 29½ minutes; the pulsator is in action during the wash and rinse periods only.

During these washing operations, the following action takes place in the dial timer circuit:

1. *Wash-fill period*—Assuming that the installation of the machine has been properly made according to the manufacturer's directions, and that hot and cold water is available at the prescribed temperatures and pressures, the wash-fill period of the washing cycle is started by pulling out the dial knob and moving it clockwise until it meets a positive stop. This is the "start" position and is so marked on the dial.

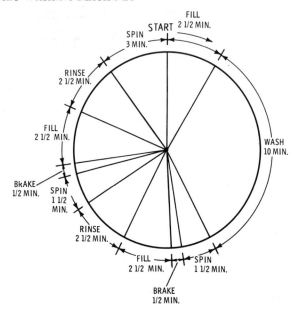

Fig. 15. *The complete washing cycle for one type of automatic washer.*

The prescribed weight of dry clothes is next deposited in the washer tub, together with the necessary amount of soap or detergent. The amount of soap or detergent varies with the temperature and with the hardness of the water. If a water softener or blueing agent is to be used, it can be added at this time.

Next, turn the water-temperature-selector switch to "hot" (except for rayons, woolens, and delicate fabrics), and push the dial in to start the washer. During this period, the No. 1 cam, Fig. 14, in the dial-timer assembly has raised the first contact arm until it has closed with the upper contact. As shown by the black line in the illustration, current is now supplied to the timer motor and to the other two contact arms. At the same time, the upper No. 2 contact is closed by the No. 2 cam, and, when the water-temperature-selector switch is in the "hot" position, the solenoid on the hot-water valve is energized through the red wire, thereby opening the valve. The open valve permits hot water to flow into the washer-tub assembly. The flow of water is controlled by the action of a metering washer in the water-control flow-washer retainer.

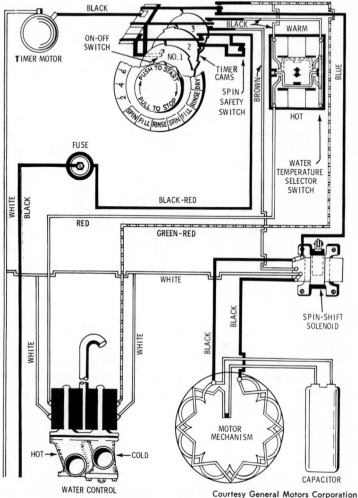

Fig. 16. The typical electrical connections during the wash and rinse operation.

Courtesy General Motors Corporation

At the end of the fill period, the No. 2 contact arm returns to the neutral position, the circuit to the hot water solenoid is opened (thus stopping the flow of water), and the timer dial moves into the "wash" position. The time consumed for the fill period is 2½ minutes.

2. *Wash period*—During this period, the No. 1 contact arm is still in the raised position, with the No. 3 arm resting on its lower contact, as shown in Fig. 16. This contact position energizes the washer motor through the blue wire, which is connected to the black motor lead at the spin-shift solenoid; the motor then drives the pulsator through the wash period. The wash period is 10 minutes.

3. *Wash-spin period*—At this part of the washing cycle, the No. 1 contact closes in the lower position, the No. 3 contact remains closed in the lower position, and the No. 2 contact closes in the lower position, thereby energizing the spin-shift solenoid through the brown wire, as shown in Fig. 17. When the spin-shift solenoid is energized, it rotates the trip-shaft lever, releases the stop that has held the clutch spring in place, and permits the entire washer mechanism to go into a spin of approximately 1100 rpm. This spinning action extracts water from the clothes by centrifugal action and forces the water from the spinning tubs into the washer cabinet. The water is then drawn off by the water pump and expelled through the drain hose. Since the water pump is connected directly to the motor shaft, it is in constant operation except during the fill and brake periods, and it expels all water entering the washer cabinet.

As noted in Fig. 17, the "hot" wire from the lower No. 1 contact is open. This is part of the safety feature on spin operations. Closing the washer cabinet lid closes this contact and permits the spinning action to take place. The wash-spin period takes approximately 1½ minutes.

4. *Brake period*—During this period, all contacts except the main line contact are open; the trip-shaft lever returns to its normal position by spring action and catches the end of the clutch spring. The clutch spring, which catches on the trip-shaft lever, comes to a stop and, in turn, stops the brake-torque plate. Two friction brake plates, which are attached directly to the spinning mechanism, act on the now stationary brake-torque plate and thus bring the spinning mechanism and tubs to a stop. This braking action consumes ½ minute.

5. *First rinse-fill period*—As in the wash-fill period, the No. 1 contact cam is in the raised position; however, since the rinse fill is to be made with 100°F. water, the solenoid for the mix side of

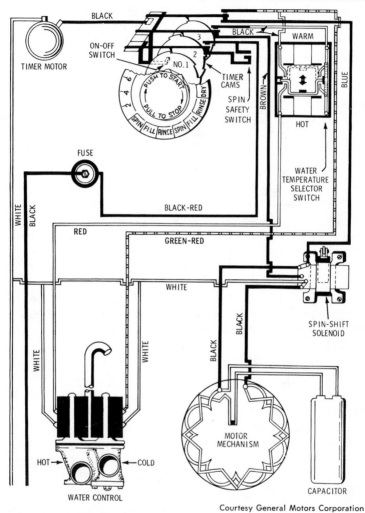

BLACK

ON-OFF
SWITCH

TIMER MOTOR

BLACK

WARM

3

NO.1

2

TIMER
CAMS

SPIN
SAFETY
SWITCH

BROWN

BLUE

HOT

WATER
TEMPERATURE
SELECTOR
SWITCH

FUSE

WHITE

BLACK

BLACK-RED

RED

GREEN-RED

WHITE

WHITE

WHITE

BLACK

BLACK

SPIN-SHIFT
SOLENOID

HOT

COLD

MOTOR
MECHANISM

CAPACITOR

WATER CONTROL

Courtesy General Motors Corporation

Fig. 17. The typical electrical connections during the spin operations.

the water-control valve must open. Therefore, the No. 3 contact arm must close with its upper contact and energize the mix solenoid through the green wire with the red tracer, as shown in Fig. 18. During this period, the cold-water valve is fully opened, and the hot-water valve is closed. However, a thermostat located in the

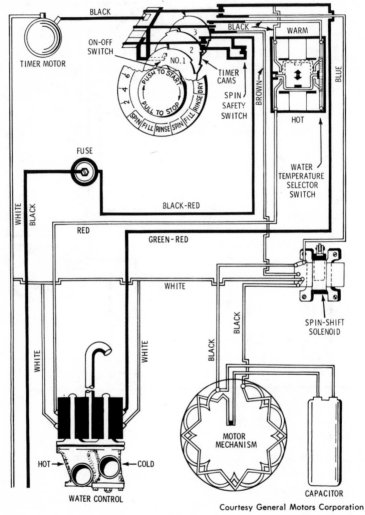

Fig. 18. The typical electrical connections during the rinse-fill operation.

Courtesy General Motors Corporation

mixing chamber of the valve allows a sufficient quantity of hot water to mix with the cold water, thereby raising the rinse water temperature to 100°F., plus or minus 5°. The time consumed for this operation is 2½ minutes.

6. *First rinse period*—The action during this period is identical to that in the wash period. The pulsator is in motion during this operation, and the time consumed is 2½ minutes.

7. *First rinse-spin period*—The action during this part of the washing cycle is the same as that of the wash-spin period. The time used for this operation is 1½ minutes.

8. *Brake period*—This action is the same as the first brake period. The time consumed is ½ minute.

9. *Second rinse-fill period*—This is the same as the first rinse-fill period. The time consumed is 2½ minutes.

10. *Second rinse period*—This is the same as the first rinse period. The time consumed is 2½ minutes.

11. *Dry-spin period*—The action of the machine during this period is the same as for other spin periods, except that this period is twice as long. At the end of this period, the washer mechanism is stopped by the brake action, and the dial mechanism moves to the "off" position. The timing dial stays in this position until it is manually reset at the "start" position. The time for the dry-spin period is 3 minutes. Any or all of the various periods may be repeated, if desired, by manually resetting the dial.

Agitator-Type Washing Machines

This unit differs considerably from the pulsator type previously described, both with respect to its construction and methods of control. In the agitator-type washer, the controls are cam-operated instead of being electrically operated by solenoids.

The approximate time for the complete washing cycle, as shown in Fig. 19, is as follows:

Wash, 15 minutes; drain, 2½ minutes; spin, 2½ minutes; brake, 2½ minutes; rinse, 2½ minutes; drain, 2½ minutes; spin, 5 minutes; and brake, 2½ minutes. These periods add up to a 35-minute cycle for the complete operation, discounting the 2½ minutes required for the original filling.

This maximum preselected period may be shortened considerably, since the wash period may be decreased to about 2½ minutes, depending on how badly soiled are the clothes being washed.

During these washing operations, the following actions take place:

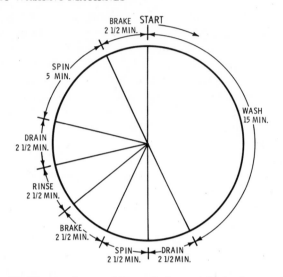

Fig. 19. The complete washing cycle for an agitator-type washer.

1. *Fill*—The water valve opens, and the tub fills at temperatures selected by the operator.

2. *Wash (15 minutes)*—The water valve closes, and the motor starts. The agitator oscillates for 15 minutes. Clothes are added after the soap dissolves. Reset the temperature dial for rinsing. Other washing periods of less than 15 minutes' duration are obtained by setting the cycle control until the desired time period registers on the indicator dial. For a longer cycle, the dial may be turned back to repeat the wash cycle.

3. *Drain (2½ minutes)*—The agitator stops, and the drain valve opens. The tub drains into the draining casing and is then pumped to the sewer.

4. *Spin (2½ minutes)*—The tub spins at approximately 550 rpm and extracts the remaining soapy wash water from the clothes by centrifugal force.

5. *Brake (2½ minutes)*—The brake is applied and gradually slows the tub rotation to a dead stop. The water valve opens, and the tub fills. The clothes are immersed in clean rinse water at a preselected temperature.

6. *Rinse (2½ minutes)*—The agitator oscillates, and a high velocity spray flushes scum from the water surface.

7. *Drain (2½ minutes)*—The agitator stops, and the drain valve opens. The tub drains into the drain casing and is then pumped to the sewer.

8. *Spin (5 minutes)*—The tub spins at approximately 550 rpm and extracts the remaining excess rinse water from the clothes.

9. *Brake (2½ minutes)*—The brake is applied to stop rotation; the power switch cuts off the current from the machine, and the wash cycle is completed.

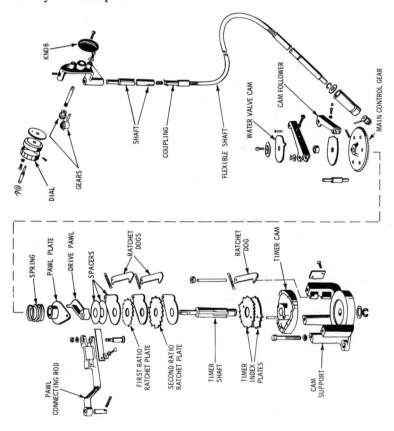

Fig. 20. The operating-control and timer assemblies of a typical automatic washing machine.

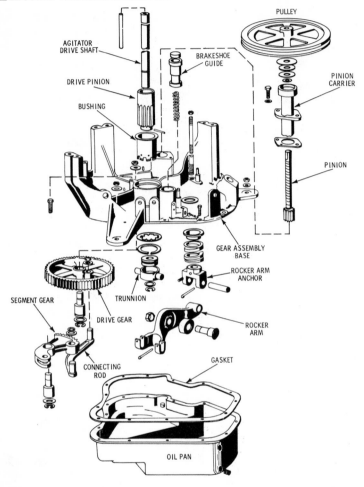

Fig. 21. The gear-case assembly of a typical agitator-type washing machine.

Timing Control Mechanism—The operating control and timer assemblies of the machine are shown in Fig. 20. The cam-controlled timing mechanism is composed of three toothed ratchet plates, which are actuated by a pawl. The pawl is driven through a connecting rod from the agitator drive mechanism (a motor-switch cam, a water-valve cam, and a main-drive cam).

410

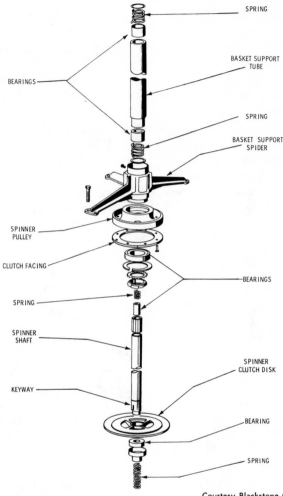

SPRING

BASKET SUPPORT
TUBE

BEARINGS

SPRING

BASKET SUPPORT
SPIDER

SPINNER
PULLEY

CLUTCH FACING

BEARINGS

SPRING

SPINNER
SHAFT

SPINNER
CLUTCH DISK

KEYWAY

BEARING

SPRING

Courtesy Blackstone Corporation

*Fig. 22. The basket-support-and-clutch-drive assembly of an agitator-type
washer.*

When the deep-cut teeth in the first and second ratchet plates
line up, the pawl drops into a tooth in the timer index plate, thereby
pushing this plate and the attached cam forward one step to actuate
the next operation cycle. The main drive cam has a contoured
surface of varying elevations, and it actuates a rocker arm by

411

means of a roller, which, in turn, raises and lowers the agitator shaft and the spinner shaft connected to the tub.

The agitator drive clutch, located at the upper end of the agitator shaft, disengages when the agitator shaft is raised. The tub is lifted off its seat for draining when the spinner shaft is raised. When the spinner shaft is lifted to a second raised position, it engages the spinner clutch with the spinner pulley for rotation. These operations are reversed when the shafts are lowered. The operating-cycle control operates through a flexible shaft to advance the timing from "stop" to "fill" and to set the length of the wash-period time.

Gear Case—The various parts of the gear case are illustrated in Fig. 21. The gear case consists of the agitator shaft and pinion

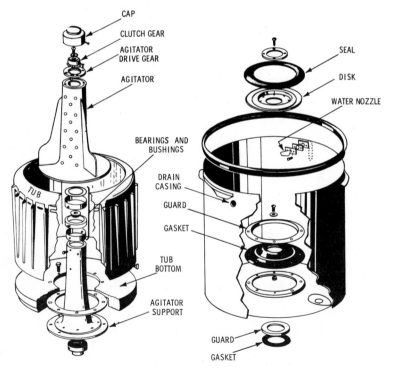

Fig. 23. The tub and collector-tank assemblies of an agitator-type washing machine.

assembly, the lower timer assembly, the brake assembly, the connecting rod assembly, and the main pulley and pinion assembly.

Basket Support and Clutch Drive—This assembly is shown in Fig. 22. It consists of a spider casting assembly mounted on the gear case that supports the tub and houses the spinner shaft assembly, together with the assemblies of the flywheel and bearings.

Tub and Collector Tank—The collector tank and tub assemblies are illustrated in Fig. 23; the agitator, drive gear, center-support gaskets, and seals compose its associated parts. The drain casing consists of a galvanized sheet-steel tank, which completely surrounds the tub. It is sealed from the top panel by means of a flexible rubber gasket. The drain-cleanout assembly is attached to the bottom. The water nozzle is attached to the side wall near the upper rim, and a flexible rubber gasket is assembled to the bottom of the casing to prevent water leakage. The tub is supported within the drain casing by means of the tub-support disc assembly, which also seals the drainage holes in the bottom of the tub during the wash and rinse cycles.

Cylinder-Type Washing Machines

The assembly view of a front-loading, cylinder-type washer is illustrated in Fig. 24. This machine, in common with other automatic washers, is time-controlled by means of a centrally located timer and a thermostatically controlled water valve, which control the action of the machine automatically through the complete washing cycle.

The cycle of operation, after filling, takes place in the following order:

Soak, drain, flush rinse, spin extraction, automatic stop, wash drain, flush rinse, spin extraction, deep rinse, drain, spin extraction, deep rinse, drain, spin extraction, tumble, automatic stop. Depending on the control setting, the total time for the complete operation is from 40 to 50 minutes.

During the washing cycle, the cylinder turns in a clockwise direction with a speed of approximately 60 rpm, and between 300 and 550 rpm during the high-speed spin or extracting cycle, depending on the particular model. The clothes are rotated by means of a series of baffles located on the inner surface of the cylinder. As the baffles pass under the garments, the garments are picked up and carried almost to the top of the cylinder. There they are thrown toward the other side and downward again to meet the baffles and be carried upward in an elliptical path, thus insuring maximum cleansing efficiency.

During the water-extraction or spin period, the clothes continue to revolve with the cylinder and are held against the cylinder wall by centrifugal force; since the speed of the cylinder is increased quite gradually, the clothes are evenly distributed around the cylinder, thus producing a balanced load with uniform drying action.

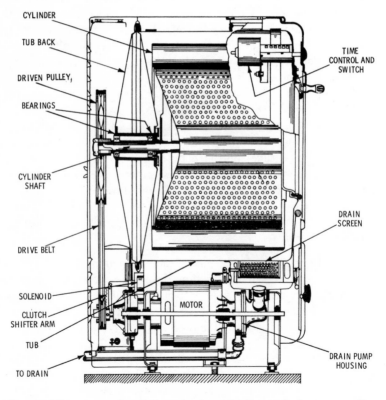

Fig. 24. The cross-sectional view of a front-loading, cylinder-type washing machine.

Controls—Washers of this type usually have three controls, one for water metering, one for water temperature, and one for timing the washing period. The electrical circuit of a cylinder-type washer is shown in Fig. 25, and its wiring harness is shown in Fig. 26. The time-control assembly, or timer, is a device that is used to coordinate the electrically controlled assemblies of the washer by opening or closing their electrical circuits, so as to

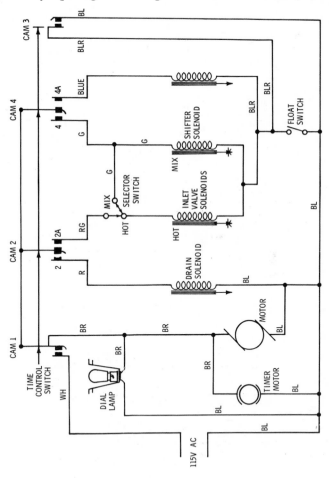

Fig. 25. *The schematic circuit of a typical front-loading, cylinder-type automatic washer.*

415

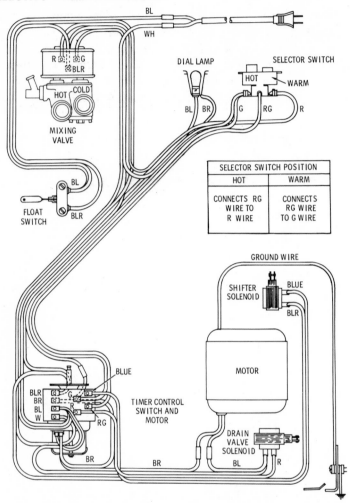

SELECTOR SWITCH POSITION	
HOT	WARM
CONNECTS RG WIRE TO R WIRE	CONNECTS RG WIRE TO G WIRE

Fig. 26. The wiring harness and electrical components of a cylinder-type washing machine.

cause them to perform their various functions at the proper time during the operation cycle.

The function of the various time-control circuits can best be understood by a careful study of the circuit diagram illustrated in Fig. 25. The timer is a set of four switches, three of which have one common terminal connected to one side of the power-supply

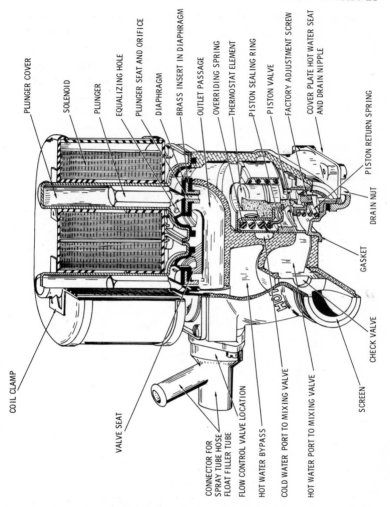

PLUNGER COVER

SOLENOID

PLUNGER

EQUALIZING HOLE

PLUNGER SEAT AND ORIFICE

DIAPHRAGM

BRASS INSERT IN DIAPHRAGM

OUTLET PASSAGE

OVERRIDING SPRING

THERMOSTAT ELEMENT

PISTON SEALING RING

PISTON VALVE

FACTORY ADJUSTMENT SCREW

COVER PLATE HOT WATER SEAT AND DRAIN NIPPLE

PISTON RETURN SPRING

DRAIN NUT

GASKET

CHECK VALVE

SCREEN

HOT WATER PORT TO MIXING VALVE

COLD WATER PORT TO MIXING VALVE

HOT WATER BYPASS

FLOW CONTROL VALVE LOCATION

FLOAT FILLER TUBE

CONNECTOR FOR SPRAY TUBE HOSE

VALVE SEAT

COIL CLAMP

Fig. 27. *The cross-sectional view of the inlet and mixing-valve assembly for an automatic washing machine.*

circuit. The switches distribute current to the various electrical parts of the unit as the cam assembly causes them to make and break contact.

From the timer motor outward, the first three switches control, respectively, the washer motor and the timer motor and the drain solenoid and mixed- or hot-water inlet valve for the wash period (depending on the setting of the selector switch). The fourth switch controls the mixed-water inlet valve and the shifter solenoid by means of a cam. When this cam allows the arm to drop into a slot, a connection is made to the contact that controls the mixed-water inlet valve. To break the circuit, the arm is raised to a neutral position midway between the two contacts, and from that position a riser on the cam pushes the arm up at the proper time to contact the shifter-solenoid terminal.

The timer motor is geared to a spring-driven escapement, and the escapement, in turn, drives the camshaft. The time interval between impulses is approximately 50 seconds, and the amount of cam-assembly rotation for each impulse is $5°$. Thus, 72 impulses are required to produce one complete rotation of the shaft, and the total time represented is one hour. The "off" positions on the dial occupy $30°$ of space on the motor cam, so actually only 55 minutes of operating time are available for any one cycle. This cycle is usually shortened further by the operator, who may set the control knob for a shorter washing time.

The camshaft extends beyond the timer case and forms the shaft for the control knob. The knob is keyed to the shaft with a setscrew spring-and-plunger arrangement that prevents backward rotation by engaging the shaft only when turned in a clockwise direction. A ratchet is provided in the escapement to permit the camshaft to be advanced clockwise at any time without interrupting the action of the escapement or the motor.

The selector switch is a single-pole, double-throw, snap switch connected in the electrical circuit of the inlet valve, and it permits selection by the operator of hot or warm water for the washing and first rinse operation. The switch is energized only during the wash and first rinse periods. At all other times it is out of the circuit. One contact on the switch is connected to the hot inlet valve solenoid, and the other contact is connected to the mix, or warm, inlet valve solenoid. Thus, depending on the setting established by the operator, either the hot- or the mixed-water inlet valve will open when the timer is turned to the wash period.

Inlet and Mixing Valve Assembly—The function of these valves is to control the admission of hot and cold water to the machine. A typical valve of this type is shown in Fig. 27. The hot- and cold-water lines are solenoid-operated, as shown in Fig. 25. To control the amount of hot and cold water necessary to produce a predetermined temperature, a sensitive thermostatic element has been incorporated in the unit to control the position of a large cylindrical piston. The hot water enters through an opening at the end of the piston chamber that is farthest from the thermostatic element; the cold water enters at the other end of the chamber. A small gasket on the piston prevents the water from mixing, except through the openings in the center of the piston, which partially closes each water vent. As hot water flows through the piston and over the element, the sensitive element expands and forces the piston to close the hot-water port and open the cold-water port. If the water becomes too cold, the element contracts and permits a coil spring to force the piston in the opposite direction, thereby decreasing the flow of cold water and increasing the flow of hot water. Once the piston has adjusted the valve openings to the proper size, it will remain motionless unless the temperature of either water supply varies.

INSTALLATION OF WASHING MACHINES

Satisfactory performance of an automatic washing machine depends to a great extent on a carefully planned and properly designed initial installation. The location where the laundry is done should be well lighted and adequately equipped with convenient electrical outlets. The plumbing connections must be made properly, and when required, the washer must be anchored to the floor to prevent movement.

Plumbing

The plumbing requirements depend on the location of the washer. In kitchen installations, the hot- and cold-water outlets are close at hand, and all that is required is to attach the hose connections to the sink or set-tub water taps. In basement installations, however, it is often necessary to provide the cold- and hot-water and drain connections. In addition to arranging for water supply and drainage, it is also necessary to make certain that the water pressure is adequate and that there is a sufficient amount of hot water available at the required temperature.

In most installations using hose, the water supply lines and drain facilities should be located within a close proximity to the

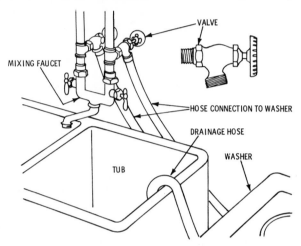

VALVE

MIXING FAUCET

HOSE CONNECTION TO WASHER

DRAINAGE HOSE

WASHER

TUB

Fig. 28. The plumbing installation for an automatic washer. The water supply is tapped off the lines by means of angle-valve and hose connections. Drainage is provided by a third hose connection that leads from a special drainage coupling in the lower part of the washer to the tub.

machine. The length of hose and the location of fittings should be the determining factors in rearranging plumbing to provide convenient water facilities. Various local code regulations apply in most communities to permanent plumbing and electrical installations. The serviceman should be familiar with these regulations and should make sure that all installations conform to the local codes. In all basement installations, the machine is usually placed adjacent to set tubs, which are equipped either with individual hot- and cold-water faucets or with a mixing faucet. Typical basement installations are shown in Figs. 28 to 31.

Fig. 32 shows an above-the-basement type of installation. The water inlet connections are made with a minimum ⅜-inch O.D. copper tubing instead of rubber hose. Pipe clamps are again used for installation of shutoff valves, although "tees" may be substituted if desired. When tubing is used, a tube coupling is necessary

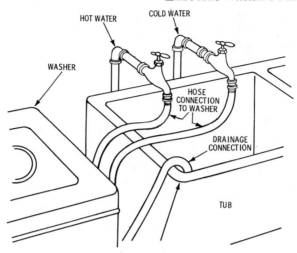

HOT WATER

COLD WATER

WASHER

HOSE
CONNECTION
TO WASHER

DRAINAGE
CONNECTION

TUB

Fig. 29. The plumbing connections for hot and cold water to an automatic washer.

at either end. The coupling screws into a reducing valve at one end and into the inlet fittings on the back of the machine on the other. Gaskets are not necessary, although they may be used.

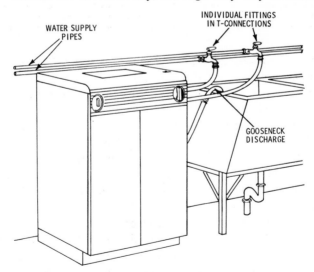

INDIVIDUAL FITTINGS
IN T-CONNECTIONS

WATER SUPPLY
PIPES

GOOSENECK
DISCHARGE

Fig. 30. Water supply and drainage connections in a basement washing-machine installation.

421

Electric Washing Machines

When copper tubing is used, make large even bends, and form adequate loops to facilitate moving the machine away from the wall for subsequent servicing. These steps will also minimize the possibility of developing leaks in the intakes as a result of vibrations. If a permanent installation is desired, and the machine is to be mounted flush against a wall, it will be necessary to expose the studding from the floor to the above-the-water intake connections in order to accommodate the copper tubing and the intakes and the drain elbow, which extends beyond the depth of the machine.

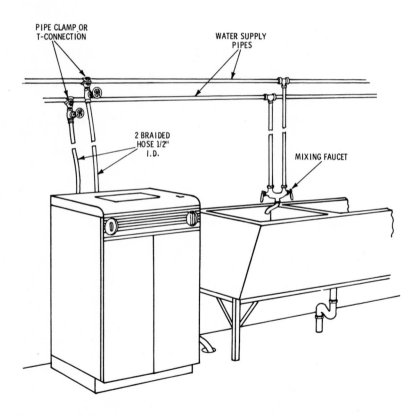

Fig. 31. An alternate type of rubber-hose installation with simple faucet connections for a basement location.

Condition of the Floors

A weak and unstable foundation may be a contributing cause of vibration. If the machine is to be installed above the basement on a weak wood flooring, the joists and flooring should be strengthened, as indicated in Figs. 33 and 34. In the case of a basement installation where a damp floor condition is encountered, or where there is no cement floor, a suitable foundation, such as that shown in Fig. 35, may be built to prevent rust and insure proper machine

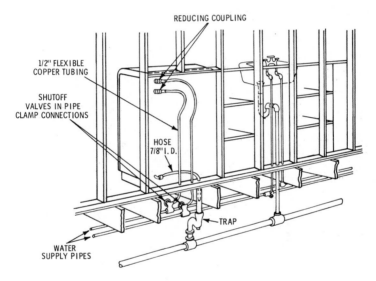

Fig. 32. An above-the-basement washing-machine installation.

operation. A solid wooden platform is usually satisfactory. In any event, it is necessary that the floor or foundation be absolutely level. This can be ascertained by placing a carpenter's level on the top rear and front of the machine; raise or lower the adjustable legs until perfect leveling in all directions is obtained, as noted on the level.

Some washers must be bolted down; this may be accomplished in various ways, depending on the thickness of the concrete floor. If the floor is in good condition and at least 2 or more inches thick, drill two ¾-inch holes at least 1½ inches deep, and insert a lead bolt anchor in each hole with a lead anchor set punch, as shown in Fig. 36. To be sure that the bolt anchors are firmly set, place a piece of ⅜-inch pipe approximately 2½ inches long over each ⅜″ × 4″ machine hold-down bolt. Screw the bolt into the anchor, and tighten it sufficiently to pull the expansion part of the lead anchor up approximately ¼ inch.

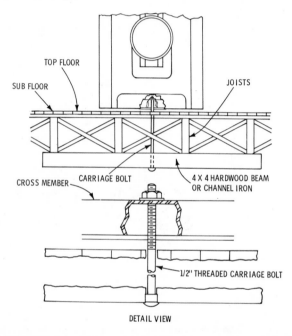

Fig. 33. The floor must be strengthened if the washing machine is installed on a weak wooden floor. The strengthening method shown in this illustration is recommended for most installations of this type.

Another excellent method of anchoring the washing machine to concrete floors is to use molten lead. After drilling the ¾-inch holes in the concrete, set the bolt upright in the hole. Be certain that the hole is thoroughly dry so the lead will not splatter; grease the hole to prevent splattering. Now pour the hole full of molten lead, and allow the lead to harden. This process will provide a firm foundation for anchoring the hold-down bolts.

If the concrete floor is less than 2 inches thick, various methods may be used to anchor the machine. One method consists of

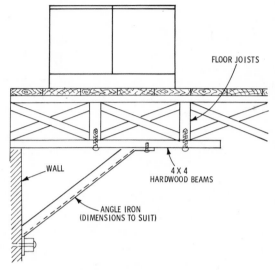

FLOOR JOISTS

WALL

4 X 4
HARDWOOD BEAMS

ANGLE IRON
(DIMENSIONS TO SUIT)

Fig. 34. An alternate method of strengthening the floor below a washing-machine installation.

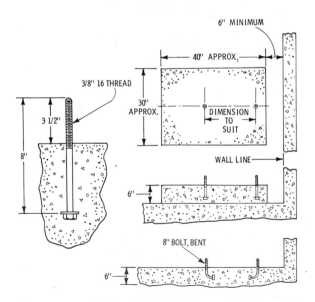

6" MINIMUM

3/8" 16 THREAD

40" APPROX.

3 1/2"

30"
APPROX.

DIMENSION
TO
SUIT

8"

WALL LINE

6"

8" BOLT, BENT

6"

Fig. 35. One recommended method of building a suitable cement foundation for the basement installation of a washing machine.

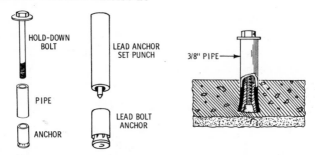

Fig. 36. *The bolt–anchor method of securing an automatic washer to a cement floor.*

making two openings in the floor 2½ inches wide and at least 7 inches long or a round hole 4 inches in diameter. Insert two 2″ × 2″ angle irons with a hole in the centers for a ⅜-inch bolt, as shown in Fig. 37. Next, place each angle iron in the correct position in the floor so that the bolts fit into the proper holes in the cross members of the frame. Be sure that the ends of the angle irons are inserted and turned so as to go underneath the edges of the old concrete. Fill the hole with new cement, and complete the installation.

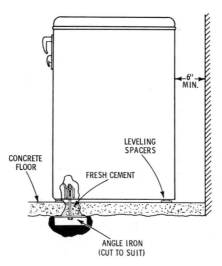

Fig. 37. *The usual method of securing a washing machine to a cement floor that has a thickness of two inches or less.*

Water Pressure and Temperature

The water volume should be carefully checked to insure the availability of an adequate pressure. Inadequate water pressure (below 20 pounds at the tap) is usually unsatisfactory for the proper operation of automatic washers. It is also important to check the water volume at the tap. Adequate pressure must be available to provide for the introduction of a sufficient volume of water into the machine. The installation of larger supply lines or a thorough cleansing of the corroded pipes is essential for increased water flow. If the water pressure exceeds 120 pounds, which is unusual, a pressure-regulating valve must be installed. Private pumping systems should be checked for correct pressure at the pump cutoff, and this pressure should not exceed 120 pounds to be efficient for washer use.

An adequate amount of hot water at a temperature of approximately 160°F. is of the utmost importance for any washing-machine installation. In private dwellings using automatic hot-water heaters, a hot-water supply of 20 to 30 gallons from a gas or oil water heater is usually sufficient. Electric hot-water heaters should have a storage capacity of approximately 50 gallons, since their heat recovery tends to be less rapid. The hot water should be available in sufficient quantity and at the required temperature at all times during the washing operation in order to prevent delay and unsatisfactory results.

Electrical Wiring

Each machine is usually supplied with an electrical cord of sufficient length to reach a convenient outlet. The washing machine should be connected to a separate electrical circuit that is designed to provide a 115-volt, 60-cycle, single-phase alternating current. The conductors should preferably be No. 12 size wire (never smaller than No. 14) and should be installed from the fuse panel to a receptacle near the permanent location of the washer. The supply circuit should be protected with a 15-ampere, time-delay-type fuse. In some installations, a rest-type circuit breaker may be part of the original wiring installation. Check to see that the correct size element is installed.

Some washers are equipped with a special polarized plug that requires a special receptacle to accommodate this plug. If such a receptacle is already installed, it is advisable to check the house wiring to make sure that the grounded wire is connected to the nickel-plated terminal of the receptacle. This precaution is a *National Electrical Code* requirement and should be adhered to in all new installations.

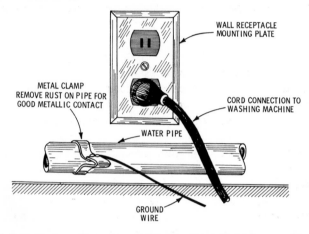

Fig. 38. For safe operation, the washing machine should be grounded to a water pipe by the method shown in this illustration.

Another grounding method is shown in Fig. 38. The internal wiring of the washer is arranged to take advantage of the safety measures thus provided. The employment of light-weight extension cords is not advisable because of the possible excessive voltage drop, which may be sufficient to cause improper washer operation or even damage the machine.

The electrical installation should conform to the requirements of the *National Electrical Code* in addition to any existing local codes.

SERVICING AND REPAIRS

Because of the timing and other associated operating mechanisms, automatic washers require that the serviceman have a more precise mechanical knowledge than that needed to diagnose and repair simple manual machines. An automatic washing machine will not cause any servicing difficulties once its fundamental principles of operation have been thoroughly understood. It is important that the serviceman have an intimate understanding of the functions of both the mechanical and the electrical systems of the various washer types before any attempt is made to render service. In particular, the serviceman should be familiar with the recommendations and specifications of the manufacturer's service manual in order to correctly repair or replace any faulty components when required.

Fast and positive diagnosis of the service problem is highly desirable; this not only prevents waste of time and expense but also creates the necessary element of good will that is so important for any successful service organization. The necessary experience will soon enable the serviceman to readily identify and correct many difficulties by a simple examination of the machine. This routine, however, can be speeded up considerably if a check list is made up of the most common failures that can cause a machine to become inoperative.

Service Tools

The tools required to properly service most washing machines may be divided into three classes; they are:

1. Standard tools,
2. Special tools,
3. Test equipment.

Standard tools are those tools that can be procured locally from any hardware store and which are usually a part of any mechanic's tool kit, whereas special tools are used to perform special services and must usually be obtained directly from the washing-machine manufacturer. Test equipment for shop and field service purposes is extremely essential. The service shop need not be a fully equipped laboratory, but certain basic requirements for handling the various washer mechanisms and performing necessary service adjustments can cut handling costs and speed the entire service process.

The minimum equipment needed to carry out the various test and disassembly operations are as follows:

1. AC voltmeter (0-150V),
2. AC wattmeter (0-2500W),

3. Water-pressure gauge (portable),
4. Test lamp (for testing circuit continuity),
5. Gear case holding fixture,
6. Hoist (small),
7. Plumbing facilities for water test,
8. Carpenter's level (about 25 inches long),
9. Arbor press (bench type).

To these should be added the special tools, as required, from the various manufacturers, and such standard tools as:

1. ¼-inch drills,
2. 5-inch spirit level in a metal case,
3. ⅛-inch punch,
4. $\frac{3}{16}$-inch punch,
5. ½-inch center punch,
6. Screwdriver—T.S. No. 4,
7. Screwdriver—Phillips No. 2,
8. Screwdriver—standard 6-inch,
9. ½-inch star drill,
10. ½-pound ball-peen hammer,
11. 1-pound rawhide hammer,
12. Soldering iron—115 volts at 50 watts,
13. 10-inch adjustable pliers,
14. Wrench—8-inch adjustable crescent,
15. Wrench—$\frac{9}{16}'' \times \frac{11}{16}''$ open-end,
16. Wrench—⅞-inch socket,
17. Chisel—$\frac{1}{2}'' \times 6''$,
18. Dowel pins—$1\frac{1}{4}'' \times \frac{1}{4}''$.

Convenient Service Charts

In most washing machines, there is a direct relationship between troubles and causes, and their remedy. Service instruction charts have been carefully designed to help the serviceman locate and repair faults that he is likely to meet in day-to-day service work. The troubleshooting chart given on the following pages should be referred to, and, although the cause of the trouble may at times differ from that given in the service chart, experience will quickly reveal the exact cause, after which the remedy in most instances will readily be found.

Washer Service Chart

Trouble	Possible Cause	Remedy
Washer will not start.	No power supply at electrical outlet.	Check condition of fuse. Replace if necessary.
	Low or incorrect voltage.	Inspect wiring from fuse block to electrical outlet. Check voltage with voltmeter, and compare with that on washer nameplate. If wiring is excessively long, it should be replaced with heavier size conductors to minimize voltage drop.
	Defective motor.	Remove service-line plate from motor, and check current with test lamp at motor terminal posts. If circuit is complete and motor does not run, remove for repair or replacement.
	Thermal-overload switch is opening motor circuit.	Permit motor to cool, then reset the overload switch. If temperature surrounding the motor is too high, provide exterior ventilation or change location of washer.
	Faulty wiring or electrical switches.	See that all plug-in connections are tight. Check cord connection at electrical outlet. Check continuity of electrical system. Check wiring for loose connection at switch terminals. Replace faulty switches.
	Motor hums without starting.	Remove belt from motor pulley. Turn agitator or cylinder by hand. If cylinder is blocked, remove obstructions. With belt removed from motor pulley, plug in cord and if motor continues to hum, it must be assumed that blocking is in transmission or motor. Dismantle transmission for check.
	Broken or loose belt.	Replace or adjust by means of belt tension screws.

Washer Service Chart (Continued)

Trouble	Possible Cause	Remedy
Motor overheats and stops.	Washer overloaded.	If dry load did not exceed that given in the instruction book, check whether towels, blankets, or other highly absorbent fabrics are causing washer to become overloaded. If machine is oversoaped, this will clog drain and impose an undue load on motor.
Noisy motor.	Pump impeller loose.	Remove pump, and check impeller for loose setscrew. Tighten if necessary.
	Transmission defects.	See that gears do not show signs of excessive wear. Replace when necessary. Check oil level. If transmission check is satisfactory, and motor noise prevails, motor bearings or other motor parts are faulty. Repair or replace motor.
Motor does not reverse.	Belt slipping.	Check tension of transmission to V-belts. See that belt is not cracked; if so, replace.
	Motor reverse relay not functioning.	See that connections are secure or that relay or timer is not loose on terminals. If relay contacts are badly burned, they should be replaced. If relay coils are faulty, replace or obtain new relay of same manufacture as that previously used.
Washer noisy.	Installation mounting loose.	Check leveling of machine. Tighten hold-down bolts, taking care not to disturb end supports or cross members.
	Pump impeller loose or faulty pump bearing.	Remove pump cover, and check impeller for loose setscrew. Clean and tighten. Check water-pump bearing for noise.

Washer Service Chart (Continued)

Trouble	Possible Cause	Remedy
	Transmission not properly lubricated.	Lubricate.
	Worn or split belt.	Check and replace if necessary.
	Gear-case mechanism out of adjustment.	Check for improper meshing of agitator drive pinion and rack or drive sector. Adjust when necessary. Check for improper worm adjustment. Adjust worm end play where necessary.
Electric shock when machine is touched.	Static electricity. Machine not properly grounded.	Install proper and effective ground connection.
Motor runs but agitator or cylinder does not move.	Broken or slipping belt. Loose pulley or motor coupling. Transmission defects.	Check and replace any defective parts, if necessary.
	Defective clutch or timer. Loose wire on control terminals or circuit defects.	Adjust and replace defective parts when necessary. Check electrical circuit, and compare with manufacturer's wiring diagram.
Water does not enter tub.	Defective hose connections.	Ascertain whether hot-water inlets and outlets are connected by same hose. Make similar checks on cold-water inlets and outlets.
	Water temperature control in "off" position (automatic units).	Turn temperature control to medium position.
	Water valve out of adjustment (automatic units).	Check valve; make sure valve cams open completely. If not, make necessary adjustments.
	Valves on water line turned off.	Turn water valves on.
	Kinked water inlet hose.	Remove kinks from inlet hose. Check water flow before attaching hose to washer.

Washer Service Chart (Continued)

Trouble	Possible Cause	Remedy
	Foreign matter lodged in water system.	Remove hoses and valves. Clean or replace parts as required.
Water does not drain from tub.	Lint trap filled with foreign matter.	Check and clean.
	Drain holes in bottom of tub clogged.	Remove agitator, and clean out foreign matter from drain holes by using small round brush or blunt end of pencil.
	Garment in drain casing.	Pull out misplaced garment.
	Water pump not functioning.	Disassemble pump-impeller housing, clean thoroughly, and reassemble. If necessary, remove pump assembly, and repair or replace as required.
	Motor running in wrong direction.	Remove panel, and check motor direction. Motor should run in counterclockwise direction, facing pulley end. Replace motor if defective.
	Worn cylinder or agitator bearings.	Check bearings. If water is getting through bearings, examine carefully for out-of-round wear, scoring, or defective seal. Replace as required.
Water temperature too high or too low.	Inlet hoses improperly connected.	If hoses are reversed at faucets or at water inlet valve, water of improper temperature will be metered into tub. Change hose connections at one end only.
	Insufficient hot water available from tank.	Check capacity and thermostat setting of water heater. If water heater tank is too small, replace.
	Inoperative thermostat in water inlet valve.	If temperature of hot water from tank is 150°F. or over, trouble may be assumed to be in water valve. Replace valve.
Tub slow in starting during spin cycle.	Loose or worn belts.	Adjust belt tension, and replace belt if necessary.

Washer Service Chart (Continued)

Trouble	Possible Cause	Remedy
	Lint trap filled with foreign matter.	Remove lint trap, and clean thoroughly.
	Overloaded electric circuit.	Check for insufficient wire capacity or low voltage at motor terminals. Washer will not operate properly if supply voltage varies more than 10% of machine rating.
	Loose clutch.	Tighten clutch-adjustment nut.
	Defective timer motor.	Check and adjust or replace as required.
Washer does not spin.	High-speed clutch requires adjustment.	Adjust or replace as required.
	High-speed solenoid defective.	Check, and replace if necessary.
Washer vibrates during spin cycle.	Washer not properly leveled.	Machine must be level and have firm, strong foundation. Check with carpenter's level, and make certain that each caster bears equally on floor or base.
	Defective rubber mounting supporting gear case.	Examine, and replace if defective.
	Weak floor.	Necessary reinforcement should be made.
Washer stops during spin cycle.	Overload switch too sensitive or trips to unbalanced load.	Reset switch or replace if required. An unbalanced load condition or improper drainage is the most common cause of motor overloading. Balance load, and check for proper drainage.
	Tight working gear unit.	Remove motor belt, and turn pump pulley with finger to check free movement. If pump runs tight, disassemble pump-impeller housing, clean, and adjust thoroughly. If this procedure does not correct difficulty, remove pump, and repair or replace as required.

Washer Service Chart (Continued)

Trouble	Possible Cause	Remedy
Agitator lifting.	Insufficient water in tub.	Observe water supply in tub during fill cycle, making certain that tub is filled to water line, which is slightly above top of agitator blades.
	Insufficient water pressure.	Water pressure should be a minimum of 20 pounds per square inch at tap for satisfactory operation, and a water flow of from 6 to 7 gallons per minute should be available.
Washer does not cycle (automatic units).	Defective timer motor.	Replace timer.
	Worn or broken transmission parts.	Check, and replace if required.
	Contact arms not closing with timer.	Replace timer.
	Defective solenoid wiring.	Check circuit, and compare with manufacturer's diagram. Install new wiring if necessary.
Oil leaks into tub.	Defective tub-mounting seal or loose nuts.	Replace seal if necessary, and tighten all nuts.
Oil leaks between mechanism housing and base.	Screws loose around base. Gasket torn or damaged. Dirt under gasket.	Tighten screws, replace gasket, and clean all metallic surfaces.
Washer tears clothes.	Rough spots on tub bottom, tub sides, or agitator.	Remove rough spots with emery cloth. Bead on underside of agitator should be smooth.
	Insufficient water supply.	Water should reach to approximately top of agitator blades. Water at tap should be at a minimum running pressure of 20 pounds per square inch and a volume flow at not less than 6 gallons per minute per water tap.

436

Washer Service Chart (Continued)

Trouble	Possible Cause	Remedy
	Tight agitator gears.	Check agitator gears for proper operation. Remove burrs and sharp edges with emery cloth.
Odor in washer cabinet.	Scum accumulation on walls of cabinet due to use of hard water or emulsification due to oil leaks.	Use stiff brush to completely clean washer to remove all deposits. Check for oil leaks.
Wringer rolls not turning.	Drive gears worn or broken. Inoperative clutch lever. Broken coupling.	Check, and replace when necessary.
Wringer rolls not running in reverse.	Worn or broken clutch lever arm or cam. Worn or damaged gears.	Check, and replace as required.
Wringer rolls sticky.	Faulty use and improper maintenance.	Clean, and release rolls after each use.
Wringer releases under load.	Excessive load, bent tie bar, worn or damaged latch.	See that clothing is put through the wringer at an even rate, thus avoiding bunching. Straighten tie bar, and replace worn or damaged parts.

CHAPTER 22

Automatic Clothes Dryers

As the name implies, the function of the automatic dryer is to dry the clothes after washing and excessive moisture removal have been completed. Depending on the type of heat energy used, clothes dryers may be divided into two general classes, electric dryers and gas dryers. The source of heat in the electric dryer is obtained electrically by means of a heating element mounted in the dryer; the heating element is controlled by a centrally located thermostat and timer. In the gas dryer, the source of heat is derived from ignited gas, which is obtained by turning on the gas flow with the thermostatic control. In order to ignite the gas flow, however, a pilot light must first be burning in the combustion chamber. This pilot ignition is automatic; the lighting is accomplished by a spark that is created by turning a knob, which is usually located on the dryer control panel.

All dryers, irrespective of heating methods used, are equipped with a suction fan; the function of this fan is to provide the movement of fresh, clean air through the clothing during the drying process.

The automatic dryer (Fig. 1) is relatively simple in operation, and it consists of the following essential parts:

1. A perforated metal drum or cylinder,
2. An electric motor,
3. A heating element,
4. A thermostatic heat control,
5. An automatic timer switch.

OPERATION

In operation, clothes are placed in the horizontally mounted drum of the dryer, usually after they have been washed. No pre-

Courtesy Whirlpool Corp.

Fig. 1. A typical automatic clothes dryer.

arrangement of the clothes to be dried is necessary. A centrally located lamp (on some dryers) serves to illuminate the dryer interior during the loading or inspection process. The interior illumination is accomplished by a lamp that is connected through the door switch, so that the bulb is energized whenever the door is opened. After the door is closed, the dryer thermostat is set to the correct heat level, and the timer is set at the desired running time. The best temperature and running time combination is determined by the operator after becoming familiar with the operation of the dryer.

When the timer is set, the drum begins to rotate at approximately 50 revolutions per minute, and the heater is then turned on. Air circulation is provided by means of a motor-driven fan, which circulates the heated air through the clothes or material to be dried. The inner surface of the drum is usually provided with a series of equally spaced metal baffles, as shown in Fig. 2, which carry the clothes to the top of the dryer drum and then permit them to drop to the bottom. In this manner, the baffles not only prevent the clothes from "lumping up" but also provide a tumbling action that speeds the drying process.

The dryer door may be opened at any time to inspect the clothes for the amount of remaining dampness or to insert more clothes. When the door is opened with the timer set, the door switch automatically stops the drying cycle, turns off the heater, and stops the drum and fan rotation. The cycle is again resumed when the door

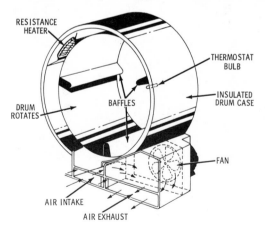

Courtesy General Electric Company

Fig. 2. The principal parts of a typical automatic dryer.

is closed; therefore, it is not necessary to reset the timer when the door has been opened. By opening the door, however, some heat will be lost. For this reason, it is not desirable to open the door too frequently during the drying cycle, or the clothes would then require a longer time to dry properly.

The automatic timer switch can be preset for any given operating time up to 60 minutes. Five minutes prior to the end of the designated time, the heating unit becomes disconnected, but the drum continues to rotate for the remaining 5 minutes. This arrangement gives the clothes a chance to cool sufficiently for comfortable removal from the dryer; it is also possible to leave the clothes in the dryer for some time after the drying process is completed. When not in use, the dryer door should be kept closed to prevent the interior light from burning.

Air Circulation

Air circulation is provided by means of a blower or fan, which may be mounted directly on the motor shaft or connected to the

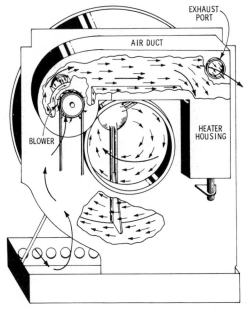

Fig. 3. The air-circulation system in a typical electric dryer.

motor by means of a pulley arrangement. Figs. 3 and 4 illustrate two common methods of air circulation. In Fig. 3, the blower draws air from the drum and forces it along the horizontal air duct to the intersection of the horizontal and vertical ducts. At this point, the air stream divides; that is, part of the air stream is allowed to escape through the exhaust tube, and another part is forced back into the drum to be recirculated through the clothes, thereby effecting the drying operation.

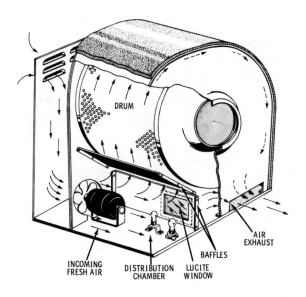

DRUM

INCOMING FRESH AIR DISTRIBUTION CHAMBER LUCITE WINDOW BAFFLES AIR EXHAUST

Courtesy General Electric Company

Fig. 4. The schematic representation of the air-flow system in a typical electric dryer.

In the unit shown in Fig. 4, the circulating air is drawn into the dryer through the louvers in the back. This air passes down across the back of the unit and is drawn into the distributing chamber by the fan. Part of the air is sent up along the side of the drum, and the remainder is sent along the bottom. In this manner, moisture-laden air is exhausted through the lint collector and out the exit port at the front of the dryer, as indicated.

443

Loading Ports

Due to their construction, most automatic dryers are of the front-loading type; that is, they are provided with a circular or rectangular front opening to admit and remove clothes from the dryer. In some dryers, the door or opening is equipped with a glass insert to permit observation while the clothes are being dried. The door is usually fitted with a positive-type door latch, similar to those used on some domestic electric refrigerators.

Dryer Drum

The horizontal dryer drum, one of which is shown in Fig. 5, is usually constructed of perforated metal to provide full air cir-

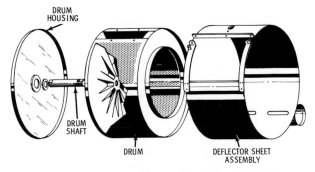

Fig. 5. A typical automatic-dryer drum assembly.

culation. The inner surface of the drum is carefully smoothed and painted with a coat of rust-resistant material, in order to eliminate the possibility of clothes being torn and stained. The rotation speed varies, depending on the type involved, but it is usually between 40 and 50 revolutions per minute.

Motor

If 60-cycle alternating current is available, a $\frac{1}{6}$- to $\frac{1}{4}$-horsepower, split-phase, 115-volt AC motor is usually used to provide power for the drum and ventilating fan. The speed reduction between the motor and drum is accomplished by means of a pulley arrangement. The drive pulley is mounted on one end of the motor shaft and rotates the dryer drum through an intermediate speed-reducing pulley. The air-circulating fan is mounted on the other end of the motor shaft.

Electrical Heating Method

The heating in electric-type automatic dryers is accomplished by means of an electric heating element, or unit, that is usually centrally located in the incoming air stream for the best drying efficiency. The power requirements of the electric-type heaters are usually approximately 5000 watts, with the dryer connected across

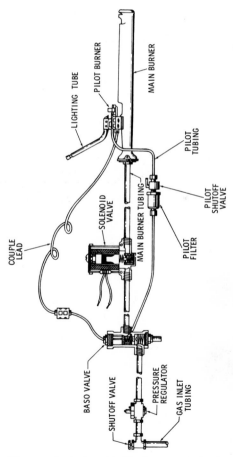

Fig. 6. The gas-control assembly in a typical gas dryer.

a 230-volt, 60-cycle AC circuit. Heaters can normally operate on any voltage in the range from 208 to 230 volts, although drying takes considerably longer at the lower voltages.

A three-wire power connection to the dryer is required. The third wire enables the power to be distributed so that a standard 115-volt, 60-cycle AC motor can be used to drive the drum and fan.

Gas Heating Method

The gas type automatic dryer is, of necessity, somewhat more complex than the electrical type, since in addition to its electrical components, the gas dryer must be piped for gas and be equipped with a burner and accompanying gas-control assembly, as illustrated in Fig. 6. The air-circulation system in a typical gas dryer is shown in Fig. 7.

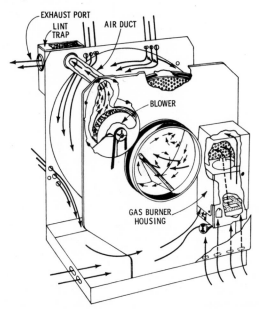

Fig. 7. The air-circulation system in a typical gas dryer.

Timer Control—The point of supply for all electrical controls is at the dryer timer. When the timer is turned to the "on" position, the electrical circuits to all the dryer controls are completed. The timer is a simple "on-off" switch that is operated by a motor-driven cam. In operation, the timer control performs as follows:

1. It starts the motor, which drives the blower and drying cylinder.
2. It opens the solenoid valve, which provides gas to the main burner.
3. It controls the length of the drying cycle required to deliver the clothes at the desired degree of dryness.
4. It rotates the drying cylinder for 5 minutes after it closes the solenoid valve (shuts off the main burner). This function allows the heat to be dissipated from all dryer parts and cools the clothes to a more comfortable handling temperature.
5. It permits the clothes to be checked at the user's convenience without the necessity of resetting the controls.

Baso Valve—The Baso valve is incorporated in most gas dryers of recent manufacture and applies the principles of thermostatic operation to the gas control assembly. As the gas enters the dryer, it goes through a manual shutoff valve, a regulator or coupling (depending on the type of gas used), and then to the Baso valve. This valve is the safety pilot, and it controls the flow of gas to the pilot and the main burner assemblies. It consists of three sections—an electromagnet, a plunger valve, and a thermocouple. A plunger in the Baso valve must be manually opened to permit the gas to flow to the pilot burner, which, when ignited, heats the thermocouple. The thermocouple generates a low voltage current to the electromagnet in the valve body. The energized magnet then electrically holds the pilot valve open. By releasing the manually operated valve, the gas is permitted to flow to the pilot burner and the solenoid valve. Opening and closing the solenoid valve then controls the flow of gas to the main burner.

If for any reason the gas supply is cut off, or the pressure drops too low, the thermocouple cools and the electromagnet is de-energized. The Baso valve then snaps shut and cuts off the gas supply to the dryer.

Thermostats

Drying temperatures are maintained by thermostatic control. The thermostats used on most automatic dryers are similar to those used on electric ranges and are of the liquid type with a capillary-tube attachment. The function of the thermostat is to control the amount of heat delivered to the dryer by breaking the heating-element circuits when the temperature of the circulating air reaches the predetermined temperature setting. In most thermostats, provisions are made for the selection of any temperature between 140° and 200°F. The bulb of the thermostat is normally inserted in the blower air duct, so that the thermostat can more efficiently control the temperature of the air.

Germicidal Lamps

Many automatic dryers of recent manufacture employ germicidal lamps because of their germ-killing power. In order to provide the necessary operating voltage, a lamp ballast or series inductor must be connected as shown in Fig. 8. The germicidal ozone lamp oper-

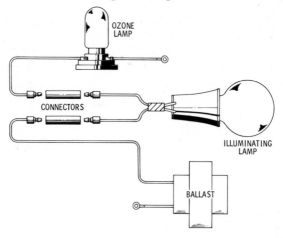

Fig. 8. The wiring diagram for the connection of germicidal and illuminating lamps, with the accompanying ballast unit.

ates automatically; that is, it is wired in such a way that it functions only with the dryer door closed.

Lint Receiver

Lint is caused by normal clothing wear and is removed in the drying process by various methods. Some dryers have a lint trap, or receiver, installed in the air-exhaust duct that is designed to catch all lint exhausted from the dryer. In other types of dryers, lint normally thrown off during the drying process is collected in the condensing-chamber water under the drum and is automatically flushed down the drain.

DRYER INSTALLATION

Since the dryer is a component part of the home laundry, the best location for the convenience of the operator is to line up the washer, dryer, and ironer in this order, so that the clothes may be moved from one appliance to the other until the laundry cycle is completed. Installation is usually made in either the basement or utility room. The dryer, in common with the washer, must be properly leveled for quiet operation and long life. A padding of paper, cloth, or cardboard should be placed under each leg to prevent damage to the floor when sliding the unit from one position to another. No bolting down is necessary.

Electric Wiring

An electrically operated dryer differs from the gas dryer mainly in that the necessary heat is derived from an electrical heating element, whereas in the gas dryer the heat is obtained from a gas burner. The electrical power required for an electric dryer is greater than that needed for a gas dryer because of the power taken by the heating element in the electric clothes dryer. Although the wattage may vary for the various types of electric dryers, the average power consumed is approximately 5000 watts. In order to limit the current flow, then, it will be necessary to employ a wiring system similar to that used for electric ranges, that is, a 115/230-volt, three-wire network system.

Two-Wire Heaters—The common method of connecting loads of this nature to a service line is illustrated in Fig. 9. It consists of the conventional 115/230-volt, three-wire network in which one of the leads (the common or neutral) has a zero voltage potential to ground and a 115-volt potential between it and each of the other leads. Thus, with reference to Fig. 9, the potential between B and N and A and N equals 115 volts, whereas the potential between A and B is 230 volts. When installing a dryer, it should be

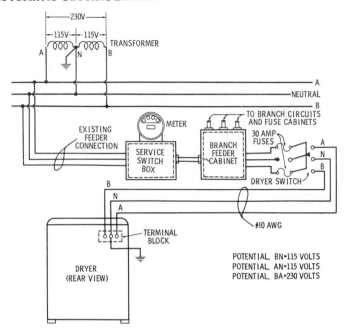

Fig. 9. The schematic wiring diagram for an automatic dryer connected to a 115/230-volt AC source.

remembered that the dryer heating element is always connected across the 230-volt circuit, while the drum motor, timer motor, and lamps are connected to the 115-volt circuit.

The dryer heating unit may also be connected to a 120/208-volt source such as is often found in apartment buildings that are serviced by elevators. A power-supply system of this type is shown in Fig. 10. In this three-phase, four-wire supply system, the common (neutral) has a zero voltage potential to ground, and there is a potential of 120 volts between ground and each of the other leads. There is also a 208-volt potential between any of the two other leads. In this system, the dryer heating unit is connected between any two of the phase leads, whereas the drum motor, timer motor, and lamps are connected between the neutral and any one of the phase leads.

Three-Wire Heaters—Three-wire heaters, as the name implies, have three wiring terminals. The wiring method is similar in

450

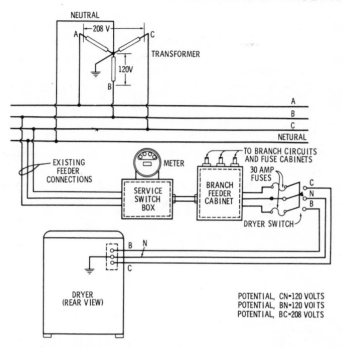

Fig. 10. **The schematic wiring diagram for an automatic dryer connected to a 120/208-volt AC source.**

every respect to that of the two-wire heaters, except that the neutral lead is connected to the center tap of the heater, and the other two leads are connected across either the 208- or 230-volt leads, depending on the type of power system used. In this type of wiring system, the potential between the neutral (heater center-tap connection) and any one of the two other leads will be approximately 115 volts.

Since the dryer is a heavy-duty appliance, it should have its own separate power line. The branch circuit should consist of three No. 10 wires fused for 30 amperes and should carry 230- or 208-volt, 60-cycle alternating current. The third wire (neutral) is a tap on the line that provides a 115-volt power source for the timer motor, the lamps, and the drum motor. The electrical installation should conform to the *National Electrical Code* and to all local codes in the particular area. The dryer cabinet should always

451

be grounded either to the neutral (white) wire of the branch circuit or by a separate ground wire attached to the cabinet. If a rigid conduit or flexible cable is to be used for connection to the electrical supply, a line switch must be installed. No. 10 wire is satisfactory if the dryer is to be located less than 15 feet from the main fuse box. At greater distances, No. 8 wire should be used.

Dryer Wiring Diagrams—The internal connections of several types of dryer heaters are shown in Fig. 11. The terminal potential may vary, depending on the power-distribution system available. A dryer heater that is intended for connection to a 230-volt source is frequently connected to a 208-volt system, but when so con-·
nected, the drying time will be slightly longer due to the effect of this lower voltage.

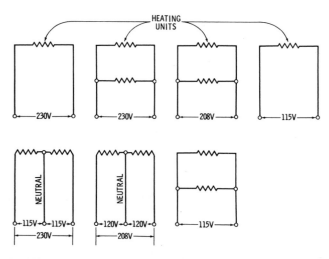

Fig. 11. The various wiring connections for two- and three-wire dryer heating elements.

Figs. 12 to 14 show the internal wiring of several dryer models, and although they all differ in certain respects, the basic principles of operation are the same. Referring to the wiring diagram of Fig. 12, the three connection terminals are marked "230V," "neutral," and "230V," respectively. This simply means that the potential between the outside terminals when connected to the line should be 230 volts, and the potential between any one of the outside

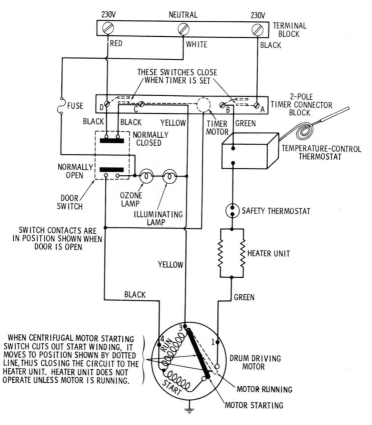

Fig. 12. The schematic representation for the internal electrical circuit of a typical automatic dryer.

terminals and the neutral should be 115 volts. With the dryer door open, the interior lamps will be illuminated but will be disconnected when the dryer door is closed, due to the position of the door switch. Because of the interlocking feature of the door switch, the dryer motor cannot be energized; that is, the timer switch cannot operate unless the dryer door is closed. With the dryer door closed, the timer setting closes a number of electrical contacts inside the timer, thereby allowing the current to pass through the main motor, heater unit, timer motor, and thermostats. The current going to the main motor returns directly to the ground,

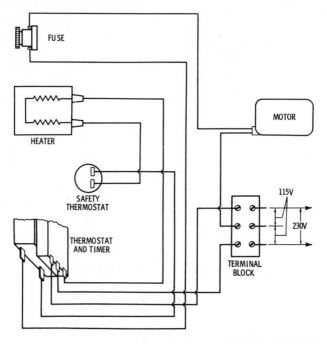

FUSE

HEATER

MOTOR

SAFETY
THERMOSTAT

THERMOSTAT
AND TIMER

115V

230V

TERMINAL
BLOCK

Courtesy Horton Manufacturing Company

Fig. 13. The internal wiring diagram of a typical automatic dryer.

or neutral, through the lower door-switch contacts, which are now in a closed position. Thus, the circuit is completed, and the dryer operates.

Some dryers employ a solenoid-operated switch in the heater circuit. In this type of system, the solenoid coil is wired in series with the thermostat, and the switch contacts are wired in series with the heater.

Ventilation

All dryers require some means of ventilation, since they evaporate well over 1 gallon of water each hour. This evaporation produces a considerable amount of moist air. If the dryer were to be located in a small closed room, the moist air from the dryer would soon saturate the air in the room, resulting in a considerable decrease in the operating efficiency of the dryer. If such a condition

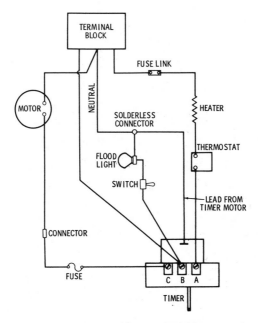

TERMINAL
BLOCK

FUSE LINK

MOTOR

NEUTRAL

SOLDERLESS
CONNECTOR

HEATER

FLOOD
LIGHT

THERMOSTAT

SWITCH

LEAD FROM
TIMER MOTOR

CONNECTOR

FUSE

C   B   A

TIMER

Courtesy General Motors Corporation

***Fig. 14. The diagrammatic representation of the internal wiring circuit
in an automatic dryer.***

is encountered, the windows and doors should be opened or the
room should be ventilated by mechanical means. If no ventilation
of this type is available, it may be necessary to install a positive
vent to exhaust the moist air through a window or wall to the
outside.

Fig. 15 illustrates some common venting systems and the adapt-
ers used to connect the exhaust duct on the dryer to the vent. The
elbows must be smooth on the inside and have at least a 2-inch
radius on the inside bend. All joints must be made so that the
exhaust end of one pipe is inside the next pipe down stream; this
arrangement prevents the accumulation of lint in the vent pipe.

The addition of a vent pipe tends to reduce the amount of air
the blower can exhaust. This reduction does not affect the dryer
operation if it is held within practical limits. No more than four
right-angle elbows should be used in venting the dryer, and no
more than 10 feet of straight pipe should be used when four elbows

455

are employed. Two feet of straight pipe may be added for each elbow when less than four are used.

If the vent passes through a wall, a metal sleeve of a slightly larger diameter should be set in the wall, and the 3-inch vent pipe should be passed through this sleeve. This practice is required by some local codes and is recommended in all cases to protect the wall from possible discoloration due to the 160°F. air passing through the vent pipe. If the vent passes through a window, the glass pane should be removed, and a sheet-metal plate should be put in its place. If at all practical, the vent should not exhaust directly below a window, since the window will have a tendency to steam under certain weather conditions. A deflector of some sort should be placed over the end of the vent to prevent rain and high winds from entering the vent when the dryer is not in use. Be sure that the deflector does not restrict the exhaust and that it terminates at least 1 foot above ground level.

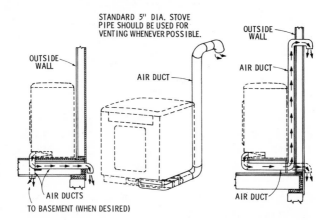

Fig. 15. Typical venting installations for automatic clothes dryers.

Venting the dryer into a chimney is not recommended, whether the chimney is used for other venting or not. Venting under a house is also not recommended. In both cases, there is a danger of lint build-up over a period of time that may prove to be a fire hazard. Venting a dryer straight up through a roof is not desirable but may be done if properly installed. The overall length of the vent has the same limits as venting through a wall. A rain cap must be placed on top of the vent and must be of such a type as to be free from clogging.

Gas Piping

In dryers equipped for gas heating, a sufficiently large gas supply line should be brought to the back of the dryer in order to prevent a pressure drop. The gas inlet connections, depending on the type of dryer, may have a ½-inch or ⅜-inch NPS female thread. The gas supply line should be ½-inch rigid pipe. Although galvanized steel pipe is the most economical material to use, there are certain localities where brass or aluminum pipe may be required, due to the corrosive actions of certain fuel gases.

At some convenient point in the gas supply line, a shutoff valve should be inserted so that if necessary, the dryer gas supply can be cut off without interrupting the service to the other appliances. As usual, all codes pertaining to the safety and installation of gas appliances apply also to the installation of dryers. Clean and adequate threading of pipes and fittings and the use of good white lead or other pipe joint compound should be employed in order to comply with the rules governing gas piping installations.

SERVICING AND REPAIRS

Servicing of automatic clothes dryers requires that the serviceman be acquainted with the operation and functioning of both the mechanical and electrical systems. Although the various types of clothes dryers may differ in appearance and location of controls, they all operate on the same principles and are fundamentally similar with regard to servicing. Clothes-dryer timers are quite similar to those used on automatic washing machines, while thermostats used in automatic dryers are of the same type as those used on electric ranges.

Since the only moving parts consist of the motor, drive, drum, and exhaust fan, the clothes dryer, when properly installed, should give years of trouble-free service. When called on, the serviceman should be familiar with the recommendations and specifications of the particular manufacturer's service manual in order to be able to correctly replace any worn-out or faulty component.

AUTOMATIC CLOTHES DRYERS

Electric Dryers

The operating test on the dryer consists of running the unit and checking for any abnormal condition, such as noise, too high or too low operating temperature, and proper timing. Before attempting to check the interior of the unit, be sure that the power is disconnected. This is particularly important, since most dryers are connected to a 230-volt source. This voltage is dangerous and

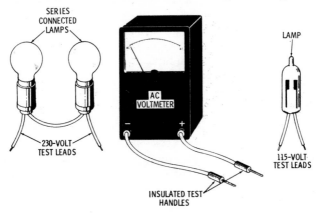

Fig. 16. Test lamps and an AC voltmeter are used to test the circuit continuity and terminal potential, respectively, in automatic dryers. Potentials of 208 or 230 volts are always checked with a voltmeter or two incandescent lamps connected in series, whereas the low-voltage circuits (110 to 120 volts) are usually tested by means of a single test lamp.

particularly so if the dryer is installed on a wet or uninsulated basement floor.

Several operating malfunctions are given below, along with their causes. In most instances, the remedies are obvious once the causes have been pinpointed. Test equipment is shown in Fig. 16.

Dryer Does Not Start—If the dryer does not start, check the electrical connection to the dryer before dismantling. If a blown fuse is found to be the cause, the dryer should be thoroughly examined for internal short circuits. A loose connection on the terminal block, motor, switch, or timer may also be the cause of an open circuit. If all these connections are found to be in proper operating order, the motor itself should be checked.

Motor Hums but Drum Does Not Rotate—This could be due to several causes, such as a loose motor pulley, a loose drum pulley, a loose or broken belt, and also because of overloading the drum. The remedy is obvious if any of these conditions exist.

Dryer Will Not Shut Off—This may be caused by one of the following:

1. The timer motor may be jammed, or the clock spring may be broken. Turn the dial to the 30-minute position, and listen for the ticking of the clock. If the clock fails to tick, pull the dial slightly away from the front panel, since it may be binding. If the clock still does not operate, replace it.

2. The timer may run but not be able to stop the dryer due to the improper positioning of the stop pin, which is located to the right of the timer shaft behind the dial. The dial has a pair of raised fins which strike this pin and thus prevent the customer from turning the dial too far. If the timer fails to shut off, remove the dial, and trim the edge of the fin behind the word "off." Trim the edge next to the larger open segment until the dial can advance enough to break the contact points built in its base. This may be corrected on some units by turning the timer shaft over in the dial.

3. If the contact points in the timer are arced closed, the unit will fail to stop, even if the timer is in the "off" position. This can best be checked by pulling the timer dial approximately ¼ inch away from the front panel and turning it to the "off" position. If the unit continues to operate, open the dryer door. If the unit then stops, replace the timer assembly. If it continues to run, see item 4 below.

4. It is possible to have a grounded motor that will not stop even with a good timer or with the door open. This condition will only occur when part of the motor winding is shorted out against the motor frame, which is grounded. The motor must be replaced if the ground cannot be corrected.

Dryer Noisy—This may be caused by one or all of the following:

1. Loose fan or motor pulley or loose or dry belt. An intermittent or persistent squeak is caused by what is known as a "dry" belt. An application of a thin coat of surface belt

dressing on the pulleys will usually correct this situation. Also, be sure that the pulleys are properly aligned.

2. A noisy suction fan is usually caused by incorrect alignment, or, where the fan is belt-driven, by a loose or damaged belt. Align the pulleys; replace defective belts.

3. Loose items between the drum and cylinder housing and rattles of various sorts indicate loose cabinet screws. Check all screws, and tighten securely.

Clothes Dry Too Slowly—This is a common complaint from people who are not familiar with automatic clothes-dryer operation. In general, however, slow drying may be caused by the following:

1. Clothes are too wet when transferred to the dryer. Since the capacity of most dryers is limited to the removal of approximately 10 pounds of water in an hour's time, any excess water contained in the clothes will tend to slow up the drying process.

2. Check for lint cloggage in the lint box; this condition can prevent proper circulation. If an outside vent is used, check the exhaust duct from the lint box to the outside for a blocked air duct.

3. The thermostat may be set too low. The correct temperature setting is determined by placing a thermometer in the lint trap. The temperature should check slightly below the thermostat setting at the cutout time of the element. However, the calibration of the thermostat is set at the factory and should not be changed unless it is certain that the control-knob setting is faulty.

4. The voltage may be too low. Check the voltage at the dryer terminals, and see that this voltage corresponds to that given on the appliance nameplate. If an excessive voltage drop is caused by the use of insufficiently heavy wire over too long a distance from the fuse box or source, the result will endanger the correct operation of the dryer and will, in addition, be a constant fire hazard.

5. Check with the customer to determine her technique of loading the dryer. Normally, not more than two double bed

sheets or similarly large pieces should be put in the dryer at the same time. Finish out the load with smaller articles. Never insert more than three sheets in one load, since there is not sufficient room to open properly and dry efficiently.

6. If nothing can be found to indicate possible trouble, request a sample load of clothes from the customer (enough to weigh between 7 and 9 pounds dry), and wet them until they weigh approximately 18 pounds. Explain to the customer that the dryer is designed to remove between 9 and 10 pounds of water in an hour's time. Preheat the dryer for 3 to 5 minutes. Note the exact weight of the wet clothes, and place them in the dryer. Set the timer for 60 minutes, and allow the dryer to run for either 30 or 60 minutes, whichever is more desirable. Remove the clothes, and reweigh them; double the weight lost in 30 minutes to determine the drying rate per hour.

7. If the weight of water removed per hour is up to standard (9 to 10 pounds), the dryer is operating as designed, and the customer is overloading the unit either in total weight or in water weight. The average wringer or spinner washer can remove enough water to leave 9 to 10 pounds or less of water in a load of clothes. If the dryer fails to come up to standard on the water-removal test, recheck item 5 to locate the trouble.

Gas Dryers

The modern gas dryer differs from its electrical counterpart mainly in the internal wiring arrangement. Because of the absence of the main heating element, gas dryers are usually connected to the standard 115-volt AC house circuit. The following list of troubleshooting pointers should aid the serviceman in determining the cause of any common gas-dryer malfunction.

Pilot Will Not Light—Check for one of the following causes:

1. Main gas valve not turned on.
2. Air in gas line; needs additional bleeding.
3. Pilot orifice or line clogged.
4. Wrong size of pilot orifice.

461

5. Pilot filter-adjustment screw closed.

6. Defective safety pilot.

Main Burner Goes Out Repeatedly—This may be caused by one of the following conditions:

1. Main burner orifice too large (overrated burner).

2. Cycling of high-limit switch; not enough air flow through unit; check for lint accumulation.

3. Loose electrical connection(s).

Main Burner and Pilot Go Out—This condition is usually caused by either a defective thermocouple or a defective safety pilot. Check both units, and replace if necessary.

Clothes Dry Too Slowly—Check for one of the following causes:

1. Lint trap clogged; cycling of high-limit switch.

2. Exhaust duct clogged; cycling of high-limit switch.

3. Loading door not sealed properly.

4. Impeller loose on motor shaft.

5. Main burner orifice too small; underrated dryer.

6. Low gas pressure.

7. Lint trap not in place.

8. Defective high-limit switch.

9. Dryer overloaded; clothes cannot tumble.

10. Defective operating switch.

11. Clothes not spun dry before placed in dryer.

12. Air-selector switch set on "Dry Air" instead of "Heated Air."

Drying Chamber Does Not Turn (with motor running)—This condition may be caused by one of the following:

1. Loose drying-chamber pulley.

2. Worn or broken belt.

3. Broken idler-pulley spring.

4. Tub locked, belt slips.

Motor Will Not Start—Check for one of the following causes:

1. Service cord disconnected.

2. Blown fuse or tripped overload relay.
3. Loose or broken electrical connection(s).
4. Loading door open.
5. Defective door switch.
6. Defective motor.
7. Locked impeller.

Noisy Operation—This condition may be due to:

1. Pulley loose on shaft.
2. Impeller loose on motor shaft.
3. Drum rubbing against drum case.
4. Foreign matter between tub and case.
5. Cabinet not secured to inner unit or base.
6. Dryer not properly level.

Dryer Will Not Shut Off—This condition is normally caused by a defective timer control. Check this control, and replace it if necessary.

Main Burner Flame Characteristics—A sharp blue flame indicates the necessity of reducing the flow of primary air; a soft yellow flame necessitates an increase in the primary air flow.

CHAPTER 23

Electric Dishwashers and Garbage Disposers

Electric dishwashers (Fig. 1) are manufactured in a number of different types to suit the various kitchen requirements. They may be furnished as single units or in combination with sink-and-drainboard units. For best results, the dishwasher should be permanently installed in combination with the ordinary sink. A cabinet underneath the sink can serve as storage space for kitchen utensils. In addition to the sink, some manufacturers supply units that include both garbage disposer and dishwasher.

DETERMINING FACTORS FOR GOOD OPERATION

Prior to a technical discussion of the operating principles of automatic dishwashers, it will be of assistance to discuss certain fundamental factors concerning the proper loading of dishes, detergents to use, water temperature, and general treatment of various stains and food soils, since a knowledge of these will assist in obtaining the desired cleaning result.

Food Soils

Food soils may be roughly divided into two groups, those that are soluble in water and those that are insoluble. The second group, which is by far the larger of the two, consists of the soils that are water dispersed, water swelled, melted by heat, or entirely unaffected by water. Uncooked eggs, sugar and syrups, as well as the juices of fruits, vegetables, and meats are water soluble. Of these, syrups present the only difficult problem of removal; although syrup can be dissolved in water, the rate of dissolution is relatively slow. A hot solution that flows swiftly over the surface of the syrup is conveniently used to achieve the most rapid rate of disolution.

465

Courtesy Whirlpool Corp.

Fig. 1. An automatic electric dishwasher.

Some food soils break up in water into fine particles, which are readily scattered through the liquid. Others may be finely divided solids, such as tomato juice, strained vegetables, and applesauce; these soils are the most difficult to remove from dishes when they are allowed to dry, and it is only after they have been thoroughly penetrated by a wash solution that they can be pried from a plate and dispersed by the mechanical action of water. This is a time-consuming operation, and none but the most efficient detergent causing rapid penetration can assure the removal of this soil in the time allowed for the dishwashing operation.

Gelatinous foods and starchy foods, such as potatoes, rice, and some breakfast cereals, swell when soaked in water and shrink when allowed to dry. These foods adhere strongly when allowed to dry on a dish and must be thoroughly wet before they can be removed. Greases and fats are melted, and fluid oils become less viscous when heated to dishwashing temperatures. The action of the hot detergent solution then breaks up fats and oils into microscopic droplets, which can then be rinsed from the dishes and flushed down the drain.

Detergents to Use

Never under any conditions should soap, soap flakes, or soap powders be used in a dishwasher. Always use the detergent recommended by the dishwasher manufacturer. These detergents, in addition to their excellent cleaning properties, should have the ability to prevent film formation in hard water. The action of the detergent in removing food soil from dishes is twofold. One is the ability to increase the attraction of water to solids and greases. This particular property is known as "wetting power," and it is by virtue of this characteristic that the wash solution is able to penetrate and soften dried-on foods in much less time than that required by water alone. Another characteristic of the wash solution provides a tendency to support or suspend greases and smaller food particles in the liquid. The detergent rapidly wets masses of small particles, such as mashed potatoes or rice, loosens the particles so that the water action scatters them, and by means of absorption, imparts to them a buoyancy that makes them practically part of the solution.

Treatment of Stains

Under certain conditions, stains or films may form on china and glassware that have been washed in a dishwasher. These stains are more noticeable in a dishwasher because the film is not rubbed off as it is when dishes are washed and wiped by hand. The most common source of stain is the water supply and rust that emanates from the piping. When iron film (rust) appears alone, it is an overall brown stain that may be deposited on dishes washed in water containing a comparatively large amount of iron. To remove iron film, wash the dishes in the usual manner using a cupful of detergent; then wash them again, using a quarter-teaspoon of oxalic acid crystals. Follow with another wash using the detergent to remove all traces of the acid. *Do not use the detergent simultaneously with the acid.*

Tea stains, which are reddish brown to dirty brown in color, result from the union of the tannic acid found in tea with hard water and are usually confined to cups and saucers. They give a dull appearance to chinaware. These stains can be readily removed by hand washing, using a mild abrasive such as baking soda. They are also readily removed by using bleach, although the strength of the bleach should be moderated so that the bleach does not damage the chinaware.

Lime film accumulates on the back of dishes, in glasses, and on dishwasher racks and tub. It may be almost any color, from creamy white to mottled brown. When the correct amount of the proper detergent is used, lime film usually disappears in the washing process.

Water Temperature

For efficient electric dishwashing, the water temperature should be between 140° and 160°F. In homes equipped with automatic hot water heaters, the thermostat setting should conform to this requirement. Also, the length of the pipe connecting the water heater with the dishwasher should be as short as possible, and its size (diameter) should be as recommended by the dishwashing-machine manufacturer.

Dishrack Loading

Most dishwashers of recent manufacture have two wire racks, as shown in Fig. 2. The lower rack is designed for platters, dinner plates, salad plates, dessert plates, and other flat pieces and also for milk bottles, pots, pans, casseroles, and pitchers. In addition, the lower rack is usually furnished with a basket in which the silverware is placed. The upper rack accommodates such items as cups, stemware, tumblers, cereal bowls, etc. All dishes should be placed in an inverted position for proper drainage. The larger dishes in the lower rack should be placed so as not to block off the passage of water to the upper rack.

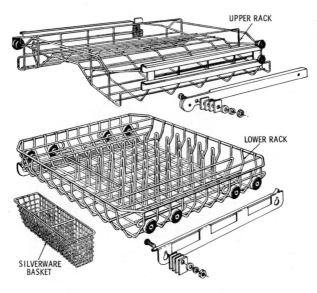

Fig. 2. The upper and lower rack assemblies, and the silverware basket, as employed in a typical automatic dishwasher.

OPERATING PRINCIPLES

In operation, the soiled dishes are loaded on special wire racks from the front or top, depending on construction. After the prescribed amount of detergent has been added to a special receptacle, the dishes are ready for washing; the racks are then pushed back into the washer, and the door is closed. The dishwasher is placed in operation by closing and latching the door and then setting the starting button. The machine then goes through its various operations to complete its washing and drying cycles. The complete assembly of a front-loading automatic dishwasher is shown in Fig. 3.

The electrical circuits of most automatic dishwashers are so arranged that they will not operate unless the door is closed and securely latched. The purpose of this precaution is to prevent water

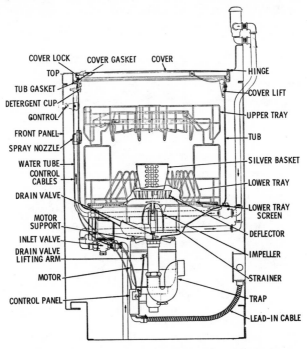

Courtesy General Electric Company

Fig. 3. The complete assembly of a typical front-loading automatic dishwasher.

from being thrown out into the room if the door is inadvertently opened during operation.

The various cycles of spraying, washing, rinsing, and drying are controlled by a timer and a small synchronous motor, much in the same manner as that of an automatic clothes washer. The timer consists of a synchronous motor that drives a set of cams, which in turn move electrical contacts to open and close the various circuits during the machine operation. When the starting switch is closed, the timer motor starts, the pilot lamp lights, and the drain valve is closed. As the timer motor runs, the contacts close or open in proper sequence to start and stop the motor, thus caus-

470

ing the solenoids to open and close the inlet and drain valves and turn the heater element on and off at the proper time during the washing operation.

Fig. 4 shows the circuit wiring diagram of a typical dishwashing machine; the complete operational cycle is as follows:

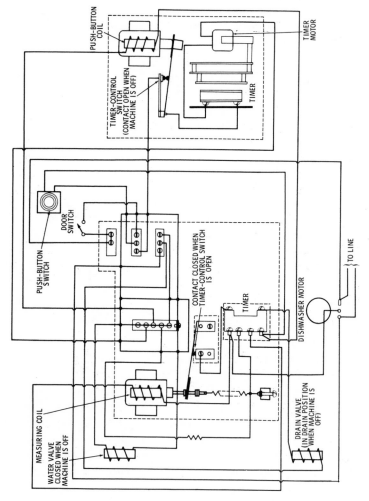

Courtesy General Electric Company

Fig. 4. The schematic wiring diagram of an automatic dishwasher.

1. The water-inlet valve, drain, water-measuring device, and the main motor are electrically operated through the three-position control switch.

2. Operation of the push-button switch energizes the push-button coil, which in turn closes the timer and control switch. The timer motor now commences operation.

3. The impeller-motor circuit is then closed, and the motor begins to revolve; the water-inlet-valve solenoid opens the valve, thus permitting water to be sprayed into the machine.

4. After a brief time interval, the drain-valve solenoid closes the drain valve, resulting in a water accumulation in the washer. When the water has reached a predetermined level, the water valve is closed by the measuring coil.

5. Since the measuring coil is connected in series with the impeller-motor winding, an accumulation of water in the machine results in an additional current draw by the motor. When the water reaches a predetermined volume, the current through the motor winding causes a relay coil to open a set of contacts in the water-inlet-valve solenoid circuit, which in turn closes this valve and prevents additional water from entering the washer.

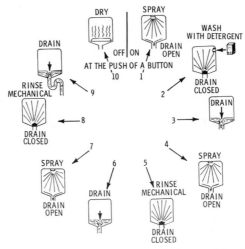

Courtesy General Electric Company

Fig. 5. *The complete cycle of operation for a typical automatic dishwasher.*

6. The drain is simultaneously closed when the water-inlet valve is closed; the detergent receptacle opens, and the motor spins the impeller, thereby completing the washing cycle.

7. At the completion of the washing cycle, the timer again opens the drain valve by means of the solenoid. As the water is drained out, the inlet valve opens; water is then sprayed into the machine for a few seconds, after which the drain valve again closes, thus allowing additional water to accumulate. The water level is determined automatically by the measuring coil.

8. The water-inlet valve is again closed, and another rinsing cycle then begins, with the motor operating.

9. After the completion of the washing cycle, which operates according to a predetermined sequence, the drain valve opens, the water-inlet valve closes, the motor stops, and a heating element is automatically connected across the input terminals; the heat produced by the element provides a thorough drying for the dishes.

The time required for a complete washing cycle, as shown in Fig. 5, varies to some degree, depending on the model and make of the machine. The average washing cycle, including the drying period, is approximately 35 minutes. After the washing cycle has been started, it can be advanced to any point by simply turning the control knob in a clockwise direction. Also, the washing cycle can be interrupted at any point by merely unlatching the door. This action opens the timing switch and, therefore, all other circuits except the one containing the pilot lamp. If the drain valve is closed at the time of interruption, its control circuit will also remain closed.

SERVICING AND REPAIRS

Although an electric dishwasher may give trouble-free operation for many years, service may be required at times because of abuse due to a lack of complete knowledge concerning the proper use of the machine. When making a service call, the serviceman should endeavor to give helpful suggestions with relation to the proper operation of the machine. For example, some dishes must be precleaned prior to being placed in the dishwasher, and they must also be placed in the machine in a certain logical sequence and order for the most effective cleaning action.

When receiving service complaints concerning unsatisfactory washing in a seemingly normally operating machine, the serviceman should instruct the customer with respect to the proper preparation

Dishwasher Service Chart

Trouble	Possible Cause	Remedy
Machine fails to operate.	Door not latched.	Latch door.
	Defective switch or timer.	Replace.
	Switch linkage out of adjustment.	Adjust.
	Open circuit.	Check circuit with test lamp until fault is found.
Machine does not wash clean.	Soap used.	Use only an approved detergent.
	Dishes not loaded properly on racks.	Load dishes as per instructions.
	Dishes not properly precleaned.	Preclean dishes.
	Incorrect water temperature.	Adjust water-heater thermostat.
	Not enough water.	See "Insufficient fill" and "Water does not remain in tank."
	Fill-valve strainer clogged.	Remove strainer and clean.
	Timer inoperative on "fill" cycle.	Replace timer unit.
	Fill-solenoid coil inoperative.	Replace fill-solenoid coil.
	Measuring coil inoperative or out of adjustment.	Replace or repair as required.
Water does not remain in tank.	Leaking drain valve.	Tighten flange on drain valve.
	Inlet valve not opening.	Adjust linkage, repair or replace solenoid as required.
Machine noisy.	Drain- or fill-solenoid core not properly centered in solenoid coil.	Realign solenoids to assure perpendicular and centered action of solenoid cores.

Dishwasher Service Chart (Continued)

Trouble	Possible Cause	Remedy
	Motor out of alignment.	Realign motor.
	Vibration.	Machine not resting solidly on floor. Check to see if machine is level and has firm foundation.
	Impeller scraping against impeller screen.	Check and adjust as necessary.
Door or cover will not close.	Door or cover seal binding inside of tank.	Loosen screws on seal retainer, and force seal inward. Reset screws to retain seal.
Insufficient fill.	Low water pressure.	Check water supply to increase pressure.
Slow draining.	Drain solenoid inoperative.	Check and replace drain solenoid.
Dishes do not dry.	Incorrect water temperature.	Adjust water-heater thermostat to approximately 150°F.
	Leaking inlet valve.	Check valve and replace valve-seat washer.
	Inoperative heating element.	Check heating-element wattage with wattmeter. Output of most dishwasher heating elements varies between 750 and 1000 watts at rated voltage. If wattmeter shows no reading, circuit is incomplete. Check and replace element if burned out. Check timer-control unit.
Tarnishing silverware.	Chemicals in water.	Local water conditioning agency should be consulted, since a water softener or mineral filter may be required. In areas having soft water, the amount of detergent may be reduced from two to one tablespoonful.

and loading of dishes, observe that the proper detergents are used, and see that water of the correct temperature is available in sufficient quantity. He should also make certain that the operating voltage is constant within 10% of the prescribed 115 volts. Run the dishwasher through several cycles in the customer's presence, and check its operation closely to see that the timer control functions according to the manufacturer's specifications.

The serviceman and the doctor have a lot in common. Usually neither is called until there is trouble. In the case of both, several instances of incorrect diagnoses or wrong remedies can result in a bad reputation. A good serviceman must first thoroughly analyze the trouble and then proceed with the correction. The job should be accomplished thoroughly and completely on the first call, thereby eliminating the need of costly and time-consuming return calls. In case of doubt, however, always check back to make certain that the system is functioning properly.

In most dishwashers, there is a direct relationship between trouble, cause, and remedy. The service chart has been worked out to help the serviceman locate and repair malfunctions that are liable to be encountered. The chart refers to automatic dishwashers only, but, since all machines have numerous components in common, the chart will be helpful for all types of machines.

DISHWASHER INSTALLATION

Every electric dishwasher requires the following essential items for its proper operation:

1. Drainage,
2. Hot water supply,
3. Electrical connections to motor.

In addition to arranging for the proper hot water supply and drainage, it is also necessary to make certain that the water pressure is adequate and that there is a sufficient amount of hot water available at the required temperature. All plumbing must be in accordance with local plumbing codes and the best sanitary practice. Local restrictions regarding traps, vents, etc. must be followed.

The electrical wiring must conform to the requirements of the *National Electrical Code*, in addition to any existing local codes, and must be of adequate capacity to supply the dishwasher motor and heating element without an appreciable voltage drop. If there is any doubt about the power requirements, a check should be made with the local power company before starting the installation.

GARBAGE DISPOSER

The function of a garbage disposer is to grind or shred the kitchen garbage into minute particles, which are then washed down the drain and disposed of in the plumbing waste line. The principle part of the disposer is a high-torque electric motor. This motor drives an impeller and fly cutters, in addition to a flywheel that is designed to provide the necessary inertia to moderate and compensate for any speed fluctuation in the disposer. The rotation of the motor shaft and fly cutters or pulverizing hammers in a stationary shredding device provide the necessary cutting action. The garbage disposer is designed for mounting under the kitchen sink in such a manner that the intake flange of the disposer is drawn up against the bottom of the sink bowl. The disposer drain outlet is fitted for a slip-joint connection to an "S" drum or "P-type" nonsyphoning trap whenever local plumbing codes permit.

DISPOSABLE GARBAGE

Disposable kitchen wastes consist of most vegetable matter, bones, and similar substances, which may readily be cut up in the shredding or pulverizing mechanism. The garbage disposer, of course, will not dispose of inorganic matter, such as tin cans, bottle caps, broken dishes, glass, etc. Paper should also not be fed into the machine, since it tends to clog the drainage system.

OPERATING PRINCIPLES

Fig. 6 shows a cross-sectional view of a typical garbage-disposer unit. It consists essentially of an upper and a lower housing in which the motor and operating mechanism are located. The lower housing contains the sealed drive motor and a stationary shredding element or ring, which contains a number of sharp cuttings edges surrounding its inner surface. A flywheel is fitted to the motor shaft below the shredding ring. Depending on the construction, the operating mechanism contains a set of retracting impellers and a pulverizing device that forces the garbage against the stationary shredding ring.

In addition, most garbage disposers depend for their operation on a water-flow interlock. This interlock is mounted in the cold water line ahead of the cold water faucet; because of its electrical connections, the interlock prevents the disposer from operating unless the correct amount of cold water flows into the sink drain to carry the food waste away. With the garbage in the disposer, and with a sufficient amount of cold water admitted from the cold water faucet, the drive motor begins to rotate. The garbage is spun

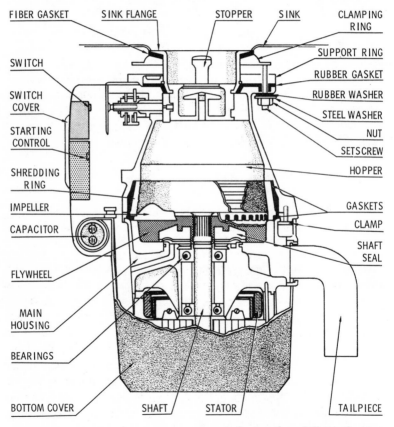

Courtesy General Electric Company

Fig. 6 The cross-sectional view of a typical kitchen garbage disposer.

outward by the rotating flywheel and shredded to small bits by the cutting edges of the shredding ring, which is aided by the centrally located impeller. As the garbage is shredded, it is washed down through the holes in the flywheel-strainer plate and into the drain.

GARBAGE DISPOSER INSTALLATION

Garbage disposers are delivered as completely assembled units, ready for installation on the customer's premises. They are designed to fit the sink and drain installation, thus permitting the mounting of the disposer unit without any extensive alteration to the plumbing system. A typical garbage-disposer installation is shown in Fig. 7.

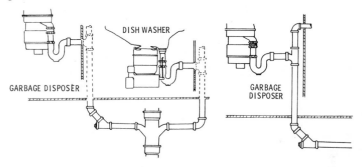

Courtesy General Electric Company

Fig. 7. The plumbing arrangements for two typical garbage-disposer installations.

As a first step in the installation procedure, disconnect the sink drain fitting, and place the main disposer unit in its approximate position directly below the sink opening, with the supporting buffer rings and flange in place. Then, block up the disposer unit until the buffer unit bears slightly against the sink bowl. In order to provide a watertight connection, lay a heavy cushion of plumber's putty around the sink opening. Place the sink mounting flange through the sink opening, and screw it into the mounting ring in the main unit. Tighten the mounting screws carefully and progressively until the unit is sealed firmly in place in the sink opening. The supporting blocks are removed at this time, and a slip-joint plumbing connection is provided to the drain outlet of the disposer.

Drainage

Since certain localities, due to drainage difficulties, do not permit garbage-disposer installations, the serviceman should find out whether municipal authorities have approved such installations for the community where the installation is to be made. Prior to the actual installation, a thorough survey of the home drainage system should be made. Generally, the system will conform to plumbing codes, but an examination of the system may disclose that certain changes and alterations may be necessary for trouble-free disposer operation. In older drainage systems, check for grease traps; unless local codes require such traps, they should be removed, since the very nature of trap construction can lead to almost certain clogging. However, if grease traps are required, the home owner should be advised to clean the trap at certain periodic and predetermined intervals.

Any old sections of drain lines, fittings, or house traps remaining at the start of the disposer installation, must be thoroughly cleaned mechanically. Always provide a separate trap for the garbage disposer. Never connect it to a trap serving other units, such as a dishwasher, laundry tub, etc. If absolutely necessary, the drain line from the disposer and the dishwasher can be connected below the floor level.

The drain pipe should be at least 1½ inches in diameter on all horizontal runs of 6 feet or less; when in excess of 6 feet, a 2-inch pipe or larger should be used. Also, the drain line should have a minimum slope of ¼ inch per foot of run, although a slope of

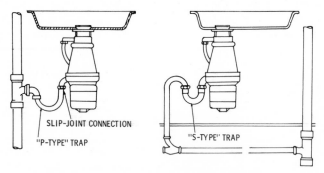

Fig. 8. Two types of disposer plumbing connections.

½ inch per foot of run is preferable. Fig. 8 shows two typical plumbing layouts for garbage-disposer installations.

Septic Tanks

In suburban areas where septic tanks are used, great care should be exercised when installing garbage disposers, since the additional load imposed on the sewage-disposal system may cause trouble in the form of overflow and backing up of sewage. Based on field experience, authorities have found that the septic tank should have a minimum capacity of 500 gallons when serving homes having two bedrooms or less that are regularly occupied; for each additional bedroom regularly occupied, 250 gallons should be added when a garbage disposer is used. Thus, in the case of a four-bedroom house, a septic tank with a capacity of at least 1000 gallons is usually required. In all cases, local codes should be diligently adhered to.

Electrical Connections

The electrical components in garbage disposers, as shown in Fig. 9, usually consist of a motor control or starting switch; an automatically reset, thermostatically controlled, motor-overload relay; a water-flow interlock; and a split-phase AC drive motor. Certain motor-control systems provide a motor-reversing switch

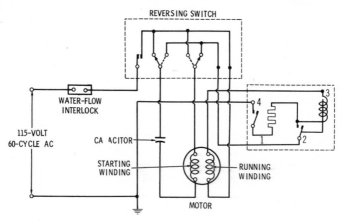

Fig. 9. The schematic wiring diagram for a garbage disposer that is de-signed for reversible-motor operation.

481

by means of which it is possible to run the drive motor in either direction. Since these various components are completely wired at the factory, all that is required at the site of the installation is to provide the necessary cable connections between the disposer junction box and the main switch cabinet in the home. It is advisable, however, to check the local codes and the *National Electrical Code* for the approved method of connecting the disposer unit to the home wiring circuit before proceeding with the installation.

SERVICING AND REPAIRS

Since garbage disposers are ruggedly built and have relatively few moving parts, they will usually operate for years without the need for servicing and repairs when correctly installed. When

Garbage Disposer Service Chart

Trouble	Possible Cause	Remedy
Motor will not start.	Fuse blown.	Check and replace.
	Flywheel jammed.	Release flywheel by removing obstructing part. Clean and adjust as necessary.
	Inoperative motor.	Check motor connections and assembly.
	Inoperative capacitor.	Replace.
	Inoperative control switch.	Replace.
	Flow interlock stuck or out of adjustment.	Check plunger, adjust or replace as required.
Motor will not stop.	Control switch stuck.	Replace or adjust.
	Flow interlock stuck or out of adjustment.	Replace or adjust plunger.
	Short around starting switch.	Check and remove short.
Slow grinding.	Stuck or badly worn impellers.	Free impellers; if necessary replace flywheel assembly.

Garbage Disposer Service Chart (Continued)

Trouble	Possible Cause	Remedy
	Badly worn shredder.	Mix food wastes, or cut into smaller pieces.
Noise or excessive vibration.	Inoperative bearing.	Check, lubricate, or replace as necessary.
Water leaks.	Mounting screws loose at sink, or putty seal broken.	Replace and putty seal, and tighten screws.
Drain stoppage.	Clogged drain.	Clean out.
	Insufficient water.	Clean and adjust flow interlock.
	Improper venting of drain line.	Provide proper vent.
	Grease trap stoppage.	Clean out.
	Septic tank filled.	Clean out.

troubles do occur, however, they usually consist of motor failure, a jammed flywheel, slow grinding, drain stoppages, and/or water leaks. The preceding chart will serve as a guide to assist the serviceman in determining the trouble, giving the possible cause and suggested method of remedy.

Check the power supply and the disposer motor rating to see that the voltage and frequency as given on the motor nameplate are the same as that of the power supply. If there is any doubt concerning the power requirements, check with the local power company before starting the installation.

CHAPTER 24

Household Refrigerators

Refrigeration may be defined as the process of removing heat from a body or substance. It is accomplished by placing a colder medium into or adjacent to the body to be refrigerated. The medium used in a household refrigerator to attract and absorb the heat rejected by the body being refrigerated is called the *refrigerant*. In the refrigeration process, the refrigerant goes through constant changes in its physical state during which an exchange of heat takes place.

When a liquid refrigerant is evaporated to a gas, the change in its physical state is always accompanied by the absorption of heat. Evaporation has a cooling effect on the surroundings of the liquid, since the liquid obtains from its surroundings the necessary heat to change its molecular structure. This action takes place in the *evaporator* unit of a refrigeration system. Conversely, when a refrigerant gas is condensed into a liquid, the change in its physical state is always accompanied by the release of heat. This action takes place in the *condenser* unit of a refrigeration system and is due to the mechanical work exerted on the gas by the compressor.

ELECTRIC REFRIGERATORS

The modern electric refrigerator, as shown in Fig. 1, is one of the most common types of household appliances. Although refrigerators may differ somewhat in general appearance and size, they all operate on the same refrigeration principles.

Cooling Methods

The compression system, which is the most common cooling method used in electric refrigerators, makes use of an electric

Fig. 1. A typical modern electric refrigerator.

motor-driven compressor to pump the heat from the refrigerator compartment. In the absorbtion system, as used in gas refrigerators, a small gas flame produces the circulation of a refrigerant medium (usually ammonia) to remove the heat.

As previously mentioned, the cooling action in an electric refrigerator is accomplished by the evaporation of a liquid refrigerant. Refrigerants are heat-carrying mediums which absorb heat at a low temperature level and are compressed by the compressor to a higher temperature level where they discharge the absorbed heat, together with that added during the compression process, to the condenser. The ideal refrigerant is one that can discharge to the condenser all the heat it is capable of absorbing in the evaporator or cooler. All refrigerant mediums, however, carry a certain portion of the heat from the condenser back to the evaporator; this characteristic reduces the heat-absorption capacity of the medium in the low side of the system.

Pressure-Temperature Relations of Liquids

The boiling point of water at pressures found at sea level is 212°F., when heated in a vessel that is open to the atmosphere. At an altitude of several thousand feet, however, water boils at a considerably lower temperature because of the lower pressure. A lower pressure may also be attained by means of a vacuum pump, in which case water may be made to boil at lower relative

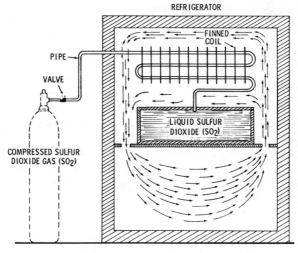

Fig. 2. A simple refrigeration system.

temperatures than the usual 212°F. This pressure-temperature relation holds true for all liquids; the boiling point rises or falls as the pressure is increased or decreased, respectively.

Some common liquids boil at a temperature below that of water, such as alcohol with a boiling-point temperature of 173°F. and ether with a boiling-point temperature of 94°F. Other substances boil at still lower temperatures; that is, they evaporate through the absorption of heat at relatively low temperatures. For example, sulfur dioxide (SO_2) boils at 14°F. at atmospheric pressure, ammonia boils at $-28°F.$, and Freon 12 (CCl_2F_2) boils at $-22°F.$ There are numerous other substances that vaporize and liquefy at comparatively low temperatures and thus have possibilities for use as a refrigerant; that is, they can be used to remove heat from a refrigerator.

A Simple Refrigeration System

The principle of using the latent heat of vaporization of a liquid, such as sulfur dioxide, for producing refrigeration can be easily illustrated by thinking of a refrigerator of extremely simple design, similar to the one shown in Fig. 2. The refrigerator consists of a box that is completely insulated on all six sides to prevent the entrance of heat by conduction, convection, and radiation. A series of finned coils with one end connected to a cylinder charged with surfur dioxide is placed in the cabinet. Two pounds of sulfur dioxide are charged into the coil, after which the compressed cylinder is again sealed and disconnected from the line, with the charging end of the pipe open to the atmosphere.

Since the liquid sulfur dioxide is exposed to the air, the only pressure to which the liquid is subjected is atmospheric pressure, which is approximately 14.7 pounds per square inch absolute pressure, or zero pounds gauge pressure. At this pressure, sulfur dioxide liquid will boil or vaporize at a temperature of 14°F. or at any higher temperature. If the temperature of the room in which the refrigerator is located is 70°F, the temperature of the cabinet at the time of adding the sulfur dioxide liquid will also be 70°F. The liquid sulfur dioxide in the coils, therefore, will immediately start boiling and vaporizing because the surrounding temperature is well above the boiling point (14°F.) of the liquid. As the liquid

488

boils away, it absorbs heat from the cabinet; for every pound of sulfur dioxide liquid that is vaporized, 168 Btu of heat will be extracted from the cabinet. As soon as the temperature of the coil is reduced to a point lower than that of the cabinet, the air in the cabinet will start circulating in the direction shown by the arrows in Fig. 2, because heat always flows from the warmer to the colder object.

With this method, however, the 2 pounds of sulfur dioxide liquid would soon be vaporized, and the evaporization of the gas would be given off to the air outside the cabinet; thus, the refrigeration process would then stop until a new charge was placed in the cooling coil. Since sulfur dioxide is expensive and difficult to handle, some means must therefore be used to reclaim the sulfur dioxide vapor so that the original charge may be used continuously. The inconvenience of recharging the coil must also be prevented, and the refrigerator must be built so that it will automatically maintain proper food-preservation temperatures at all times, with absolutely no inconvenience to the customer. This is accomplished by using a compressor to pull the warm sulfur dioxide gas from the cooling unit and pump it into the condenser, where the gas is condensed and is made ready to return to the cooling unit.

Refrigerator Operation

Practical methods of refrigeration provide a closed refrigerating circuit; that is, the refrigerant is kept in a closed metallic system where its vapor is compressed and condensed back into liquid form and used over again many times to cool the refrigerator compartment. The evaporator, as shown in Fig. 3, is mounted inside the refrigerator and is connected by two metallic refrigerant-carrying tubes to the compressor, which is driven by an electric motor. The evaporator temperature is controlled by a switch-and-bellows device, which is generally referred to as the temperature-control switch. The function of the temperature-control switch is to automatically start and stop the motor and compressor as often as is necessary to maintain the desired temperature in the refrigerator.

A temperature-control switch, shown schematically in Fig. 4, consists primarily of a thermostatic bulb that is fastened to the bellows by means of a capillary tube. The bulb and tube are

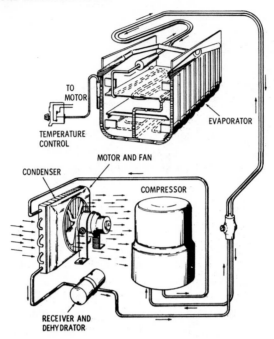

TO
MOTOR

TEMPERATURE
CONTROL

EVAPORATOR

MOTOR AND FAN

CONDENSER

COMPRESSOR

RECEIVER AND
DEHYDRATOR

Fig. 3. The major components of a compressor-type electric refrigerator.

charged with a highly volatile liquid. After a certain temperature is reached, depending on the setting of the switch-control knob, the gas pressure in the bulb-bellows assembly increases with a consequent expansion of the flexible metal bellows. This action, in turn, forces the moveable contact arm against the spring, and the switch snaps closed to start the motor and compressor. As the motor runs, the control bulb is cooled, thereby gradually reducing the pressure in the bulb-bellows system. This reduction in bellows pressure allows the spring to push the shaft slowly downward until it has finally traveled far enough to push the toggle mechanism off center in the opposite direction, thus snapping the switch open and stopping the motor. The control bulb then slowly warms up until the motor again starts, and the cycle repeats itself.

Motor Controls

Most electrical household refrigerators employ, in addition to the previously described temperature-control switch, an overload-

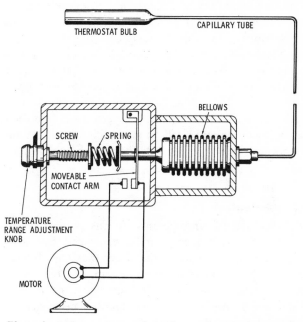

Fig. 4. The schematic representation of a temperature-control switch.

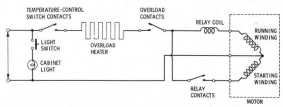

Fig. 5. A common means of motor control, as employed in many modern refrigerators.

safety control and a motor-starting relay. As shown in the wiring diagram of Fig. 5, the coil elements of both the overload-control switch and starting relay are mounted in series with the motor running winding. The function of the overload control is to remove the motor from the source if the motor becomes overloaded. The overload control generally consists of a heater coil, through which the motor current passes, and a pair of bimetallic contact blades, which open the circuit when the heater current exceeds a predetermined value.

The starting relay facilitates the starting of the split-phase motor. With the temperature and overload-control contacts closed, the circuit is completed through the relay coil and the motor running winding. The heavy current flow through the relay coil causes the relay contacts to close; this action, in turn, puts the starting winding in the circuit, and the motor starts to rotate. As the speed of the motor approaches its normal value, the current flow through the relay coil decreases. The relay contacts then open, and the starting winding becomes disconnected from the circuit. The motor now resumes normal operation with only the running winding connected in the circuit.

Refrigerant Control

In the discussion of refrigerants, it was pointed out that the temperature of a refrigerant can be controlled by changing its vapor pressure. It is apparent, therefore, that two different pressures are required in a refrigeration system, one to permit boiling and the other (sufficiently high) to stimulate condensation in the condenser. Some type of control is therefore necessary to reduce the high-temperature, high-pressure liquid in the receiver to the desired low temperature and pressure in the evaporator. This function is accomplished by what is generally called the refrigerant control. There are two principal types of refrigerant controls used in household refrigerators; they are the restrictor, or capillary-tube, control and the pressure, temperature, or refrigerant-level controls; both types of systems are employed equally.

The restrictor, or capillary-tube, system is by far the simplest in operation, since it contains no valves or adjustments; however, because of its nature, it requires more accurate designs to meet particular requirements. The restrictor, in a sense, is a fixed control that has no moveable elements responsive to load variations. Its element of variable control lays only in the natural variation of the factors affecting the flow rate of the refrigerant. The positive force to push the refrigerant through the restrictor or capillary tube is the pressure differential between the inlet and outlet; the inlet is the condenser pressure and the outlet is the evaporator pressure. Acting against this positive force is the resistance offered by the friction within the restrictor. Because of this friction factor, the

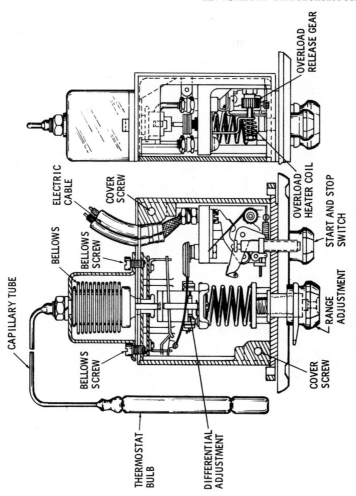

Fig. 6. The construction details of a typical temperature-control switch.

diameter and length of the restrictor are closely fixed quantities in any refrigerator unit. With this system, there is no valve to separate the high-pressure zone of the condensing unit from the low-pressure zone of the evaporator unit. Therefore, the pressures through the system tend to equalize during the "off" cycle and are retarded only by the length of time required for the gas to pass through the small opening of the restrictor.

The pressure-refrigerant control employs a pressure valve that is responsive to the evaporator pressure. This valve opens when the pressure goes down and closes when it goes up. Because of its spring balance, the valve will operate only at a predetermined pressure.

The thermostatic-refrigerant control, as shown in Fig. 6, employs a tube and bellows similar to the previously discussed system of temperature-control methods. The tube and bellows are connected to a refrigerant-charged bulb in such a manner as to exert the necessary force to close or open the needle valve as the gas falls below or rises above a predetermined temperature value of the coolant.

The refrigerant-level control employs a buoyant ball that floats on the surface of the liquid refrigerant. Depending on the location of the float in the refrigeration system, the float is called the *high-side float* or the *low-side float*. The high-side float is located in the receiver or in some chamber where the liquid refrigerant collects after leaving the condenser. The float ball is connected to a needle valve in such a way as to open when the liquid level rises, thereby allowing the liquid to pass on to the evaporator. The low-side float is located in a reservoir in the evaporator and is so connected to the needle valve that as the liquid is evaporated, and the refrigerant level is lowered, the valve is opened to admit more refrigerant from the receiver.

Defrosting Control

Frequent opening of the refrigerator door permits warm air to enter the refrigerator, which causes a consequent and rapid increase of frost or ice on the evaporator. This coating of ice, if not occasionally removed, can seriously impair the efficiency of the refrigerator. Modern refrigerators, therefore, are furnished with a defrosting control that permits the evaporator temperature to increase considerably above its normal value for a brief period of time to allow the ice to melt. The defrosting control is simply an additional feature built into the thermostatic-control switch that applies additional spring pressure on the metallic bellows. When the defrosting cycle is completed, the accompanying temperature increase actuates a tripping mechanism, which automatically returns the refrigerator to its former temperature setting.

Two-Temperature Refrigerators

This type of refrigerator has two temperature zones with two evaporators, one for normal refrigerating temperature food storage and a second (usually contained in the upper part of the unit) for temperatures well below freezing for storage of frozen foods. There are several methods used to obtain two different temperature zones within the same refrigerator cabinet. Fig. 7 illustrates the principles of operation of this type of unit; the system functions as follows:

After the refrigerant is liquefied in the condenser, it passes through the dehydrator and capillary tube, where it is reduced in pressure to conform to the normal temperature requirement

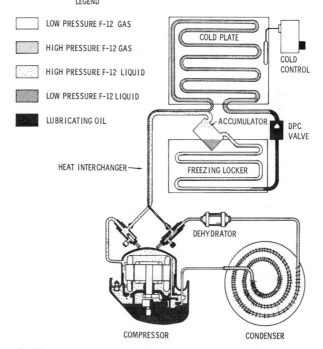

LEGEND

☐ LOW PRESSURE F-12 GAS

▨ HIGH PRESSURE F-12 GAS

▨ HIGH PRESSURE F-12 LIQUID

▨ LOW PRESSURE F-12 LIQUID

■ LUBRICATING OIL

COLD PLATE

COLD CONTROL

ACCUMULATOR

D.P.C. VALVE

HEAT INTERCHANGER

FREEZING LOCKER

DEHYDRATOR

COMPRESSOR

CONDENSER

Fig. 7. The principal components of a two-temperature refrigerator.

of the evaporator. Part of the liquid refrigerant evaporates in this evaporator to maintain the food-compartment temperature. The remainder of the liquid and the low-pressure gas pass through a differential pressure-control valve (D.P.C.) and then into the freezing compartment evaporator. This differential pressure-control valve is constructed like a spring-loaded check valve; it further restricts the flow of the refrigerant and produces a considerable pressure drop. The liquid in the second evaporator is consequently under lower pressure. Its boiling point is reduced to approximately −5°F., thus maintaining a lower evaporator temperature. It is in this manner that two different temperatures can be obtained in the same refrigeration system. From the second evaporator where the remaining liquid is evaporated, the low-pressure gas passes through the accumulator and suction line to the compressor. The accumulator is located at the outlet of the second evaporator; this accumulator traps any liquid that may be carried through with the gas and thus prevents the liquid refrigerant from entering the suction line until it is completely evaporated.

GAS REFRIGERATORS

Refrigeration by means of a gas flame differs from the conventional electric method, mainly in that the heat from a small gas fllame is substituted for an electric motor to produce the necessary circulation of the refrigerant. The common method of utilizing heat for refrigeration purposes is called an *absorption system*. The heat from a gas flame is most frequently used (although a kerosene flame or an electric heater may also be used) to produce the necessary energy for refrigerant-circulation purposes. The circulation component used in an absorption system is called the *generator*. In the absorption type of refrigeration, the refrigerant (usually ammonia) circulates between the inside and the outside of the refrigeration compartment; it absorbs heat from the inside and discharges it to the outside, thus maintaining refrigeration temperatures inside the refrigerator. The principal advantage in the absorption system of refrigeration lies in the fact that since there are no moving parts, the repairs and maintenance cost over a long period of time will be low.

Operation Fundamentals

In the absorption system of refrigeration, the generator acts as a distiller and a pump. To obtain efficient operation, the heat input must be correct and controllable. After the proper heat has been applied to the generator, the ammonia will evaporate from the water. The vapor bubbles, in trying to escape, will carry water up the percolator tube. The vapor and water are allowed to separate so that the vapor is free to continue upward into the condenser. With proper air circulation, the ammonia vapor is then condensed

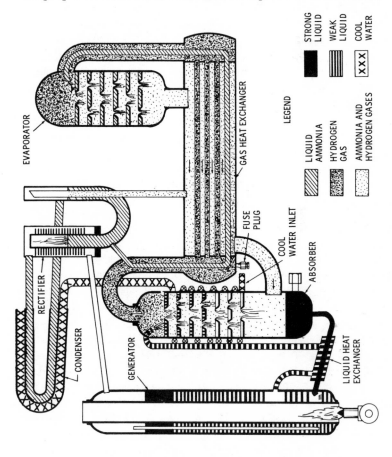

Fig. 8. The operating principles of an absorption-type refrigerator.

497

into a liquid; it then flows through a liquid trap into the evaporator. When the evaporator shelf is level, the proper slope is established in all coils to induce a gravity flow downward. As soon as the unit is charged, a small amount of hydrogen is introduced. At this point in the cycle, hydrogen flows upward into the evaporator and tends to mix with the ammonia vapor to encourage more evaporation. It is this evaporation process that produces refrigeration.

Since the mixture of hydrogen and ammonia vapor is considerably heavier than hydrogen alone, the normal tendency for this mixture is to flow downward. It is encouraged to do this, and in so doing is forced to pass upward through the absorber.

Water, which has been separated from the ammonia by heat, is flowing downward through the absorber. The water temperature has been reduced so that it will again absorb the ammonia quite readily; the water and ammonia solution then flows back to the generator for recirculation. Since the hydrogen has been washed free of ammonia, and has thus been lightened, it flows upward again through the evaporator.

When the absorption system is working normally, all of these actions are continuous, as shown in Fig. 8. A thermostatically controlled gas valve, with a feeler attached to the evaporator coil, varies the heat input and consequently varies the amount of refrigeration that the load of the refrigerator requires.

Gas Controls

Before proceeding with gas-control adjustment, be sure that the refrigerator is properly installed, that it has proper air circulation, and that it is in a level position. The gas controls normally consists of a burner assembly, a pressure regulator, an automatic shutoff valve, a gas thermostat, and a defroster.

Burner Flame—The energy that operates the refrigerating unit is supplied from the burner flame; the correct size of the burner flame supplies the right amount of heat to the refrigerating unit. The burner flame, therefore, has an important bearing on refrigerator performance. The flame is controlled between the maximum and minimum limits by means of a thermostatic valve. This valve is opened and closed automatically by the temperature of the evaporator. The burner flame should be centered within the generator flue to prevent the flame from contacting the flue walls; if the flame contacts the flue, it will cause the production of an odor outside the refrigerator, and the accompanying carbon deposits will ultimately cause flue stoppage.

Position of Heat Conductor—For proper operation, the heat conductor must just touch the minimum flame. Normally, this will occur when the concave surface of the heat conductor is lined up with the inside rim of the burner cap. When installing the burner, be careful not to alter the correct position of the heat conductor. The burner should be installed on the unit in a fixed position; it is usually held in place with a burner bracket by means of a setscrew. The distance from the end of the generator-flue opening to the end of the air-shutter barrel should be exactly as given in the manufacturer's specifications.

Gas-Pressure Regulator—All refrigerators that are equipped for use on gas should also be equipped with a gas-pressure regulator, as shown in Fig. 9. The gas-pressure regulator is designed to maintain a constant gas pressure at the burner. A gas-pressure regulator, however, cannot provide a gas pressure at the outlet in excess of the gas pressure at its inlet. The gas-pressure regulator

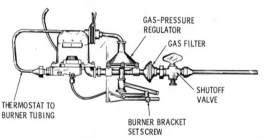

Fig. 9. The typical installation of a gas-pressure regulator.

may be installed in the burner compartment at the gas-thermostat inlet, and the gas filter may be installed in the inlet of the pressure regulator. The regulator must be installed so that the gas will flow through it in the right direction; it should also be in an upright, level position, so that its inlet and outlet will be horizontal.

Thermostat—The function of the thermostat is to control the size of the burner flame. This control depends on the evaporator temperature. The thermostat is also used to cause defrosting of the refrigerator. The various types of thermostats used are called, according to their principles of operation, *manual, semi-automatic,* and *dual.*

With the manual-type thermostat, defrosting is controlled by turning the temperature control to the "defrost" position; when the defrosting operation is completed, the temperature-control pointer is returned to the opposite position. In the semi-automatic type, defrosting is accomplished by turning the temperature control to the "defrost" position and then turning it immediately back to the operating position. When the evaporator reaches a predetermined temperature value, the thermostat valve will be opened automatically and refrigeration will be resumed. In the dual-type thermostat, defrosting can be accomplished either manually or semi-automatically, since the features of both the preceding types have been incorporated into this type of thermostat.

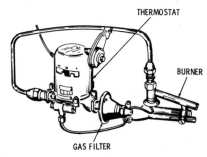

THERMOSTAT

BURNER

GAS FILTER

Fig. 10. Thermostat and gas-filter installations in a typical gas refrigerator.

Refrigerator Installation

All refrigerators must have proper air circulation for proper operation. When the refrigerator is in operation, air enters from the bottom, travels upward through the rear section of the cabinet, and is expelled at the top. This air circulation takes place naturally, unless it is prevented from doing so by insufficient clearance or blocked air passages. Proper air circulation will usually result when the refrigerator is installed indoors directly in front of a wall if a minimum clearance of 2 inches from the back of the refrigerator to the wall and at least a 12-inch clearance above the refrigerator is allowed. When the recommended top clearance cannot be obtained, an air duct should be installed, as shown in Fig. 11.

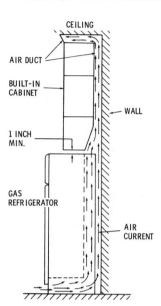

Fig. 11. The installation of an air duct to provide proper ventilation for a gas refrigerator.

The air duct should have approximately the same width as the refrigerator and should be approximately 7 inches deep. This duct may return the air to the room at ceiling height, or it may exhaust the air through a suitable vent in the roof.

Leveling—The equal distribution of the liquid within the freezing compartment requires the unit to be installed and maintained in the level position, both front and back and side to side. Leveling of the refrigerator can conveniently be accomplished by shimming the bottom supports with small wooden strips or other available material, although most late-model refrigerators are equipped with adjustable supports.

Gas-Line Connection—When connecting the gas line, use tubing and fittings as prescribed by local codes. Install the gas line so that the refrigerator can be disconnected at the inlet valve without damage to the controls. On liquid-petroleum (LP) gas installations, a gas-pressure regulator in the gas line is not needed, since the regulator at the gas-supply tanks should maintain a constant gas pressure at the burner.

SERVICING AND REPAIRS

The household refrigerator, in common with any other appliance, requires a certain amount of attention for maximum operating efficiency at a minimum cost. Here are a few important facts that the serviceman should pass on to a customer:

1. Door openings and duration of opening should be kept to a minimum. Constant opening and closing of the door will cause the unit to operate longer, more frequently and also result in more wear on the door gasket, hinges, strike, and catch.
2. Defrost regularly. A heavy frost deposit causes the unit to work harder and longer to maintain proper temperature. Frost also absorbs odors.
3. Clean interior regularly; wash cabinet inside and out; also wash shelves, containers, etc.
4. Maintain a good door-gasket seal to increase the efficiency of the refrigerator and reduce operating costs.
5. Keep the forced-draft condenser clean. Dust and dirt accumulations on the fins result in lower operating efficiency and higher operating cost.

Electric Refrigerator Service

The household refrigerator has been greatly simplified and improved over the years. These improvements have been directed into two general channels, (1) to simplify the operating mechanism and at the same time make it more compact, thereby saving space; and (2) to improve the interior and exterior of the cabinet in order to enable the customer to store a larger quantity of foods in a refrigerator unit of truly distinctive appearance.

The majority of today's household refrigerators are of the hermetically sealed type, in which the entire mechanism, including the compressor, condenser, evaporator, and connecting tubing, is manufactured in one compact unit, as noted in Fig. 12. Because of its construction, therefore, the entire assembly does not permit local servicing but must be removed from the cabinet and shipped back to the manufacturer's service depot for servicing when trouble occurs.

It is necessary, however, for the serviceman to be able to properly diagnose the various complaints, and in order to do so, he must be thoroughly acquainted with the general operational details common to all refrigerators. As a rule, the common servicing complaints fall under five general headings:

1. Unit does not run.
2. Unit runs but does not refrigerate.
3. Unit does not refrigerate properly.
4. Leaks in the system.
4. Unit is noisy.

Unit Does Not Run—The failure of a unit to run when plugged into the proper electrical outlet may be due to one of

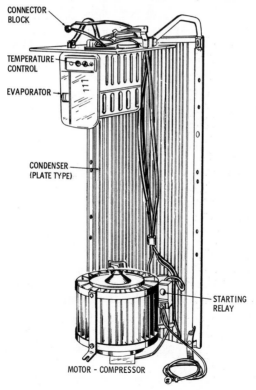

Fig. 12. The assembly details of a modern electric refrigerator with a plate-type condenser and a hermetically sealed motor-compressor unit.

several reasons; these must be determined by the serviceman before replacing the unit.

1. The power supply should be checked with a test cord.
2. The proper type of power must be used (usually 110-120 volt, 60-cycle, AC).
3. Check for broken wires in the lead-in cord.
4. Check the "on-off" switch to make sure the switch is in the "on" position.
5. Check for a defective thermostatic-control switch, which may prevent the thermostat from making contact at this point in the circuit. This switch can be temporarily eliminated from the circuit by shorting it out, in which case the motor will be connected directly to the circuit.
6. Check the protective relay of the motor.
7. On units equipped with a capacitor, test the capacitor to determine if it is functioning properly.
8. If all these points have been carefully checked, and the unit still does not run, the unit will have to be disassembled to locate the trouble and will probably have to be sent back to the manufacturer.

Unit Runs But Does Not Refrigerate—This condition is liable to occur immediately after installation, when a new cabinet has been stored in an extremely cold place or has been exposed to low temperatures during delivery, with insufficient time to warm up after being installed. Allow some time for the refrigerator unit to reach the approximate temperature of the room, and then plug it in. It should now function properly.

Unit Does Not Refrigerate Properly—Improper refrigeration may result from factors that are external to the refrigerating unit or from trouble within the unit itself. Frost on the evaporator is usually a good indication of whether the fault is within the unit or elsewhere. When checking a refrigerator for improper refrigeration, make sure that the unit has operated for a sufficient period of time to create normal operating conditions. The time required to reach normal operating conditions depends on how long the unit has been shut down.

1. *Unsatisfactory cabinet temperature*—Complaints of this kind, especially when the evaporator frosts and freezes ice satisfactorily, may be divided into two types, cabinet temperature too high or cabinet temperature too low.

a. *Cabinet temperature too high*—Since the evaporator frosts satisfactorily, the complaint probably has no bearing on the sealed unit itself and may be caused by any one of the following:

1. Improper thermostatic-switch setting,
2. Defective thermostatic switch,
3. Blocked air circulation in cabinet,
4. Improper sealing of door gasket,
5. Excessive food load in cabinet.

Because the thermostatic-switch setting closely regulates the evaporator temperature, it may be that the switch-control knob is set in too warm a position. Often the customer belives that a warmer control-knob setting produces more economical operation, but he does not understand that these warmer settings, when used under certain food-load conditions, might not provide a sufficiently low cabinet temperature.

A defective thermostatic switch is generally the result of burned contacts, partially discharged thermal elements, or an incorrect setting. Burned contacts cause erratic operation of the unit. For example, cabinet temperatures may be exceptionally low at one time, and then the unit may fail to start, thus resulting in an increase in cabinet temperature. Switches with badly burned contacts should be replaced. A partially discharged thermostatic-switch bulb can prevent the switch from cutting in properly when set on "defrost." It can also cause excessive defrosting in the evaporator during the "off" cycle because of the higher evaporator temperature required to cut in the switch. This defrosting condition will increase steadily over a period of time. Improper switch operation, which indicates a faulty charge in the thermostat bulb, should be remedied by replacing the bulb.

Blocked air circulation in the cabinet is usually caused by the improper placement of foods. When the air circulation becomes blocked or restricted due to excessive crowding of foods or coverings placed on shelves, the cabinet temperature in these blocked or restricted areas will rise, and food spoilage may result. The only remedy for this problem is the proper distribution of the foods to be refrigerated; the manufacturer can furnish specifications for this purpose.

The improper sealing of the door gasket against the outside cabinet shell causes an excessive leakage of warm air into the cabinet; this leakage usually results in a certain amount of sweating on the outside cabinet shell adjacent to where the door gasket is improperly sealed. An improperly sealed gasket can also cause excessive ice accumulation on the evaporator, which necessitates more frequent defrosting. If the door gasket is found to seal improperly, the door latch may be adjusted, or the hinge butts may be reshimmed. If after these adjustments are made, certain spots are still found to seal improperly, a piece of nonoxidizing tape may be placed between the door gasket and the door panel in order to back up the gasket sufficiently and make it seal properly.

An excessive food load causes the air temperature of the cabinet to rise. The air temperature will usually continue to be higher than normal until the food is cooled. Frequent opening of the cabinet door causes complete air changes in the cabinet and also places an excessive load on the refrigerating mechanism. This condition may cause, when combined with a warm food load, not only the cabinet temperature to rise above normal but may also considerably retard ice freezing.

b. *Cabinet temperature too low*—If the cabinet temperature is too low, the unit is evidently refrigerating too much. The most obvious cause of the condition is a low setting on the temperature-control switch. Since there is a definite relationship between the room temperature and the cabinet air temperature, different temperature-control settings may be required for the same degree of cooling. Thus, in some locations, low room temperatures combined with a colder control setting may cause the cabinet air temperature to go below freezing. Also in high altitudes, the lower barometric pressure will lower the range of the switch. This necessitates resetting the switch to a warmer position, so that the cabinet temperature is not held too low when installations are made at an altitude higher than 1000 feet above sea level.

2. *No refrigeration*—On a complaint of this type, the compressor is usually found to run continuously while the evaporator just

feels cool or is approximately the temperature of the cabinet. The reason for the absence of refrigeration may be caused by either a full or floating restriction within the unit, such as a closed capillary tube, a permanently closed refrigerant-control valve, internal compressor trouble, or a lost refrigerant charge due to a leak in the system. A condition of no refrigeration usually shows an extremely low wattage consumption when checked with a wattmeter. If any of these conditions do exist, the defective unit, or units, should be repaired or replaced.

3. *Poor refrigeration*—This condition is usually recognized by a comparatively short "off" cycle and a running cycle that is longer than normal. The trouble may be due to a low refrigerant charge in the unit, a restriction in the evaporator, or thermostatic switch trouble. This complaint may also be caused by frequent and/or abnormal food-load changes in the cabinet.

Leaks in the System—In the event that any part of the unit in a hermitically sealed system should develop a leak, the refrigeration unit should be removed from its cabinet and shipped back to the manufacturer's service depot for repair or replacement. The leak is usually discovered by the presence of oil around the point at which the leak developed. It must not be assumed, however, that the presence of oil on any part of the unit is an indication of a leak.

An actual refrigerant leak can be found by the use of a leak detector, of which there are several on the market. A shortage of refrigerant in the system may result in partial frosting of the freezer; this condition should be checked with the cold control in the coldest position. If the entire freezer does not frost when the cold control is set in this position, the unit may be partially discharged. The leak must first be found and repaired. The refrigeration unit can then be recharged with the same refrigerant (usually Freon) as was in the system originally.

Noisy Refrigerator—Refrigerators of recent manufacturer seldom exceed a noise level that is sufficiently high to attract attention. However, noisy refrigerator complaints are difficult to properly diagnose, since there are no recognized standards by which a unit can be judged in the field.

Motor hum originates in the magnetic circuit due to the passage of an alternating current. A continuous hum during the compressor-operation period, however, may be due to vibration, either in the motor or in the tubing. Check the motor mounts and the tubing holders; if either is loose, tighten securely.

A high internal knock, which continues until the compressor stops, is commonly caused by a stuck divider block. In a hermetically sealed unit, it will be necessary to replace the unit in order to remedy this type of complaint.

A vibration complaint is one of the most difficult types of noise to analyze. These complaints, however, can usually be traced to loose internal cabinet parts or some external cause, such as vibrating walls, floors, etc., either during the running cycle or when the compressor stops. Floor strength and location of the cabinet in the room are often big factors in causing vibration complaints. If the floor under the cabinet is not solid, there may be a natural freuency of vibration in it that is sufficiently close to that of the unit, when running, to cause a vibration, quiver, or hum in the floor. Usually, a complaint of this type may be corrected by strengthening or adding support to the floor directly under the cabinet. It may also be necessary to move the cabinet to that part of the floor which is adequately supported in order to remedy the vibration trouble.

ICE MAKER

This particular ice maker (sold by Western Auto Stores) is small, compact in design, and controlled by a remote sensor thermostat that starts the motorized mechanism to harvest the ice cubes. The ice maker incorporates the use of a full 360° rotation of the tray with a twist action to dump the ice cubes. See Fig. 13. The various electrical switching is performed by a commutator and brush assembly which is contained within the mechanism. When installed properly in a freezer compartment of 8° F. or less, the sensor cools slowly until the fixed temperature thermostat closes and starts the drive motor and the water in the sensor. The motor drives the commutator and the tray assembly.

The following information is provided to enable the serviceman to understand the functioning of the ice maker in normal operation and as a guide to correct diagnosis in service. The ice container should be removed from the ice maker if the refrigerator is run before the water supply is connected. This will stop the operation of the mechanism.

Ice Maker Components

The major components of the ice maker are as follows:

Drive Mechanism Assembly—The drive mechanism assembly consists of a 115 volt AC drive motor and control system, preset to perform the entire ice harvesting operation. All component

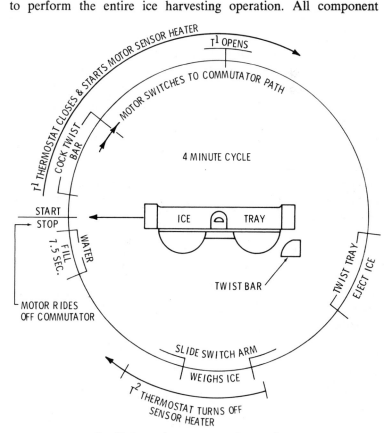

Fig. 13. Ice maker sequence of operation.

parts of the mechanism are replaceable and are accessible from the front of the ice maker without necessitating removal of the ice maker from the appliance.

Control Circuit—The control circuit, built into the same housing as the drive mechanism, works in conjunction with a temperature sensing device called a sensor which physically senses the amount of time required to freeze the ice cubes and start the harvest cycle. A commutator and brush assembly sequences the complete harvest and re-fill operation.

Sensor—The sensor is an epoxy mold assembly, containing an *on* thermostat, and a heater *off* thermostat, wired in series to a special heater coil.

Starting with a warm freezer, the cold thermostat (T_1 in Fig. 14) in the sensor will gradually be cooled down to a temperature of 11°F. ± 3°F. This thermostat will close and start the harvest cycle. The drive motor will start and drive the tray at a speed of ¼ rpm. A 22 watt heater in the sensor will be energized and heat the sensor for approximately 2½ minutes until the circuit is opened by a high temperature thermostat (T_2) at approximately 59°F. (See Fig. 14).

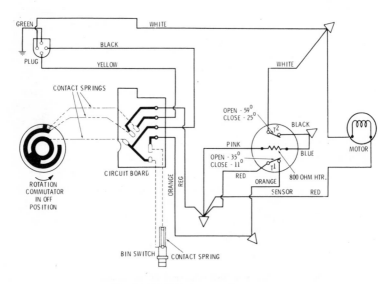

Fig. 14. Wiring circuit of ice maker.

The sensor being subject to the same conditions as the ice tray, will cool at the same rate the water cools. The sensor will cool to

510

11°F. and start the harvest cycle, at which time the ice cubes are solid and ready for harvest.

Ice Tray and Twist Bar—The ice tray is of the flexible type and the cycle functions are obtained in a single revolution of the tray. The twist bar is a slide mechanism in a bracket mounted to the right side of the frame assembly.

The brackets on each end of the ice tray are designed so the front bracket will be stopped by the twist bar, the tray will continue to rotate for approximately 18 seconds, when the cam action on the front tray bracket will slide the twist bar and release the rear bracket to dump the ice cubes. The ice tray will continue to rotate 15° before the level position where the solenoid valve will open for 7.5 seconds, and the tray will fill with water. The cycle will stop with the tray in the level position.

Ice Container—The ice container assembly has storage capacity of approximately 4 pounds. When the bin is full, the weight of the ice depresses the bin weighing springs, lowering the ice container and opening the slide switch.

At 270° of rotation of the tray, a cam on the drive gear, in the mechanism, trips the slide switch arm which forces the slide switch down to weigh the ice.

Water Valve Assembly—A controlled water flow valve, activated by a solenoid coil, meters water to the ice tray. The water valve will operate satisfactorily at pressures from 15 psi to 120 psi.

Slide Switch—The normally open slide switch, mounted in the bottom of the drive mechanism assembly stops the ice harvest when the ice container is full. It will also suspend ice maker operation any time the container is removed or not correctly positioned in the guides. If the ice maker operation does not stop when the ice container is removed, a defective cut-off switch is indicated.

IDENTIFYING TROUBLES

It is important for the serviceman to remember that the ice maker normally performs the harvest operation only after the ice in the tray is completely frozen and the temperature at the sensor reaches 11°F. ± 3°F. Therefore, in the case of a suspected malfunction, it is unlikely the serviceman will be able to observe the operation of the ice maker without first making sure all the conditions for operation are met.

To save time, it is suggested the serviceman make sure of the exact nature of the customers complaint by asking questions. Four crucial questions are:

1. When did the ice maker stop performing properly?
2. How is the ice maker malfunctioning?
3. If "not operating" or "too long between loads" has the freezer been used more than usual—that is with heavier loads of fresh foods or more frequent opening of the freezer or refrigerator door?
4. At what point does the customer set the cold control knob?

The answer to these and/or other questions, depending on the nature of the complaint, will help to pin-point the cause of the condition and indicate the steps necessary to make the proper correction. In many cases, a service call will be unnecessary if the condition can be corrected by the customer's setting the cold control knob a little colder or simply waiting until a particularly large load of fresh food has frozen.

SERVICING PROCEDURES AND CHECKS

Wiring Procedure—All wiring on the circuit board is done with the use of spades and spade connectors. All wiring on the outside of the motor plate is accomplished with the use of 4 wire nuts. Check the wiring diagram inside the facia. Be sure of good contact within the wire-nuts.

The ice maker proper is equipped with a 4 prong male connector to accept the ice maker service cord.

To remove the facia:

1. Disconnect the service cord from the power supply.
2. Remove the screw from the lower right underside of the facia.
3. Pull the bottom of the facia out and raise it off the mechanism housing.

To remove the drive motor:

1. Disconnect the service cord from the power supply.

2. Remove the facia.
3. Disconnect the 2 wire nuts on the motor leads.
4. Remove the 2 screws securing the motor to the motor mounting plate.
5. Replace the motor by reversing this procedure. (Check wiring procedure and diagram.)

To remove the sensor:
1. Disconnect the service cord from the power supply.
2. Remove the facia.
3. Remove the wire nuts from the sensor leads.
4. Slide the sensor out of the insulator.
5. Replace sensor by reversing this procedure. (Check wiring procedure and diagram.)

To remove gear and shaft, commutator and slide switch arm:
1. Disconnect the service cord from the power supply.
2. Remove the facia.
3. Remove the 4 screws securing the motor plate to the housing.
4. Lift motor plate, sensor and sensor insulation from housing.
5. Remove slide switch arm and spacer from housing.
6. Remove gear and shaft from housing.
7. Remove washer and commutator from shaft.
8. Reassemble by reversing. (Check wiring).

To remove circuit board and slide switch:
1. Disconnect the service cord from the power supply.
2. Remove the gear and shaft, commutator and slide switch arm.
3. Remove the 2 screws securing the circuit board to the housing.
4. Lift the circuit board out of the housing, taking care not to lose the slide switch spring.
5. Remove the slide switch from the housing.
6. Reassemble by reversing the procedure. Be sure the slide switch, slide switch spring, and slide switch arm and spacer are properly installed.

To replace the tray assembly:
1. Disconnect the service cord from the power supply.
2. Remove the ice container.
3. Remove the ice maker shield.
4. Disengage the spring retainer from the rear pivot, lower the rear of the tray and remove the rear pivot bearing.
5. Raise the rear of the tray assembly and slip the front of the tray assembly from the drive shaft.
6. Remount the tray assembly by reversing this procedure.

To remove the ice maker for service:
1. Disconnect the service cord from the power supply.
2. Remove the ice container.
3. Remove the ice maker shield.
4. Loosen the two screws securing the ice maker frame to the liner.
5. Remove the one screw securing the mounting bracket to the liner.
6. Slide the ice maker forward and then to the left or right to clear the screws loosened in Step 5.
7. Disconnect ice maker service cord. The individual parts may now be replaced.
8. Remount the ice maker by reversing the above procedure.

To remove the twist bar:
1. Remove the ice maker from the appliance.
2. Remove the 3 screws securing the twist bar bracket to the right side of the frame.
3. Remove the twist bar and bracket.
4. Replace by reversing this procedure.

To replace the frame assembly:
1. Disconnect the service cord from the power supply.
2. Remove the ice maker from the appliance.
3. Remove the ice tray assembly.
4. Remove the 4 screws from the rear of the housing.
5. Remove the fill spout.
6. Reinstall the fill spout on new frame.
7. Reinstall frame and ice maker by reversing this procedure.

To replace the fill spout:
1. Disconnect the service cord from the power supply.
2. Remove the ice maker from the appliance.
3. Drill the top off the 2 pop rivets and remove fill spout.
4. Reinstall new spout using 2⅛ × ⅜ aluminum pop rivets.
5. Reinstall ice maker by reversing this procedure.

To check the drive motor:
1. Remove the facia.
2. Disconnect the 2 wire nuts on the drive motor leads.
3. With a test cord check the motor operation.

To check the sensor:
1. Remove the facia.
2. Remove the wire nuts on the sensor leads.
3. With an ohmmeter, check sensor heater across the pink and blue leads. Your reading should be 600 ohm.
4. At room temperature, check T_2 across black and white leads. It should be open.
5. At room temperature, check T_1 across red and orange leads. It should be open.
6. Cool the sensor to 8°F. T_1, should close. Sensor must be heated to 35°F. for T_1, to open.

Continuity check of the commutator and circuit board:
1. Unplug the ice maker service cord.
2. Remove the facia.
3. Remove the drive motor.
4. Remove the wire nut from the red lead, and isolate the red lead to the circuit board.
5. With a low voltage test light or ohmmeter, check from the red circuit board lead to the black lead on the ice maker service cord.
6. Turn the shaft 30° clockwise and the circuit should close. The circuit should remain closed clockwise around to the level position, then open. (See Fig. 13).
7. Check from the black lead to the yellow lead of the ice maker service cord during the last 15° of commutator cycle. The water fill circuit should close.

515

To check the commutator:

1. Remove the commutator from mechanism.
2. Check for dirt. If dirty, clean with a fine abrasive pad.
3. Check for break or burned spots. If burned or broken, replace.
4. With a low voltage test light or ohmmeter, check from the orange circuit board lead to the black lead on the ice maker service cord. The circuit should be closed with the empty ice container in position, and open with the ice container removed.

To check circuit board and slide switch:

1. Remove circuit board and slide switch from mechanism.
2. Check brushes for pits. Determine if brushes can be cleaned, if not replace.
3. Check printed circuit for breaks. If broken, replace.
4. Check slide switch for dirt, or burned spots. If dirty, clean with fine emery wool. If burned, replace.

Oversize Ice Cubes—Stuck or "Hung-Up" Ice Cubes

Ice cubes not being ejected during the harvest may hang up due to mineral or other deposits in the tray. When this occurs, the water entering the tray during the next fill cycle, will overfill the tray and, possibly, spill into the ice container. In these cases, the tray should be removed and washed in vinegar. Rinse and reinstall. Check the twist bar.

CAUTION: Do not use any abrasive material to clean ice tray.

Oversize ice cubes may also be caused by too much water entering the tray during the fill or a leaking water valve. The fill time should not exceed 8.2 seconds. If it does, the commutator must be replaced. If the fill time is within limits, replace the flow washer in the valve. If the valve leaks, replace the valve.

To check ice tray:

1. Visually check tray for cracks or loose brackets.
2. If tray is cracked—replace tray assembly.
3. If brackets are loose, tighten rivets, if rivets can not be tightened or if brackets are broken—replace tray assembly.

4. Check rear pivot bearing position.

5. Check rear pivot bearing retainer spring.

To check twist bar:

1. Visually check for broken bar or bracket. If broken, replace.

2. Check for loose mounting screws. If loose, tighten. The 2 screws in the front slots are for timing the amount of twist on the tray. The twist should be from 14.8 seconds to 21.2 seconds.

In some cases it may be necessary to adjust more twist on the ice tray. This can be done by lowering the front of the twist bar bracket.

ICE MAKER TROUBLESHOOTING CHART

TROUBLE	POSSIBLE CAUSE	CORRECTIVE ACTION
1. No water no ice	Freezer too warm	Freezer must operate below 8° F.
	Ice Maker not plugged in or improper wiring	Check power to ice maker
		Remove facia and check all twist on connectors
	Slide switch stuck or not making contact	See "to continuity check slide switch"
	No water supply	Water line not connected
		No electric connection to water valve
		Plugged Screen
	T_1 Thermostat not making contact	Make sure freezer is cold enough
		See "to check sensor"
	Motor will not start when cold	Replace motor
	Motor gears stripped	Replace motor
2. Dumps water	T_2 not closed or not making contact	Replace sensor
	Sensor heater open	Replace sensor

517

ICE MAKER TROUBLESHOOTING CHART (Cont'd)

TROUBLE	POSSIBLE CAUSE	CORRECTIVE ACTION
	Fill tube not installed properly	Install fill tube correctly
	Wires to sensor loose	Be sure all wires connected properly
	Tray retainer spring un-hooked. Rear bearing dropped down.	Re-install retainer spring
		Install new spring
3. Intermittent oper-ation, water mixed with good cubes	Burned or dirty commutator	Clean commutator
		Replace commutator if necessary
	Intermittent contact of T_2 thermostat	Replace sensor
4. Hollow cubes or cubes frozen together	Fill spout not correctly positioned	Reposition fill spout
	Air flow over ice maker obstructed	Check for plastic bags etc. in air stream
	Ice maker not level causing larger cubes on one side or end	Level ice maker
5. Not enough ice	Freezer too warm	Freezer must operate below 8° F.
	Ice container binds in frame or frame bent	Adjust frame or straighten frame to prevent binding
	Slide switch sticking	Remove slide switch ass'y and adjust contacts. File opening if necessary to obtain free movement
	Slide switch contacts out of position	Replace slide switch assembly if necessary
	Ice container springs broken or weak	Replace ice container
	Frame low on housing	Loosen screws and adjust as far up as possible
	Ice container springs broken or weak	Replace ice container

ICE MAKER TROUBLESHOOTING CHART (Cont'd)

TROUBLE	POSSIBLE CAUSE	CORRECTIVE ACTION
6. Won't shut off, ice container too full	Ice on rail	Align fill spout
	Ice maker not level	Level ice maker
	Sticking slide switch, slide switch contacts out of place	See "to continuity check the slide switch"
	Ice bucket binds in frame, frame bent	Adjust or straighten to prevent binding
7. Fill tube plugged with ice	Water valve leaks	Replace valve
	Fill tube improperly installed	Install with proper drain angle

Gas Refrigerator Service

The following points should be observed for proper adjustment and maintenance of gas burning refrigerators:

Lighting the Burner—This is a rather simple operation, provided the manufacturer's instructions accompanying the unit are followed. The lighting procedure is generally as follows:

1. Be sure the gas valve is turned on.
2. Push the lighter button.
3. Ignite the gas at the end of the burner tube.
4. Continue to push the lighter button until the burner valve clicks open, and the burner flame ignites. If the burner flame goes out, wait 5 minutes before attempting to relight the flame.

Do not allow the burner to be ignited unless all final adjustments have been made. The burner flame should burn with a blue color and must sufficiently enter the flue opening.

Cold Control—The position of the thermostat dial depends on the refrigeration load. If the food load is heavy, turn the dial toward a colder position; if the food load is light, turn the dial to a warmer position. A colder setting is usually required in the summer than in winter. Experience and observation of the effect of

519

the various thermostat settings will usually provide the operator with the required knowledge after a short period of time.

Refrigeration Stoppage—If it is necessary to discontinue refrigeration for any length of time, turn off the gas supply. Remove the ice cube trays, and empty them. Dry the interior of the refrigerator. Leave the door partly open to ventilate the cabinet interior and to keep it fresh.

Improper Thermostat Operation—Normally, the thermostat has been properly adjusted by the manufacturer. This setting enables the refrigerator to operate satisfactorily under most conditions; therefore, these settings should not be changed. If the factory setting of the thermostat is not changed, the usual causes for improper thermostat operation are:

Gas Refrigerator Service Chart

Trouble	Possible Cause	Remedy
Refrigerator too cold.	Thermostat improperly adjusted.	Adjust thermostat to proper setting.
	Low room temperature.	Will correct itself with rise in room temperature.
	Minimum flame too large.	Flame can be corrected by proper setting of adjustment screw.
Refrigerator not cold enough.	Insufficient gas.	Check maximum heat input of thermostat adjustment, gas regulator, shutoff valve, and defrosting valve. Replace defective unit.
	Heat spreader on vertical flue is missing.	Install new spreader, as specified by the manufacturer.
Gas odor.	Improper air-vent adjustment.	Correct the air-vent adjustment for the best position possible.
	Flame impinges flue.	Relocate burner.
	Insufficient ventilation.	Provide proper ventilation.
	Flue dirty.	Clean and adjust if necessary.
	Gas leaking.	Locate and repair leak.
	Flue cover not installed or improperly installed.	Install flue cover to meet manufacturer's instructions.

1. Dirt in the bypass orifice,
2. Incomplete thermostat-bulb contact with bulb sleeve,
3. Lost thermostat charge.

To properly clean the bypass orifice, wash it out with a solvent recommended by the manufacturer, and blow it out with air. Before reassembling, make sure the orifice hole is clear. If the thermostat bulb makes an incomplete contact with the bulb sleeve, the bulb temperature will be higher than that of the freezing compartment, and the burner will operate continuously at a maximum flame, thereby lowering the temperature of the refrigerant by an excessive amount. The thermostat, if defective, should be replaced. If the thermostat loses its charge, it will become inactive; the power element is then no longer able to expand and contract. The gas valve, therefore, remains closed when the temperature control is set within the operating range. In this case, the burner will burn continuously at a minimum flame, which, in turn, will result in a high cabinet temperature and continuous defrosting. Again, the thermostat must be replaced for efficient operation.

CHAPTER 25

Room Air Conditioners

A room air conditioner is generally defined as a unit air conditioner that is suitable for placement in the particular room or area to be air conditioned. The room in question may be a room in an office or a residential room, such as a bedroom, living room, etc.

Room air conditioners are usually grouped according to their design and method of installation as window units and console units. As the name implies, window units are installed on the window sill, whereas console units are placed on the floor in front of the window with a suitable air duct projecting through the window to provide the necessary air circulation. The advantage in the use of air conditioners of the console type is its relatively low cost in providing summer cooling in the room selected; it is also portable and easy to install, thus permitting it to be moved from one room to another as conditions and desires dictate. When properly installed, the room air conditioner will give a large measure of comfort, providing air that is free from the irritating exhaustion that comes with sweltering hot spells and high humidity; the installation of a typical air conditioner is shown in Fig. 1.

CAPACITY

Room air conditioning units should be large enough for the room or rooms to be cooled. The capacity of air conditioning units is calculated in Btu per hour, under certain specific conditions of Dry and Wet Bulb temperatures. The term "one ton of refrigeration" is equivalent to 12,000 Btu per hour. Therefore, a room air conditioner that is rated at 10,000 Btu, for example, will supply 10/12, or approximately 0.83, ton of refrigeration.

The human body generates considerable heat; this heat is dissipated by radiation, convection, and by the evaporation of perspiration on the skin surface. An increase in the air motion greatly increases the removal of heat by convection and increases the evaporation rate of perspiration.

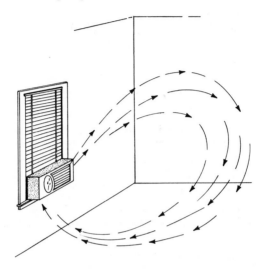

Fig. 1. The air circulation and cooling provided by a typical window-mounted air conditioner.

The important variables to keep in mind when estimating the Btu requirements for a room air-conditioning installation are:

1. Room size, in square feet of floor area,
2. Wall construction, whether light or heavy in weight,
3. Heat gain through the ceiling,
4. The percentage of outside wall area that is glass,
5. Whether the room is to be occupied during the day or at night only,
6. The amount of wall exposure to the sun in the room to be air conditioned.

Additional factors to be taken into consideration are: room ceiling height, number of persons using the room, and miscellaneous heat loads (such as wattage of lamps, number of radio and television sets in use in the room, etc.).

OPERATION

Although the various types of room air conditioners may vary in cabinet design as well as in the arrangement of components, they all operate on the same principles. In operation, the evaporator fan draws the recirculated air into the unit through louvers, which are usually located on the side of the unit. The air passes through the air filter and the evaporator and is discharged through the grille on the front of the unit into the room. The part of the unit that extends into the room is insulated to reduce the transfer of heat and noise.

The condenser-compressor compartment extends outside of the room and is separated from the evaporator compartment by an insulated partition. The condenser air is drawn through the condenser-coil sections on each side of the condenser-fan housing. The air then passes through the compressor compartment and is discharged through the center section of the condenser, which is covered by the fan housing. The fan circulates the air and also disposes of the condensed water (condensate) from the evaporator; this water drops into the base of the air conditioner and flows to the condenser end of the unit. The cooling principles of room air conditioners are the same as those discussed in the preceding chapter under the section on electric refrigerators.

INSTALLATION OF WINDOW UNITS

After the unit is removed from the crate, the mounting frame must be located on the side of the desired window. The window selected should be on the shady side of the house. If this is not possible, and the unit must be exposed to the sun, then some shading of the unit should be used for greater operating efficiency. Awnings are most effective for this purpose, since they shade both the unit and the window at the same time; the awning, however, must not restrict the free flow of air to and from the unit. The top of the awning must be held away from the building side, so that the hot discharge of air can escape. Venetian blinds or shades are a second choice in cutting down the great amount of heat transmitted by the sun to the room through the windows, but such devices are better than no shading at all.

Installation in Double-Hung Windows

Most window units are manufactured for installation in sliding windows, with free openings from 27 to 40 inches in width. The proper installation of these units is extremely important to their continued satisfactory performance. As with all other appliance installations, the manufacturer's instructions should be carefully studied and adhered to.

Installation of Window Angles—The purpose of window angles, Fig. 2, is to hold the side panels securely in the window channel. Place the window angle outside the panel. Adjust it for height; then outline and drill pilot holes. Screw the angle firmly in place. Repeat this operation with another angle on the opposite side of the window frame.

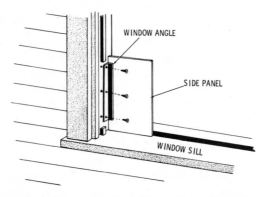

Fig. 2. The method of mounting a window angle to the window frame. The side panel may be cut to suit the particular window width encountered.

Measurement—Measure and mark the center of the window sill with a pencil line, as shown in Fig. 3. Extend this line to the outside sill for use in installing the mounting-frame support. Place the outer casing on the sill, and locate the line squarely through the center hole. Pull the casing toward you until the locating flange, which extends down from the front cross member, is tight against the sill edge. Hold the casing firmly in place. Outline the screw holes, and drill them with an undersized drill to avoid splitting the sill when the screws are inserted.

Mounting-Frame Support—Place the mounting-frame support, Fig. 4, on the outside sill with its inside angle toward you and with the center line squarely through the center hole. Outline and drill the holes; insert the screws, and secure them tightly. For stone or brick sills, expansion plugs and bolts will be necessary.

Adjusting Angles—These angles, as shown in Fig. 5, are to be bolted between the mounting-frame support and the mounting-frame tracks. The vertical portion of the adjusting angles should

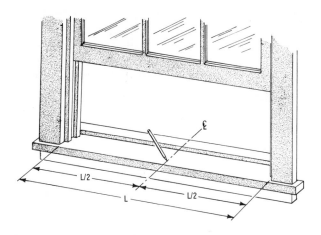

Fig. 3. The method of measuring the window sill to establish the mounting position of the air conditioner.

be located inside the mounting-frame support for safety; if the vertical portion is too long, cut it off. The horizontal portion may be turned in either direction. Be sure to use the nuts, washers, and lock washers that are furnished with the unit. Tighten up on the bolts through the mounting-frame support. Tighten up securely on the bolts through the mounting-frame tracks. Be certain that the heads of the bolts through the tracks are up and have washers under them.

Window Sill Felt Seal—The installed position of this seal is under the mounting-frame front member, as illustrated in Fig. 6. The seal is cut to fit the recess that is formed where the tracks join the front member. Stretch the felt to fill this recess completely, and push it in as far as it will go. Place the center screw through the felt; then insert the remaining screws, and tighten them all up securely.

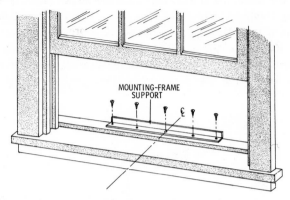

Fig. 4. The method of fastening the mounting-frame support to the window sill.

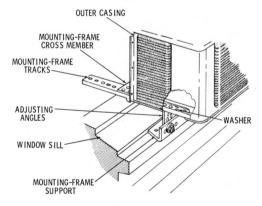

Fig. 5. Installation of adjusting angles to frame support and mounting.

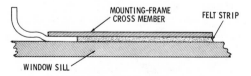

Fig. 6. The method of fitting the felt strip underneath the mounting-frame cross members.

Leveling—Use the carpenter's level, as illustrated in Fig. 7, to make sure that the mounting frame is level, front to rear and side to side. Adjustment can be made by tapping or prying the back edge of the mounting frame; the adjusting angles will support the mounting frame until the final tightening. A level-mounted frame assures proper condensate disposal and also prevents dripping.

Side Panels—Measure the side panels, shown in Fig. 8, to fit the space remaining between the outer casing and the window channel. To cut the side panels, scribe a deep line, and place the panel so that the scribed line rests on the edge of a flat surface. Hold

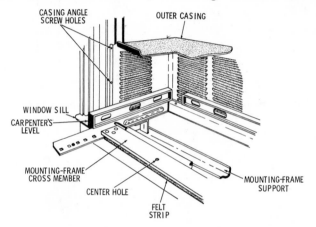

Fig. 7. The supporting frame is leveled by means of a carpenter's level.

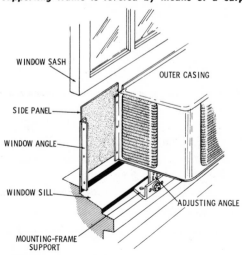

Fig. 8. The method of fitting the side panel to the outer casing.

529

the panel securely with the palm of the hand, and strike the over-lapping section of the panel a sharp blow with the heel of the other hand. Put the panels in place; make sure that the inside surface of the panel (the smooth surface) is on the room side. Put the casing angles in place, align the holes and slots, insert the screws, and tighten them securely.

Top Molding—Cut the top molding to the proper length, and install it over the top flange of the casing and the top of the side panels, as shown in Fig. 9A. The rounded section of the top molding only fits on the top flange of the casing; the square portion only fits on the side panels. Lower the window, and press it down tightly behind the molding until the bottom edge of the sash rests on the ledge of the molding.

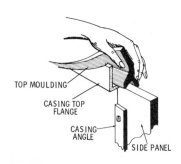

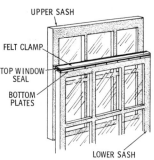

A. Fitting the top moulding to the top flange.

B. Installing the top sash seal.

Fig. 9. Moulding and seal installations.

Top Sash Seal—Remove the window lock, and seal the opening between the upper and lower sashes with the assembled felt clamp, top window seal, and bottom plates furnished for this purpose, as illustrated in Fig. 9B. Cut or notch the sashes as required, and fasten in position.

Sealing Compound—Sealing compound is usually furnished to fill the small openings that exist where the filler panel meets the outer casing at the window sill.

Chassis—The chassis is now ready to be installed in the outer casing. Check the fans by hand spinning them, and plug the unit into an electrical outlet for a trial test before installing the unit in the outer casing. Inspect the bottom pan for foreign matter, and remove all such matter, if present. Slide the chassis into the casing, and hold the front of the unit up slightly to allow sufficient clearance for the felt sealing strip, which is attached to the bottom pan. Push the chassis until the bulkhead flange is tight against the casing flanges.

Window Lock—Since the window lock can no longer be used when the air conditioner is installed, it is necessary to provide other means to prevent the window from being opened from the outside of the building. For this purpose, a suitable lock should be obtained from a local hardware store and should be securely installed.

Installation in Casement Windows

Casement windows are available in so many different types and varieties that it would be impossible to set up a single procedure to cover them all. Each job requires special treatment, thought, and planning. Installation is accomplished by removing a sufficient amount of glass and mullions to allow for passage of the unit, as illustrated in Figs. 10 to 12.

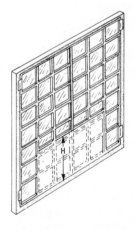

Fig. 10. The necessary changes required to permit the installation of a window-mounted air conditioner in a casement-type window.

It is necessary to build up the room-side window sill until the top is above the horizontal cross member that forms the bottom of the frame on the metal window. If the outer cabinet is allowed to rest on this cross member, any vibration will be transmitted to the window frame and wall and will be greatly amplified. This procedure is basic to all casement windows.

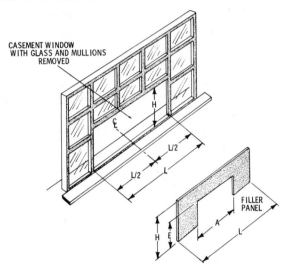

Fig. 11. Filler panel installation in a casement window. These panels are usually furnished in standard dimensions but may also be cut to suit any particular requirement. Dimensions L and H are the exact dimensions of the opening measured on the outside of the window; dimensions E and A are the height and width of the cabinet at the point of insertion.

Measure the height and width of the opening left by the removed glass, and cut a piece of ¼-inch Masonite or equivalent material to fit this opening. This board is referred to as a *filler panel*. Cut out the center of the filler panel to the exact outside dimensions of the outer cabinet. When cutting the height dimension in the board,

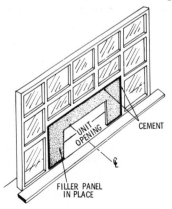

Fig. 12. The method of mounting the filler panel in a casement-type window.

allow for the height of the bottom cross member on the window frame. Install the outer cabinet as described previously under the section on double-hung windows. Install the filler panel in the opening, and seal the edges to the window frame with putty or caulking compound. The horizontal cross member of the supporting frame may then be screwed to the temporary sill by using the holes provided for this purpose; the clamping assembly may also be used if it is practical for the installation. Finally, install the chassis and the inside cabinet.

Alternate Method—Installation in Casement Windows

Measure the height of the opening, and from it subtract the height of the unit. Cut a piece of plywood, Masonite, or plexiglass equal in height to fit the measurement obtained as a result of the above subtraction, with a width equal to the opening. Install the outer cabinet in the exact center of the opening, and secure it rigidly. Install the filler board across the top of the unit, and cement it in place. Cut the regular side panels supplied with the unit to fill in at the sides. Cement these in place at the window frame.

Other casement windows, where the glass panes are small, will require the removal of one or more horizontal cross members as well as several vertical mullions. In this case, it may be advisable to reinstall the cross member at a height equal to the height of the outer cabinet. Cut and reinstall the glass above the cross members. The regular side panels supplied with the unit may be used to fill in the sides, or plexiglass may be used for this purpose. French windows will require that the lower portion of the center upright on which the doors lock be cut away, and a sufficient number of mullions and cross members, with the appropriate amount of glass, will also have to be removed.

INSTALLATION OF CONSOLE UNITS

A careful survey of the room should be made prior to the actual installation of the unit. Determine the most favorable location by taking into account the desirable location in the room, the exposure of the window, the width of the window, the height of the window from the floor, and the location of the electrical supply outlet. Since these units, as a rule, have a considerably larger cooling capacity than the window-sill units, it is usually necessary to install a special electrical connection from the meter or distribution panel directly to the location of the unit.

The installation of a typical console room air conditioner in a double-hung window is illustrated in Figs. 13 and 14. A normal installation of this type allows the window to be opened or closed without interference from the duct or window filler panels. To completely close the window, the rain hood, which protects the air duct, must be retracted. In order to adjust the height of the unit to obtain the necessary height for the duct outlet (window sill height may differ by several inches), special wooden bases that are normally made up of several laminations, as illustrated in Fig. 15, are usually employed.

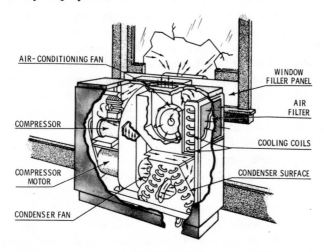

Fig. 13. Arrangement of components in a typical console-type air conditioner. The refrigerant circuit is of the conventional dry-coil, direct-expansion type and consists of a condensing unit, a liquid-to-suction-line heat exchanger, a thermostatic expansion valve, and a finned-type evaporator. Room and outdoor air passes into the unit through the inlet grille; this air is cooled as it is drawn by the fan through the evaporator, and it cools the room as it is blown from the plenum chamber and discharge grille at the top of the unit.

The standard duct usually furnished with the unit is approximately 8 inches deep, as shown in Fig. 16. The unit end has a removable flange, which slides in vertical tracks that are attached to the back of the unit. The window end of the duct has a rain hood attached to it; when the window is up, the rain hood is pushed out manually and secured in the open position by inserting a screw in each side of the duct, after the holes in the duct and rain hood are aligned.

The distance between the window and the nearest permissible location of the unit (dimension C in Fig. 17) is measured from the window end and is laid out on the duct. Remove the screws holding the removable fitting to the duct. Scribe and cut the duct; file off burrs, and smooth out the sharp edges. If the depth dimension is 8 inches or less, the standard duct is used. If this dimension is

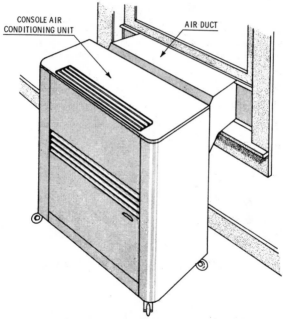

CONSOLE AIR
CONDITIONING UNIT

AIR DUCT

Fig. 14. A console-type air-conditioner installation in a standard double-hung window.

greater than 8 inches, the standard and accessory ducts can be fitted together to give a total distance of 20 inches. When cutting the accessory duct, be sure to take the measurement from the flared end to allow the extension piece to fit over the standard section. Use the unit fitting to locate new holes in the cut end. Drill the holes with a ⅛-inch drill, and reassemble the complete duct unit.

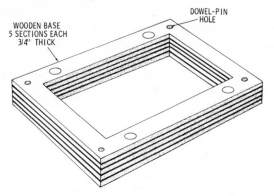

Fig. 15. A wooden base may be used to increase the height of the console unit to permit the air duct to rest on the window sill.

Installation of a console-type air conditioner in casement windows does not differ to any appreciable degree from that previously outlined for the installation of the window-sill type. An alternate method for the installation of console-type air conditioners in French windows is to install shorter windows that leave enough space below to accommodate the height of the unit.

THE ELECTRICAL SYSTEM

For the air-conditioning unit to operate properly, it is necessary that the power supply be the same as that given on the nameplate of the unit. Prior to the actual operation of the unit, determine if any other electrical appliances are connected to the selected circuit by removing the circuit fuse. Do not use an outlet that is used for any other appliance. Check the voltage of the outlet while the unit is connected and operating. The voltage drop must not exceed 10% of the voltage specified on the nameplate of the unit.

The ⅓- and ½-horsepower units, under ordinary conditions, may be connected to almost any type of standard appliance outlet. If an outlet is not available near the unit, it will be necessary to install one. Larger units, such as one that uses a ¾-horsepower motor, should be installed in a 20-ampere branch circuit that uses a 20-ampere time-delay fuse. The external supply wiring to any air-conditioning unit must comply with the requirements of the

National Electrical Code, in addition to any existing local codes where such local codes are in effect.

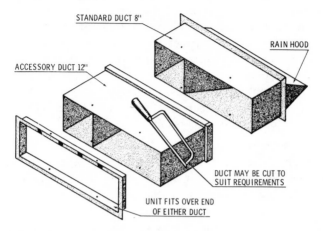

Fig. 16. **Typical air ducts that are employed in the installation of console-type air conditioners.**

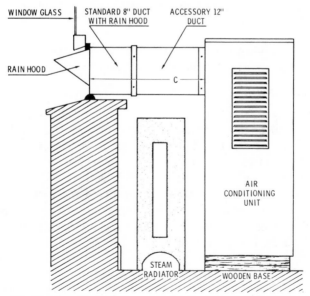

Fig. 17. **The installation of a console-type air conditioner when an obstruction necessitates the use of an extra duct unit.**

537

Description of Controls

Depending on the design of the unit, the location of the controls may be either on top or on one of the sides. These controls usually consist of an electrical control switch ("on-off" switch for the motor) and the various damper controls. The electrical control switch is usually of the knob or rotary-control dial type, with the standard four positions marked "off," "fan," "cool," and "exhaust." The damper controls are usually marked "shut," "vent," and "open." The various combinations of the control switch and damper controls provide the following results:

1. To provide cooling, the control switch is turned to the "cool" position, and the damper dial is set on "shut" or "vent," depending on whether or not outside air is desired.
2. To operate as a ventilator, the control switch is turned to the "fan" position, and the damper dial is opened as far as desired; in the "open" position, the unit only brings in and circulates the outside air.
3. To exhaust the room air, the control switch is turned to the "exhaust" position, and the damper dial is turned to the "vent" position.

With a thermostat installed for automatic cooling, the compressor and fans will cycle according to load requirements.

Thermostat Control

If automatic cooling control is desired, an approved thermostat may be installed in the electrical circuit of the compressor. The thermostat must be capable of handling the motor current, since it is normally connected in series with the compressor switch. The thermostat control can be mounted on the side of the air-conditioner cabinet at the recirculated-air intake or on the wall of the room being cooled. However, if a wall installation is made, the thermostat must not be placed in or near the direct path of the air being discharged from the unit.

SERVICING AND REPAIRS

Since most portable air-conditioning units of recent manufacture contain compressors of the hermetically sealed type, the only parts that can be serviced in the field are the relay, control switch, fans, fan motor, starting and running capacitors, air filters, and cabinet parts. The refrigeration system, which consists of the cooling units—condenser, compressor, and connecting lines—as a rule, cannot be serviced in the field.

Dismantling of the Air Conditioner

The following procedure lists the steps that are necessary to dismantle the air conditioner. This procedure can be used in its entirety or in part, according to the service required; it covers only those parts of the unit that can be serviced in the field and is not a complete teardown. By presenting a dismantling procedure, it is possible to eliminate repetition of certain steps common to many service operations. When reassembling the unit, these steps should be reversed. Be sure to place all components, except those found to be defective, in the position in which they were originally.

1. Disconnect the air conditioner from the source of electrical supply. For minor repairs, pull the service cord plug out of the supply socket; for dismantling, pull the service cord, and remove the plug from the end of the cord.
2. Remove the cabinet from the unit by first removing the screws from the base of the unit at the front, sides, and rear or outside end of the cabinet.
3. Remove the electrical control boxes by removing the screws that secure each box to the partition. Remove the control-box covers, disconnect the motor leads, and remove the control-box assemblies from the unit.
4. Loosen both fans and remove them from the shaft. Remove the sliding access panel, which fits down over the cooling unit fan shaft. Loosen and remove the two motor-cradle supports at each end of the motor, and lift the motor up and out.

Electrical Tests

In case of operating trouble, a thorough check of the electrical system is often necessary. By checking the electrical system, a great deal of time may be saved, since experience has proved that such a check can reveal the more obvious troubles. To make a complete electrical test of the unit and its controls, a test-lamp circuit, as shown in Fig. 18, can be used for both voltage and continuity checks. When checking the electrical system, refer to the wiring diagrams provided by the manufacturer. Two such wiring diagrams are shown in Figs. 19 and 20 and should be employed for the following tests. These wiring diagrams are only typical; the arrangement and wiring of components vary with the different makes of air-conditioning units.

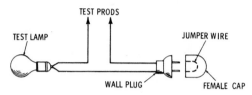

Fig. 18. The test lamp, as shown, will greatly facilitate the testing of the internal wiring in an air-conditioning unit.

Testing for Current Supply—With the electrical cord plugged into the electrical outlet, place one prod of the test-lamp circuit on terminal N and the other on terminal L. If the test lamp lights, current is being supplied to the relay.

Testing Fan Motors—Depending on the size of the unit, the air conditioner may be equipped with one or two fan motors. In some smaller units, only one fan motor is used, in which case the shaft of the motor operates two fans, one for the evaporator and one for the condenser.

To check the evaporator-fan motor, turn the switch to the "vent" position. If the evaporator-fan motor does not operate, remove the switch junction box cover and switch. Remove the connector from the four wires tied together in the switch box. Place a test lamp between terminal 1 of the switch and the wire ends from which the connector was removed. If the lamp lights, the trouble is in the motor. These motors are equipped with internal overload protection. Permit the motor to cool for several minutes (with the switch off), then recheck before replacing the motor. If the lamp did not light, a defective switch is indicated (assuming the correct voltage is available at the switch).

To check the condenser-fan motor, turn the switch to the "exhaust" position. If the condenser-fan motor does not operate,

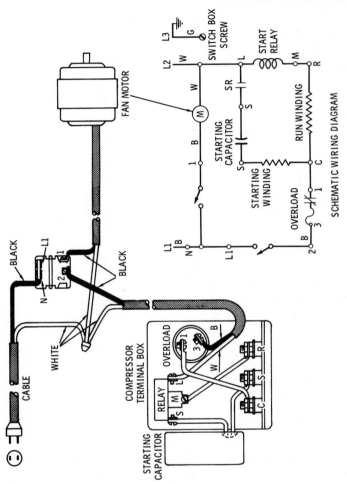

Fig. 19. The wiring diagram of a typical air conditioner with one fan motor and a hermetically sealed compressor unit.

remove the switch junction box cover and switch. Remove the connector from the four wires, and place a test lamp from these wires to terminal 3 of the switch. If the test lamp lights, the trouble is in the motor. Wait several minutes to allow the motor to cool (with the switch off), and recheck before replacing the motor. If the lamp did not light a defective switch is indicated (assuming there is a correct voltage available at the switch).

To check the evaporator- and condenser-fan motor where only one fan motor is used, turn the switch to the "vent" position. If the fan motor does not operate, remove the switch junction box cover and switch. Remove the connector from the four wires. Connect a test lamp from terminal 1 of the switch to the wire ends from which the connector was removed. If the lamp lights, the motor is faulty. These motors are also equipped with internal overload protection; therefore, permit the motor to cool for several minutes (with the switch off), then recheck before replacing the motor. If the lamp did not light, again a defective switch is indicated (assuming there is a proper voltage available at the electrical control switch).

Testing the Starting Capacitor—An electrolytic starting capacitor is used on most units in the starting-winding circuit to effect an increase in starting torque. If this capacitor becomes shorted internally, starting trouble and possibly blown fuses may result; if the capacitor develops an open circuit, the compressor will not start. The simplest way to check a capacitor is to install a new one. If the trouble is corrected, discard the old capacitor. If a replacement is not readily available, a check can be performed as follows:

Disconnect the capacitor leads. Make certain that the capacitor is not retaining a charge by placing the blade of an insulated screwdriver across the terminals. Touch the capacitor leads or terminals momentarily with the test prods of an ohmmeter. A satisfactory capacitor will cause a slight and instantaneous deflection of the ohmmeter pointer. A shorted capacitor will cause the ohmmeter pointed to indicate a continuous low resistance. The pointer will move to the zero end of the scale and will remain there as long as the prods are in contact with the terminals. The capacitor, therefore, must be replaced. If the capacitor is open, there will not be any movement of the pointer, and the capcitor must be replaced.

An additional check to further assure that the capacitor is satisfactory can be made. After the ohmmeter prods have been touched to the capacitor terminals, and a slight momentary deflection of the pointer has resulted, reverse the prods. A momentary deflection

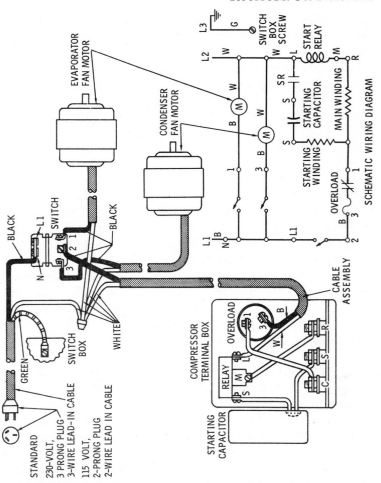

Fig. 20. The wiring diagram of a typical air conditioner with two fan motors and a hermetically sealed compressor unit.

of the pointer, approximately two times that of the first check, should result if the capacitor is satisfactory.

Testing the Compressor-Motor Relay—The relay opens the circuit to the compressor-motor starting winding when the compressor is started. The duration of this start period, or of the time that the starting winding is energized, is, under normal conditions, quite short—usually less than 5 seconds. A defective relay may fail to close, which would result in starting trouble, or it may fail to open, which could result in overload trip-outs, capacitor failure, or blown fuses.

Check the capacitor before making the following relay-operation checks, since a shorted capacitor may upset the accuracy of these checks. Remove the compressor junction box cover. With the switch in the "off" position, connect a test lamp between terminal S of the relay and motor terminal S. Turn the switch to the "cool" position; the lamp should light for only an instant if the compressor starts. If the compressor does not start, the lamp should remain lit until the switch is turned off or until the overload trips. Do not permit the switch to remain in the "on" position if the compressor does not start. If the lamp did not light, check for an opened overload. If the overload is not opened, replace the relay.

To replace the relay, remove the screws securing it to the compressor shell bracket, and install a new relay unit.

Testing the Compressor-Motor Overload—The overload is a protective device used in conjunction with the relay to open the compressor-motor circuit at abnormally high currents or dangerously high motor temperatures. The overload consists of a heater and a snap-action bimetal disc on which contacts are mounted. The heater is connected in series with the common motor winding terminal. In the case of overloads, failure of the compressor to start, or unusual voltage conditions, the current through the heater is high, thereby increasing its temperature. This heats the bimetal disc, thus causing it to automatically snap open and open the motor winding circuit. Because of its direct contact to the compressor body, as the motor temperature rises, the heat of the compressor body increases the overload-disc temperature and lessens the amount of current required to open the circuit. If the overload trips for any reason, the circuit will remain open until the compressor body, or shell, cools sufficiently to cause the disc to snap closed and again close the circuit to the motor. This protector is not adjustable, and it must be replaced if it fails to function properly.

To check the overload, remove the compressor junction box cover. Place a test lamp from terminal L of the relay to terminal 3 of the overload. Turn the switch to the "cool" position. The lamp should light if the proper voltage is available to the overload. Now, move the test lamp from terminal 3 of the overload to terminal 1. If the lamp lights from L to 3, but does not light from L to 1, the overload is opened. If the overload does not close within 10 to 15 minutes (time for the motor to cool), the overload should be replaced. If the overload does close within this time, check for the cause of overloading or overheating—low voltage, shorted capacitor, failure of the compressor to start, and/or excessive operating pressures or temperatures.

Testing the Compressor Motor—The compressor motor normally used on most units is a split-phase motor that employs a capacitor on start to increase the starting torque. Since these units use the capillary-tube method of refrigeration, the starting load is normally quite low. After a certain predetermined speed has been obtained on start, the starting winding and capacitor are shorted out of the circuit by the relay, and the full load is then carried by the running winding.

To check the compressor motor, a continuity test for an open circuit within the motor can be performed by disconnecting the wires from the three motor leads (C, S, and R). Plug the test set, as shown in Fig. 18, into a wall outlet. Place the test prods on terminal C and on terminal S. If the lamp lights, continuity through the starting winding is satisfactory. Next, place the test-lamp prods on terminals R and C for a simliar test of the running winding. If the lamp does not light on either test, the motor must be replaced, since one or both windings are open.

A test for a burned out or grounded motor can be performed by checking the resistance of the motor windings to the compressor shell. This check can best be made with a megger instrument or an ohmmeter that is capable of indicating sufficiently high resistance values; however, a rough check can be performed with the test-lamp set as follows:

With all wires removed from the motor terminals, place one prod on terminal C and touch the other prod to the compressor shell. If the lamp lights, the motor is grounded and must be replaced.

It should be remembered that faults other than motor trouble may cause the failure of a compressor to run or cause a motor to draw high current. A stuck compressor, high head pressure, low voltage, and a plugged capillary are some of the causes of compressor starting failure or unsatisfactory operation.

Oiling of Motors—The fan motors should be oiled at the start of each cooling season, or every six months if the unit is operated throughout the year. Use a good grade of electric motor oil or SAE No. 20 automobile oil; a few drops in each oil hole will usually be sufficient.

Leaks in the System

In the event that any welded part of the condensing unit develops a leak, it will be necessary to replace the condensing unit itself. A leak will usually be indicated by the presence of oil around the point at which the leak developed. It must not be assumed, however, that the presence of oil on any part of the unit is a positive indication of a leak. Always check the suspected leak with an approved leak detector.

Filters

These should be inspected at regular intervals. They are easily removed by simply lifting them up through the slots provided in the control panel. When necessary, clean them with the proper vacuum-cleaner attachment. Periodic cleaning of filters will assure maximum air delivery by the air conditioner at all times. Best results can be obtained if the filters are replaced every year and cleaned between replacements as often as is necessary.

Interior Cleaning

The interior of the unit should be cleaned periodically of all dust, grease, and foreign matter. Special attention should be given to the condenser coils and the evaporator coils. Regular cleaning will assure continuous good service from the unit.

Winter Care

In many parts of the country, the cooling unit will not need to be used during the winter months. These units may be readily removed from their location in the window and stored in a convenient place; this will prevent the build-up of moisture condensation in the unit and will also make available a greater window area. Before the unit is stored away, however, the evaporator and the condenser should be checked for dirt and should be cleaned where necessary. Cleaning the evaporator once a year is recommended to avoid the development of objectionable odors. The coils can be cleaned by the use of a stiff brush and a strong solution of soap and water or by the careful use of a garden house. All parts should be rinsed off after cleaning. It is also advisable to wash out the drain pan and retouch it with an asphalt-base paint.

When storing the air conditioner, the unit should be blocked up to take the weight off the sponge-rubber mounting. At the beginning of the cooling season when the unit is reinstalled, the fan motors should be oiled, as explained under a preceding section. After the unit has been reinstalled and started, it should be checked to make certain that the temperature drop across the evaporator is 10°F. or more. Allow the unit to operate for several minutes before making this test.

CHAPTER 26

Electric Dehumidifiers

By definition, the function of an electric dehumdifier, as shown in Fig. 1, is to remove moisture from the air and thus protect the home from excessive humidity. High humidity is particularly annoying during months of excessive heat and results in the mildewing of stored valuables, the rusting of tools and metal objects, and the swelling of floors, panel walls, drawers, doors, etc.

OPERATION

An electric dehumidifier operates on the refrigeration principle, as discussed previously. It removes moisture from the air by passing the air over a cooling coil; the moisture in the air condenses to form water, which then runs off the coil into a collecting tray or bucket. The amount of water removed from the air varies, depending on the relative humidity and volume of the area to be dehumidified. In locations with high temperature and humidity conditions, 3 to 4 gallons of water per day can usually be extracted from the air in an average size home. When the dehumidifier is first put into operation, it will remove relatively large amounts of moisture until the relative humidity in the area to be dried is reduced to the value where moisture damage will not occur. After this point has been reached, the amount of moisture removed from the air will be considerably less than that removed when the dehumidifier is first placed in operation. This reduction in the amount of moisture removal indicates that the dehumidifier is operating normally and that it has reduced the relative humidity in the room or area to a safe value. The performance of the dehumidifier should be judged by the elimination of dampness and accompanying odors rather than by the amount of moisture that

Courtesy General Electric Company

Fig. 1. A typical automatic dehumidifier.

is removed and deposited in the bucket. A dehumidifier cannot act as an air conditioner to cool the room or area to be dehumidified. In operation, the air that is dried when passed over the coil absorbs heat from the condenser; this heat is then added to the heat of compression, which raises the temperature of the surrounding air, which further reduces the relative humidity of the air.

INSTALLATION

The dehumidifying unit must be operated in an enclosed area in order to be effective. It is most effective when the home is closed, although this condition is generally unattainable when there are a number of people using the living quarters going in and out of doors regularly. The dehumidifier is also quite effective in the basement. Here, too, the windows and doors must be kept closed; the most effective dehumidification occurs when the air from the basement is not drawn by the furnace blower into other parts of the home. A dehumidifier that is operated in the basement will have little or no effect in drying a storage area unless there is adequate air circulation in and out of the enclosed area. It may be found necessary to install a second dehumidifier inside the enclosed storage area for satisfactory drying action.

For best results, the dehumidifier should be located near the center of the area to be dehumidified. However, when it is desirable to locate the dehumidifier elsewhere so as to utilize a hose connection from the collecting tray to a drain, the dehumidifier may be located some distance from the center of the area. It should be remembered that good air circulation is essential for any part of an enclosed area requiring dehumidification.

CONTROLS

The dehumidifier, as previously mentioned, operates on the principles of the conventional household refrigerator, and as such it contains a motor-operated compressor, a condenser, and a receiver. In a dehumidifier, the cooling coil takes the place of the evaporator, or chilling unit, in a refrigerator. The refrigerant (usually Freon) is circulated through the dehumidifier in the same manner as in a refrigerator. The refrigerant flow is controlled by a capillary-tube circuit. The moisture-laden air is drawn over the refrigerated coil by means of a motor-operated fan or blower.

The dehumidifier operates by means of a humidistat, which starts and stops the unit to maintain a selected humidity level. In a typical dehumidifier, the control settings range from "dry" to "extra dry" to "continuous" to "off." For best operation, the humidistat control knob is normally set at "extra dry" for initial operation over a period of three to four weeks. After this period of time, careful consideration should be given to the dampness in the area being dried. If sweating on cold surfaces has discontinued and the damp odors are gone, the humidistat control should be reset to "dry." At this setting, more economical operation is obtained, but the relative humidity probably will be higher than at the "extra dry" setting. After three or four weeks of operation at the "dry" setting, if the moisture condition in the area being dried is still satisfactory, the operation of the dehumidifier should be continued with the control set at "dry." However, if at this setting, the dampness condition is not completely corrected, the control should be returned to the "extra dry" setting. Minor adjustments will usually be required from time to time, but it should be remembered that the control must be set near enough to "extra dry" to correct the dampness condition and also as close to "dry" as possible to obtain the most economical operation.

MAINTENANCE AND SERVICE

A dehumidifier ordinarily will not operate satisfactorily below 65°F. At this temperature it becomes necessary to operate the cooling coil below freezing temperatures in order to reduce the relative humidity to a reasonable level. Although the dehumidifier can be operated with an ice formation on the cooling coil, it will be necessary to defrost the coil at least once an hour to obtain satisfactory performance; this defrosting operation requires the use of an auxiliary timer to provide the defrosting cycle. Generally, the operation of a dehumidifier at an ambient temperature below 65°F. is not recommended.

In some cases, a collection of fungus will become attached to the bottom of the dehumidifying coil after an extended period of operation. The fungus is an air-borne spore that collects on the cooling coil. Since it is air-borne and is peculiar to the particular location involved, there is nothing that can be done about it; this condition is worse in some areas than in others. Where the fungus is present, it tends to collect the air-borne dust and aggravate the situation. The best way to clean this material from the cooling coil is by the use of a soft brush and an adequate amount of clean water. As the material is loosened from the coil with the brush, it should be flushed away with clean water. It may be necessary to use a pipe cleaner or other such tool to clean out the drain of the collecting tray.

Since the principle operating parts of the dehumidifier (the motor-compressor unit) are hermetically sealed and permanently lubricated, there is little that can be done in the way of service, aside from the normal cleaning of the unit and cooling coil. However, due to the similarity that exists in the operating principles of the dehumidifier and the compressor-type refrigerator, the same methods may be used to make any mechanical corrections on the dehumidifier as were described for the refrigerator. The motor is normally provided with thermal-overload protection and operates on the conventional 115-volt, 60-cycle alternating current.

CHAPTER 27

Humidifiers and Electronic Air Cleaners

During the heating season, cold air infiltrates your home through cracks, around doors and windows causing air changes as often as twice an hour. Since cold air contains very little moisture, this infiltrated air becomes thirsty and dry when it's heated to room temperature. It soaks up moisture from your home, your furnishings, and from you, just like a sponge.

Additional moisture to your home assists in preserving that bright, shiny lustrous finish on beautiful furniture. Moisture is even good for your furniture. It keeps the joints firm and tight and prevents drying. Moisture also aids in eliminating squeaky floors. Proper moisture in the air will also eliminate static electricity plus keeping your house more comfortable in the winter at a lower temperature setting—thereby cutting fuel cost. Two models of humidifiers are shown in Figs. 1 and 2.

The humidifier is installed on the furnace plenum chamber, as shown in Fig. 3, where hot air passes through it picking up moisture from a wet evaporative pad. This moisture laden air passes through a duct and is injected into the returning air stream.

The water flows through the humidifier at right angles to the air flow. The supply water, under pressure, flows through the solenoid valve and jet pump into a distributor where it spreads out and drips onto an evaporative pad. The air passing through the pad becomes saturated with moisture. Water, not evaporated, collects in a sump where part of it is siphoned to the distributor by the jet pump. The mineral laden water remaining overflows into a drain.

When the weatherman tells you that the relative humidity is 68%, he is talking about outdoor humidity, not indoor. Air at

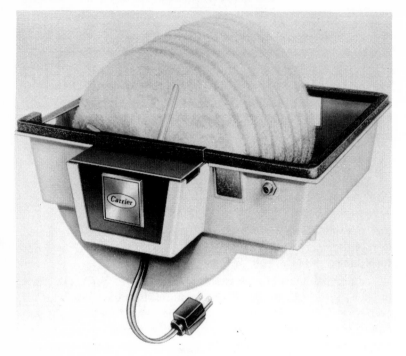

Fig. 1. A rotary-type humidifier. This unit adds 15½ gallons of moisture per day to dry indoor air.

75°F. holds 5 times the moisture of air at 32°F. Table 1 gives you a comparison between outdoor and indoor relative humidity with the recommended indoor relative humidity.

ELECTRONIC AIR FILTERS

Billions of invisible particles of smoke, grease, ash, fibers, pollen, industrial wastes and exhaust hydrocarbons float in the air of your home. These tiny particles pass through the ordinary air filters and vacuum cleaners to fade bright colors and turn white ceilings off color. These particles also spread a dulling film over windows and mirrors, shorten the life of paint and wallpaper and burrow into rugs to grind and speed wear. These tiny particles will

Fig. 2. An evaporative-type humidifier. This unit will add up to 24 gallons of moisture a day to the dry winter air in homes and small commercial buildings.

increase dry-cleaning and upholstery bills and cause countless hours to be spent cleaning and washing.

The electronic air cleaners shown in Figs. 4 and 5 may be installed in the ductwork of any forced-air heating or cooling system. The cleaner in Fig. 5 can be installed either vertically or horizontally wherever the control return air grille is found.

The three-step purification process of the electronic air cleaner is simple. First, air entering the air cleaner passes through a protective filter (1) which screens large airborne particles from the air. Second, an electrically-charged grid (2) imparts an electrical

557

Recommended Indoor Relative Humidity

OUTDOOR TEMP.	OUTDOOR REL. HUMIDITY	INDOOR REL. HUMIDITY* WITHOUT HUMIDIFIER	MAXIMUM RECOMMENDED INDOOR REL. HUMIDITY**
—10°F	30%	1%	20%
	50	1	
	70	2	
0°F	30	2	25
	50	3	
	70	4	
10°F	30	3	30
	50	4	
	70	6	
20°F	30	4	35
	50	7	
	70	10	
30°F	30	6	35
	50	10	
	70	15	

*Indoor relative humidity when outdoor air is heated to 72°.
**As stipulated by the National Environmental Systems Contractors Association.

Courtesy Carrier Corporation

Fig. 3. Installation of the rotary-type humidifier unit in the hot-air line of the heating unit.

Fig. 4. The media-type electronic air cleaner designed for homes and small commercial building. It is available in capacities of 1200, 1600 and 2000 cfm.

charge to the smaller airborne dust particles in the air, and third, the disposable filter pads act as a magnet and attract the charged particles which have passed through the grid. Airborne dust and foreign particles are trapped in the pads (4). The air cleaner design ensures a continual filtering operation even if electronic control is interrupted.

559

Courtesy Carrier Corporation

Fig. 5. Electronic air cleaner for homes and small commercial establishments, available in capacities of 800, 1200 and 1600 cfm.

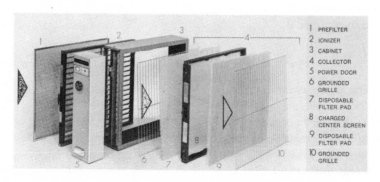

1 PREFILTER
2 IONIZER
3 CABINET
4 COLLECTOR
5 POWER DOOR
6 GROUNDED GRILLE
7 DISPOSABLE FILTER PAD
8 CHARGED CENTER SCREEN
9 DISPOSABLE FILTER PAD
10 GROUNDED GRILLE

Courtesy Carrier Corporation

Fig. 6. Component parts of the electronic air cleaner.

CHAPTER 28

Cellar Drainage Pumps

The purpose of a cellar, or basement, drainage pump is to automatically pump out seepage water as fast as it accumulates. Because it is necessary to provide a reservoir, or sump, for water storage, a drain pump is commonly referred to as a sump pump. Sump pumps are employed in many localities, particularly in areas located outside the municipal sewage system where water accumulates by seepage overflow, flood, or backing up. This water that cannot be drained off directly, either because the flow line of the sewer is not deep enough or because obstacles intervene between the sump and the sewer. While ordinarily this seepage may amount to only a small trickle, its flow is usually continuous, and if neglected, the accumulated water soon becomes a real menace to health and property.

To keep such locations dry, the incoming water is most economically disposed of as it accumulates by an automatic, electrically operated sump pump. The essential function of a sump drainage pump, therefore, is to begin pumping automatically when the accumulated water in the sump pit reaches a certain upper level, and to continue the pumping action until the pit is almost empty.

OPERATION

The pump consists essentially of a vertically mounted electric motor and shaft extension to which an impeller and housing are fitted. When the motor and impeller rotate, water is thrown outward by centrifugal force and is forced through the pump discharge outlet to the drain.

Automatic control of the pumping action is usually provided by means of a float and rod assembly, as shown in Fig. 1. The rod assembly is attached to an electric switch mechanism at its top.

561

The position of the float controls the operation of the motor and pump. Thus, the motor circuit will be completed (the motor-switch mechanism will close its contacts) when the water in the sump has accumulated sufficiently to lift the float to a certain predetermined maximum level, which depends on the setting of the upper float stop, as shown in Fig. 2. As the pump begins to discharge, the water level recedes, and the float moves downward. When the float reaches the lower float stop, the float switch will again open the motor circuit, and the motor and pump will stop.

The duration of each pumping cycle depends on such factors as pump capacity in gallons per minute (gpm), the volume of water seepage, the capacity and area of the sump pit, and the spacing of the float stops. For domestic purposes, with a water in-flow of

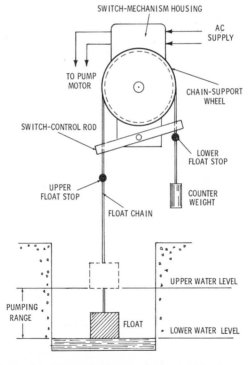

Fig. 1. The diagrammatic representation of a typical drainage-pump control system. The position of the float actuates the motor by means of its mechanical connection to the motor-switch mechanism.

from 10 to 25 gpm and with a head of up to 20 feet, a ¼-horse-power-motor pump is usually sufficient.

One unique method of level control has recently been introduced that employs neither floats nor stuffing boxes. Two electrodes are the only elements that enter the tank to make contact with the fluid contents whose level it is desired to control. In this system, shown in Fig. 3, the fluid makes or breaks contact with the electrodes and passes a minute electrical current on to the control; a power circuit amplifies this current to operate a switch, which controls signals, valves, or pumps. Accuracy is independent of temperature and pressure. When used in connection with a pump, two electrodes are suspended into the tank from a standard electrode fitting, which has been attached to the amplifier. The electrode rods project into the tank to the level that corresponds to the low point at which pumping is to stop. The electrodes are wired

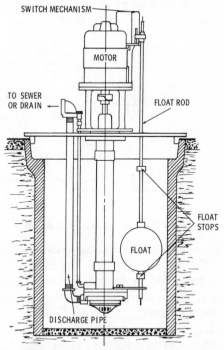

Fig. 2. *A typical drainage-pump installation, with float control. A standard sewer pipe can be used to provide a suitable drainage sump.*

to the level control. When the liquid level in the tank falls below the lower electrode, the level control closes the electrical circuit, thereby starting the pump motor and filling the tank. When the liquid rises to the level of the upper electrode, the fluid itself acts as a conductor of the minute current required for the operation of the level control. The level control then opens the electrical circuit controlling the pump motor, and the pumping operation stops.

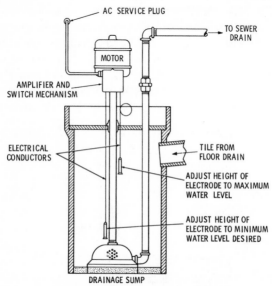

Fig. 3. A drainage-pump installation where water-level control is furnished by means of two electrodes suspended in the sump chamber.

SERVICING AND REPAIRS

When properly installed and sized, cellar drainage pumps will provide years of trouble-free service. Since it is often necessary to operate the motor pump in damp locations, corrosion should be guarded against; whenever possible, motors of the hermetically sealed type should be used.

Before starting the pump for the first time, a final inspection should be made of all unit parts to make sure that each part is ready for operation. The pump bearings should be inspected occasionally to determine whether or not there is sufficient lubrication available. The bearings should also be drained and washed out, and new lubrication should be provided as often as conditions require. This is true particularly for the lower bearing, which is immersed in water for at least part of the time.

The pump should be disassembled only when necessary. If it is found necessary at any time to renew the gaskets, use only gaskets of the original thickness. One side of the gasket should be painted with shellac; apply a mixture of oil and graphite on the other side. Make certain that all parts are replaced in their original position when reassembling the unit.

Troubleshooting

If the pump does not operate satisfactorily, the trouble may be due to one or more of the following causes:

Misalignment distortion due to pipe strains, bearings badly worn, or foundation not sufficiently rigid.	Replace defective pipe; replace worn bearings; strengthen foundation.
Low pump capacity may be due to clogged strainer, choked up impellers, insufficient speed, excessive discharge heads, air leaks, or worn impeller or impeller rings.	Remove all foreign matter from strainer and impeller; lubricate bearings; decrease discharge head; repair leaks; replace impeller and/or impeller rings.

The Discharge Head

To determine the size of the discharge head for a drainage pump, measure the distance from the bottom of the sump or catch basin to the highest point in the discharge line to the sewer or drain. Then add 1 foot to compensate for friction loss for each 20 feet of discharge pipe. For example, if the distance between the bottom of the sump to the highest point in the discharge line is 18 feet, and the total length of the discharge pipe is 40 feet, the total discharge head is 18 plus 2, or 20 feet.

Appendix—Tables and Data

Ohm's Law (DC)

$$I = E/R \qquad E = I \times R \qquad R = E/I$$

where,

I = current in amperes,
E = potential in volts,
R = resistance in ohms.

DC Power

The power expended in a load resistance when a current is caused to flow by a potential can be determined by the following formulas:

$$P = E \times I \qquad P = E^2/R \qquad P = I^2 \times R$$

where,

P = power in watts,
E = potential in volts,
I = current in amperes,
R = resistance in ohms.

Resistance

1. In series.

$$R_T = R_1 + R_2 + R_3 + \ldots + R_N,$$

where R_T is the total resistance of the circuit, and N is the number of resistors or loads in the circuit.

2. In parallel.

$$R_T = \cfrac{1}{\cfrac{1}{R_1} + \cfrac{1}{R_2} + \cfrac{1}{R_3} + \ldots + \cfrac{1}{R_N}}$$

3. Two resistors in parallel.

$$R_T = \frac{R_1 \times R_2}{R_1 + R_2}$$

Appendix

Capacitance
1. In series.

$$C_T = \frac{1}{\dfrac{1}{C_1} + \dfrac{1}{C_2} + \dfrac{1}{C_3} + \ldots + \dfrac{1}{C_N}} \,,$$

where C_T is the total capacitance of the circuit, and N is the number of capacitors in the circuit.

2. In parallel.

$$C_T = C_1 + C_2 + C_3 + \ldots + C_N$$

Inductance
1. In series.

$$L_T = L_1 + L_2 + L_3 + \ldots + L_N,$$

where L_T is the total inductance in the circuit, and N is the number of inductors in the circuit.

2. In parallel.

$$L_T = \frac{1}{\dfrac{1}{L_1} + \dfrac{1}{L_2} + \dfrac{1}{L_3} + \ldots + \dfrac{1}{L_N}}$$

Frequency
Fundamental formula.

$$f = \frac{1}{2\pi\sqrt{LC}}$$

where,

f = frequency in cycles per second,
L = inductance of the circuit in henrys,
C = capacitance of the circuit in farads,
π = a constant—3.14159. . . .

Reactance
1. *Inductive reactance.*

$$X_L = 2\pi f L$$

where,

X_L = inductive reactance in ohms,
f = frequency in cycles per second,
L = inductance in henrys.

2. *Capacitive reactance.*

$$X_C = \frac{1}{2\pi f C}$$

where,

X_C = capacitive reactance in ohms,
f = frequency in cycles per second,
C = capacitance in farads.

Ohm's Law (AC)

$$I = E/Z \quad \text{or} \quad I = \frac{E}{R^2 + \left[(2\pi f L) - \left(\frac{1}{2\pi f C} \right) \right]^2}$$

where,

I = current in amperes,
E = potential in volts,
Z = impedance in ohms.

Current, Power, and Horsepower Calculations

Table 1, on page 570, gives the electrical formulas for computing current, power, and horsepower with different types of common power-distribution systems. In the table,

I = current in amperes,
E = potential in volts,
eff = efficiency (expressed as a decimal),
hp = horsepower,
PF = power factor,
KW = power in kilowatts,
KVA = power in kilovolt-amperes.

Table 1. Electrical Formulas

Required	Direct Current	Alternating Current		
		Single-Phase	Two-Phase 4-Wire	Three-Phase
Current when horsepower is known	$\dfrac{hp \times 746}{E \times eff}$	$\dfrac{hp \times 746}{E \times eff \times PF}$	$\dfrac{hp \times 746}{2 \times E \times eff \times PF}$	$\dfrac{hp \times 746}{1.73 \times E \times eff \times PF}$
Current when kilowatts are known	$\dfrac{KW \times 1000}{E}$	$\dfrac{KW \times 1000}{E \times PF}$	$\dfrac{KW \times 1000}{2 \times E \times PF}$	$\dfrac{KW \times 1000}{1.73 \times E \times PF}$
Current when KVA is known		$\dfrac{KVA \times 1000}{E}$	$\dfrac{KVA \times 1000}{2 \times E}$	$\dfrac{KVA \times 1000}{1.73 \times E}$
Power in kilowatts	$\dfrac{I \times E}{1000}$	$\dfrac{I \times E \times PF}{1000}$	$\dfrac{2 \times I \times E \times PF}{1000}$	$\dfrac{1.73 \times I \times E \times PF}{1000}$
Power in KVA		$\dfrac{I \times E}{1000}$	$\dfrac{2 \times I \times E}{1000}$	$\dfrac{1.73 \times I \times E}{1000}$
Horsepower output of a motor	$\dfrac{I \times E \times eff}{746}$	$\dfrac{I \times E \times eff \times PF}{746}$	$\dfrac{2 \times I \times E \times eff \times PF}{746}$	$\dfrac{1.73 \times I \times E \times eff \times PF}{746}$

Table 2. Allowable Current-Carrying Capacities of Insulated Aluminum Conductors, in Amperes

Size AWG MCM	Not more than three conductors in raceway or cable or direct burial (based on room temperature of 30° C., 86° F.)					
	Rubber Type R, RW, RU, RUW (12-2) Type RH-RW Thermoplastic Type T, TW	Rubber Type RH RUH (14-2) Type RH-RW Type RHW Thermoplastic Type THW	Thermoplastic Asbestos Type TA SA Var-Cam Type V Asbestos Var-Cam Type AVB MI Cable RHH†	Asbestos Var-Cam Type AVA Type AVL	Impregnated Asbestos Type AI (14-8) Type AIA	Asbestos Type A (14-8) Type AA
12	15	15	25	25	30	30
10	25	25	30	35	40	45
8	30	40	40	45	50	55
6	40	50	50	60	65	75
4	55	65	70	80	90	95
3	65	75	80	95	100	115
*2	75	90	95	105	115	130
*1	85	100	110	125	135	150
*0	100	120	125	150	160	180
*00	115	135	145	170	180	200
*000	130	155	165	195	210	225
*0000	155	180	185	215	245	270

Table 2. (continued)

250	170	205	215	250	270	....	
300	190	230	240	275	305	....	
350	210	250	260	310	335	....	
400	225	270	290	335	360	....	
500	260	310	330	380	405	....	
600	285	340	370	425	440	....	
700	310	375	395	455	485	....	
750	320	385	405	470	500	....	
800	330	395	415	485	520	....	
900	355	425	455	....	....	....	
1000	375	445	480	560	600	....	
1250	405	485	530	....	....	....	
1500	435	520	580	650	....		
1750	455	545	615	....	....		
2000	470	560	650	705	....		

CORRECTION FACTORS, ROOM TEMPS. OVER 30° C., 86° F.							
C.	F.						
40	104	.82	.88	.90	.94	.95	....
45	113	.71	.82	.85	.90	.92	....
50	122	.58	.75	.80	.87	.89	....
55	131	.41	.67	.74	.83	.86	....
60	140	....	.58	.67	.79	.83	.91
70	158	....	.35	.52	.71	.76	.87
75	167	....	....	.43	.66	.72	.86
80	176	....	....	.30	.61	.69	.84
90	194	....	....	....	.50	.61	.80
100	212	....	....	....	....	.51	.77
120	248	....	....	....	....	....	.69
140	284	....	....	....	....	....	.59

*For three-wire, single-phase service and subservice circuits, the allowable current-carrying capacity of RH, RH-RW, RHH, RHW, and THW aluminum conductors are for sizes #2-100 amp, #1-110 amp, #1/0-125 amp, #2/0-150 amp, #3/0-170 amp, and #4/0-200 amp.

†The current-carrying capacities for Type RHH conductors for sizes AWG 12, 10, and 8 are the same as designated for Type RH conductors in this table.

Table 3. Overcurrent Protection for Motors

Col. No. 1	2	3	4		5		6		7	
	For Running Protection of Motors		Maximum Allowable Rating or Setting of Branch-Circuit Protective Devices							
Full-load current rating of motor (amperes)	Maximum rating of nonadjustable protective devices	Maximum setting of adjustable protective devices	With Code Letters Single-phase, squirrel-cage, and synchronous. Full voltage, resistor or reactor starting, Code letters F to V inclusive. Without Code Letters Same as above.		With Code Letters Single-phase, squirrel-cage, and synchronous. Full voltage, resistor or reactor start, Code letters B to E inclusive. Autotransformer start, Code letters F to V inclusive. Without Code Letters (Not more than 30 amperes) Squirrel-cage and synchronous, autotransformer start, high reactance squirrel cage.*		With Code Letters Squirrel-cage and synchronous autotransformer start, Code letters B to E inclusive. Without Code Letters (More than 30 amperes) Squirrel-cage and synchronous autotransformer start, high reactance squirrel cage.*		With Code Letters All motors code letter A. Without Code Letters DC and wound-rotor motors.	
			fuses	Circuit Breakers (Nonadjustable Overload Trip)	fuses	Circuit Breakers (Nonadjustable Overload Trip)	fuses	Circuit Breakers (Nonadjustable Overload Trip)	fuses	Circuit Breakers (Nonadjustable Overload Trip)
	(amperes)	(amperes)								
1	2	1.25	15	15	15	15	15	15	15	15
2	3	2.50	15	15	15	15	15	15	15	15
3	4	3.75	15	15	15	15	15	15	15	15
4	6	5.0	15	15	15	15	15	15	15	15
5	8	6.25	15	15	15	15	15	15	15	15
6	8	7.50	20	15	15	15	15	15	15	15
7	10	8.75	25	20	20	15	15	15	15	15
8	10	10.0	25	20	20	20	20	20	15	15
9	12	11.25	30	30	25	20	20	20	15	15
10	15	12.50	30	30	25	20	20	20	15	15
11	15	13.75	35	30	30	30	25	30	20	20
12	15	15.00	40	30	30	30	25	30	20	20

Appendix

30	35	40	40	40	35	40	40	16.25	20	13
30	35	40	40	45	35	40	45	17.50	20	14
30	40	40	40	45	40	40	45	18.75	20	15
30	40	40	40	50	40	40	50	20.00	20	16
30	45	50	50	60	45	50	60	21.25	25	17
30	45	50	50	60	45	50	60	22.50	25	18
30	45	50	50	60	45	50	60	23.75	25	19
30	50	50	50	60	50	50	60	25.00	25	20
40	60	70	70	70	60	70	70	27.50	30	22
40	60	70	70	80	60	70	80	30.00	30	24
40	60	70	70	80	60	70	80	32.50	35	26
50	70	70	70	90	70	70	90	35.00	35	28
50	80	100	100	90	80	100	90	37.50	40	30
50	80	100	100	100	80	100	100	40.00	40	32
50	80	100	100	110	80	100	110	42.50	45	34
70	90	100	100	110	90	100	110	45.00	45	36
70	100	100	100	125	100	100	125	47.50	50	38
70	100	100	100	125	100	100	125	50.00	50	40
70	110	125	125	125	110	125	125	52.50	50	42
70	110	125	125	125	110	125	125	55.00	60	44
70	125	125	125	150	125	125	150	57.50	60	46
100	125	125	125	150	125	125	150	60.00	60	48
100	125	125	125	150	125	125	150	62.50	60	50
100	150	150	150	175	150	150	175	65.00	70	52
100	150	150	150	175	150	150	175	67.50	70	54
100	150	150	150	175	150	150	175	70.00	70	56
100	150	150	150	175	150	150	175	72.50	70	58
100	150	150	150	200	150	150	200	75.00	80	60
100	175	175	175	200	175	175	200	77.50	80	62
100	175	175	175	200	175	175	200	80.00	80	64
125	175	175	175	200	175	175	200	82.50	80	66
125	175	175	175	225	175	175	225	85.00	90	68
125	175	175	175	225	175	175	225	87.50	90	70
125	175	175	175	225	175	175	225	90.00	90	72
125	200	200	200	225	200	200	225	92.50	90	74
125	200	200	200	250	200	200	250	95.00	100	76
125	200	200	200	250	200	200	250	97.50	100	78
125	200	200	200	250	200	200	250	100.00	100	80
125	225	225	225	250	225	225	250	102.50	110	82
150	225	225	225	250	225	225	250	105.00	110	84

574

Table 3. Overcurrent Protection for Motors (continued)

Col. No. 1	2	3	4		5		6		7	
	For Running Protection of motors		Maximum Allowable Rating or Setting of Branch-Circuit Protective Devices							
	Maximum rating of nonadjustable protective devices	Maximum setting of adjustable protective devices	**With Code Letters** Single-phase, squirrel-cage, and synchronous. Full voltage, resistor or reactor starting, Code letters F to V inclusive. **Without Code Letters** Same as above.		**With Code Letters** Single-phase, squirrel-cage, and synchronous. Full voltage, resistor or reactor start, Code letters B to E inclusive. Autotransformer start, Code letters F to V inclusive. **Without Code Letters** (Not more than 30 amperes) Squirrel-cage and synchronous, autotransformer start, high reactance squirrel cage.*		**With Code Letters** Squirrel-cage and synchronous autotransformer start, Code letters B to E inclusive. **Without Code Letters** (More than 30 amperes) Squirrel-cage and synchronous autotransformer start, high reactance squirrel cage.*		**With Code Letters** All motors code letter A. **Without Code Letters** DC and wound-rotor motors.	
Full-load current rating of motor (amperes)	(amperes)	(amperes)	fuses	Circuit Breakers (Nonadjustable Overload Trip)	fuses	Circuit Breakers (Nonadjustable Overload Trip)	fuses	Circuit Breakers (Nonadjustable Overload Trip)	fuses	Circuit Breakers (Nonadjustable Overload Trip)
86	110	107.50	300	225	225	175	175	175	150	150
88	110	110.00	300	225	225	200	200	200	150	150
90	110	112.50	300	225	225	200	200	200	150	150
92	125	115.00	300	250	250	200	200	200	150	150
94	125	117.50	300	250	250	200	200	200	150	150
96	125	120.00	300	250	250	200	200	200	150	150
98	125	122.50	300	250	250	200	200	200	150	150
100	125	125.00	300	250	250	200	200	200	150	150
105	150	131.50	350	300	300	225	225	225	175	175
110	150	137.50	350	300	300	225	225	225	175	175
115	150	144.00	350	300	300	250	250	250	175	175
120	150	150.00	400	300	300	250	250	250	200	200

125	156.50	175	400	350	350	250	250	250	200	200
130	162.50	175	400	350	350	300	300	300	200	200
135	169.00	175	450	350	350	300	300	300	225	225
140	175.00	175	450	350	350	300	300	300	225	225
145	181.50	200	450	400	400	300	300	300	225	225
150	187.50	200	450	400	400	300	300	300	225	225
155	194.50	200	500	400	400	350	350	350	250	250
160	200.00	200	500	400	400	350	350	350	250	250
165	206.00	225	500	500	450	350	350	350	250	250
170	213.00	225	500	500	450	350	350	350	300	300
175	219.00	225	600	500	450	350	350	350	300	300
180	225.00	225	600	500	450	400	400	400	300	300
185	231.00	250	600	500	500	400	400	400	300	300
190	238.00	250	600	500	500	400	400	400	300	300
195	244.00	250	600	500	500	400	400	400	300	300
200	250.00	250	600	500	500	400	400	400	300	300
210	263.00	250	800	600	600	500	450	500	350	350
220	275.00	300	800	600	600	500	450	500	350	350
230	288.00	300	800	600	600	500	500	500	350	350
240	300.00	300	800	600	600	500	500	500	400	400
250	313.00	300	800	700	800	500	500	500	400	400
260	325.00	350	800	700	800	600	600	600	400	400
270	338.00	350	1000	700	800	600	600	600	450	500
280	350.00	350	1000	700	800	600	600	600	450	500
290	363.00	350	1000	800	800	600	600	600	450	500
300	375.00	400	1000	800	800	600	600	600	450	500
320	400.00	400	1000	800	800	700	800	700	500	500
340	425.00	450	1200	800	1000	700	800	700	600	600
360	450.00	450	1200		1000	800	800	800	600	600
380	475.00	500	1200		1000	800	800	800	600	600
400	500.00	500	1200		1000	800	800	800	600	600
420	525.00	600	1600		1200		1000		800	700
440	550.00	600	1600		1200		1000		800	700
460	575.00	600	1600		1200		1000		800	700
480	600.00	600	1600		1200		1000		800	800
500	625.00		1600		1600		1000		800	800

*High-reactance squirrel-cage motors are those designed to limit the starting current by means of deepslot secondaries or double-wound secondaries and are generally started on full voltage.

Table 4. Overcurrent Protection for Motors

Kind of Motor	Supply System	Number and location of overcurrent units, such as trip coils, relays, or thermal cutouts
1-phase AC or DC	2-wire, 1-phase AC or DC ungrounded	1 in either conductor
1-phase AC or DC	2-wide, 1-phase AC or DC, one conductor grounded	1 in ungrounded conductor
1-phase AC or DC	3-wire, 1-phase AC or DC, grounded-neutral	1 in either ungrounded conductor
2-phase AC	3-wire, 2-phase AC, ungrounded	2, one in each phase
2-phase AC	3-wire, 2-phase AC, one conductor grounded	2 in ungrounded conductors
2-phase AC	4-wire, 2-phase AC grounded or ungrounded	2, one per phase in ungrounded conductors
2-phase AC	5-wire, 2-phase AC, grounded neutral or ungrounded	2, one per phase in any ungrounded phase wire
3-phase AC	3-wire, 3-phase AC ungrounded	2 in any 2 conductors
3-phase AC	3-wire, 3-phase AC, one conductor grounded	2 in ungrounded conductors
3-phase AC	3-wire, 3-phase AC grounded-neutral	2 in any 2 conductors
3-phase AC	4-wire, 3-phase AC grounded-neutral or ungrounded	2 in any 2 conductors except the neutral

SWITCHES S_1 SINGLE-POLE	S_2 DOUBLE-POLE	S_3 3-WAY	S_4 4-WAY
WEATHERPROOF CIRCUIT BREAKER S_{WCB} REMOTE-CONTROL S_{RC} DOOR S_D WITH PILOT S_P WEATHERPROOF S_{WP} ELECTROLIER S_E PULLS		K KEY-OPERATED	NORMALLY OPEN NORMALLY CLOSED
CONTACTOR NORMALLY OPEN NORMALLY CLOSED	SPDT	INTERCONNECTED	FUSED S_F WEATHERPROOF FUSED S_{WF}
CIRCUIT BREAKER		NORMALLY OPEN NORMALLY CLOSED PUSH BUTTON	INTERCONNECTED NORMALLY OPEN ONE NORMALLY CLOSED AND ONE NORMALLY OPEN

KNIFE	FIELD-DISCHARGE	DEAD-FRONT	FLOAT
SPST / DPDT			NORMALLY OPEN / NORMALLY CLOSED
RECEPTACLES / DUPLEX / DUPLEX (GROUNDING TYPE)	DUPLEX / (SPLIT-CIRCUIT-GROUNDING TYPE) G	(SPLIT-CIRCUIT TYPE)	ROTARY-SELECTOR
WEATHERPROOF WP / CONVENIENCE 1,3 (OTHER THAN DUPLEX)	RANGE	SPECIAL-PURPOSE / DISHWASHER DW / CLOTHES DRYER CD	CEILING / FAN (F) / EXIT-LIGHT (E)
TELEVISION TV / RADIO R	SIGNAL PLUG / MAID'S M / NURSE'S N	ALARMS / BELL / BUZZER / CHIME / HORN OR SIREN / ANNUNCIATOR	

Appendix

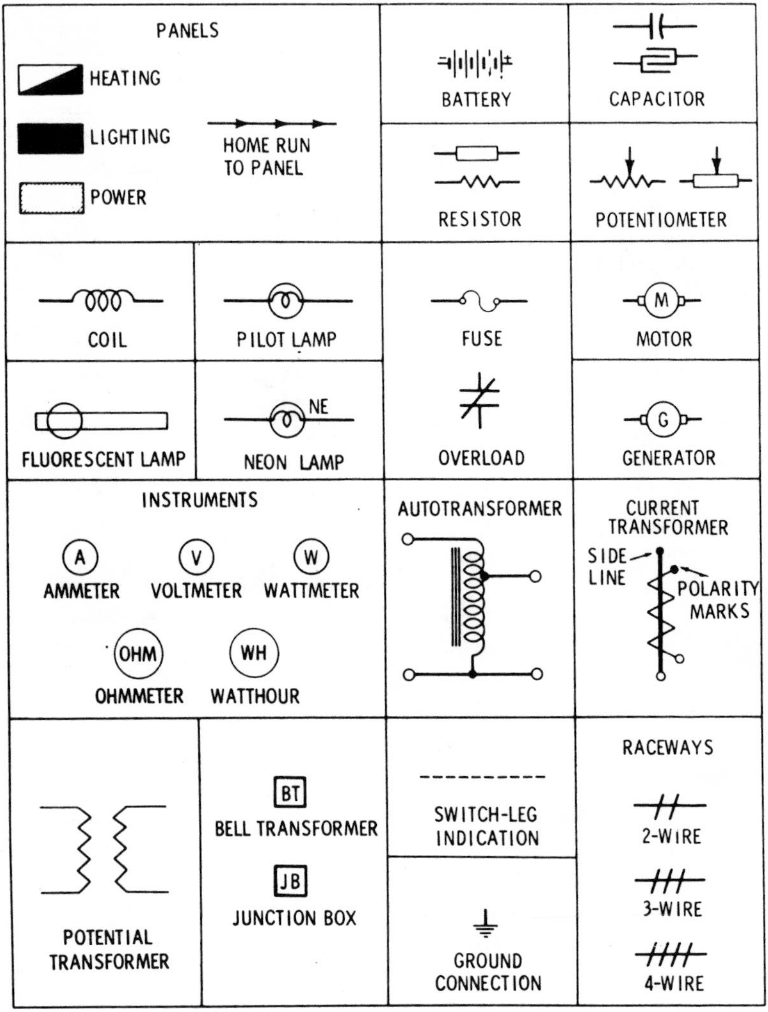
PANELS
HEATING
LIGHTING
POWER
HOME RUN TO PANEL
BATTERY
CAPACITOR
RESISTOR
POTENTIOMETER
COIL
PILOT LAMP
FUSE
MOTOR
M
FLUORESCENT LAMP
NE
NEON LAMP
OVERLOAD
G
GENERATOR
INSTRUMENTS
A
AMMETER
V
VOLTMETER
W
WATTMETER
OHM
OHMMETER
WH
WATTHOUR
AUTOTRANSFORMER
CURRENT TRANSFORMER
SIDE LINE
POLARITY MARKS
POTENTIAL TRANSFORMER
BT
BELL TRANSFORMER
JB
JUNCTION BOX
SWITCH-LEG INDICATION
GROUND CONNECTION
RACEWAYS
2-WIRE
3-WIRE
4-WIRE

TELEPHONES

INTERCOM   OUTSIDE   SWITCHBOARD

ELECTRIC DOOR OPENER
STATIONS

FIRE-ALARM
STATION

D

F

FIRE-ALARM
(CENTRAL)

FA

FS

FIRE-ALARM
(CITY)

FIRE-ALARM
(AUTOMATIC)

W

WATCHMAN

W

WATCHMAN
(CENTRAL)

DC MOTOR

ARMATURE   SHUNT FIELD

SERIES FIELD   COMMUTATOR
FIELD

VARIAC   RHEOSTAT

SYNCHRONOUS MOTOR
(AC GENERATOR)

SQUIRREL-CAGE
MOTOR

TEST LINK

Table 5. Gas Input to the Burner in Cubic Feet per Hour

Seconds For One Revolution	Size of Test Meter Dial			
	One-Half Cu. Ft.	One Cu. Ft.	Two Cu. Ft.	Five Cu. Ft.
	Cubic Feet Per Hour			
10	180	360	720	1800
11	164	327	655	1636
12	150	300	600	1500
13	138	277	555	1385
14	129	257	514	1286
15	120	240	480	1200
16	112	225	450	1125
17	106	212	424	1059
18	100	200	400	1000
19	95	189	379	947
20	90	180	360	900
21	86	171	343	857
22	82	164	327	818
23	78	157	313	783
24	75	150	300	750
25	72	144	288	720
26	69	138	277	692
27	67	133	267	667
28	64	129	257	643
29	62	124	248	621
30	60	120	240	600
31	58	116	232	581
32	56	113	225	563
33	55	109	218	545
34	53	106	212	529
35	51	103	206	514
36	50	100	200	500
37	49	97	195	486
38	47	95	189	474
39	46	92	185	462
40	45	90	180	450

Table 5. Gas Input to the Burner in Cu. Ft. per Hour (Cont'd)

Seconds For One Revolution	Size of Test Meter Dial			
	One-Half Cu. Ft.	One Cu. Ft.	Two Cu. Ft.	Five Cu. Ft.
	Cubic Feet Per Hour			
41	44	88	176	440
42	43	86	172	430
43	42	84	167	420
44	41	82	164	410
45	40	80	160	400
46	39	78	157	391
47	38	77	153	383
48	37	75	150	375
49	37	73	147	367
50	36	72	144	360
51	35	71	141	353
52	35	69	138	346
53	34	68	136	340
54	33	67	133	333
55	33	65	131	327
56	32	64	129	321
57	32	63	126	316
58	31	62	124	310
59	30	61	122	305
60	30	60	120	300
62	29	58	116	290
64	29	56	112	281
66	29	54	109	273
68	28	53	106	265
70	26	51	103	257
72	25	50	100	250
74	24	48	97	243
76	24	47	95	237
78	23	46	92	231
80	22	45	90	225
82	22	44	88	220

Table 5. Gas Input to the Burner in Cu. Ft. per Hour (Cont'd.)

Seconds For One Revolution	Size of Test Meter Dial			
	One-Half Cu. Ft.	One Cu. Ft.	Two Cu. Ft.	Five Cu. Ft.
	Cubic Feet Per Hour			
84	21	43	86	214
86	21	42	84	209
88	20	41	82	205
90	20	40	80	200
94	19	38	76	192
98	18	37	74	184
100	18	36	72	180
104	17	35	69	173
108	17	33	67	167
112	16	32	64	161
116	15	31	62	155
120	15	30	60	150
130	14	28	55	138
140	13	26	51	129
150	12	24	48	120
160	11	22	45	113
170	11	21	42	106
180	10	20	40	100

Table 6. Gas Consumption for Gas Appliances

Appliance	Input Btu per hour (approx.)
Range (free-standing, domestic)	65,000
Built-in oven or broiler unit (domestic) ...	25,000
Built-in top unit (domestic)	40,000
Water heater, automatic storage (50-gal. tank)	55,000
Water heater, automatic instantaneous, 2 gal. per minute	142,800
4 gal. per minute	285,000
6 gal. per minute	428,400
Water heater, domestic, circulating or side-arm	35,000
Refrigerator	3,000
Clothes dryer, domestic	35,000

Table 7. Power Consumption of Home Electrical Appliances

Item	Approx. Kwh per Month	Remarks
Blanket (automatic)	15	8 hr. per day (used 7 mo.)
Clock	1½	
Coffee Maker	15	25 hr. per mo.
Dishwasher	25	1½ washings per day
Dryer (clothes)	50	10 hr. per mo. (family of 4)
Fan (10-inch)	1	25 hr. per mo.
Food Freezer	40	8 cu. ft.
Garbage Disposal Unit	¾	4 min. per day
Iron	6	12 hr. per mo.
Ironer	10	10 hr. per mo. (family of 4)
Lighting	65	
Mixer	¾	5 hr. per mo.
Oil Furnace (not including circulator fan)	30	(200-500 KW-hours per year)
Radio	10	130 hr. per mo.
Range	90	(Family of 4)
Refrigerator	22	8 cu. ft.
Roaster	12	16 hr. per mo.
Sandwich Grill	4	5 hr. per mo.
Sewing Machine	1	
Television	18	90 hr. per mo.
Toaster	3	3 hr. per mo.
Vacuum Cleaner (upright)	2¼	6 hr. per mo.
Vacuum Cleaner (tank)	3¼	6 hr. per mo.
Washer (wringer-type)	2	12 hr. per mo. (family of 4)
Washer (automatic)	3	12 hr. per mo. (family of 4)
Water Heater	350	(Family of 4)

Table 8. Conversion Factors

To Convert	Into	Multiply by	Conversely, Multiply by
Acres	Square feet	4.356×10^4	2.296×10^{-5}
Acres	Square meters	4047	2.471×10^{-4}
Acres	Square miles	1.5625×10^{-3}	640
Amperes	Microamperes	10^6	10^{-6}
Amperes	Micromicroamperes	10^{12}	10^{-12}
Amperes	Milliamperes	10^3	10^{-3}
Ampere-hours	Coulombs	3600	2.778×10^{-4}
Ampere-turns	Gilberts	1.257	0.7958
Ampere-turns per cm.	Ampere-turns per in.	2.54	0.3937

Table 8. (continued)

To Convert	Into	Multiply by	Conversely, Multiply by
Angstrom units	Inches	3.937×10^{-9}	2.54×10^8
Angstrom units	Meters	10^{-10}	10^{10}
Bars	Atmospheres	9.870×10^{-7}	1.0133
Bars	Dynes per sq. cm.	10^6	10^{-6}
Bars	Pounds per sq. in.	14.504	6.8947×10^{-2}
Btu	Ergs	1.0548×10^{10}	9.486×10^{-11}
Btu	Foot-pounds	778.3	1.285×10^{-3}
Btu	Joules	1054.8	9.480×10^{-4}
Btu	Kilogram-calories	0.252	3.969
Btu per hour	Horsepower-hours	3.929×10^{-4}	2545
Bushels	Cubic feet	1.2445	0.8036
Calories, gram	Joules	4.185	0.2389
Centigrade	Celsius	1	1
Centigrade	Fahrenheit	$(°C \times 9/5)$ $+ 32 = °F$	$(°F - 32)$ $\times 5/9 = °C$
Centigrade	Kelvin	$°C + 273.1 = °K$	$°K - 273.1$ $= °C$
Chains (surveyor's)	Feet	66	1.515×10^{-2}
Circular mils	Square centimeters	5.067×10^{-6}	1.973×10^5
Circular mils	Square mils	0.7854	1.273
Cubic feet	Gallons (liq. U.S.)	7.481	0.1337
Cubic feet	Liters	28.32	3.531×10^{-2}
Cubic inches	Cubic centimeters	16.39	6.102×10^{-2}
Cubic inches	Cubic feet	5.787×10^{-4}	1728
Cubic inches	Cubic meters	1.639×10^{-5}	6.102×10^4
Cubic inches	Gallons (liq. U.S.)	4.329×10^{-3}	231
Cubic meters	Cubic feet	35.31	2.832×10^{-2}
Cubic meters	Cubic yards	1.308	0.7646
Cycles	Kilocycles	10^{-3}	10^3
Cycles	Megacycles	10^{-6}	10^6
Degrees (angle)	Mils	17.45	5.73×10^{-2}
Degrees (angle)	Radians	1.745×10^{-2}	57.3
Dynes	Pounds	2.248×10^{-6}	4.448×10^5
Ergs	Foot-pounds	7.376×10^{-8}	1.356×10^7
Fahrenheit	Rankine	$°F + 459.58 = °R$	$°R - 459.58 = °F$
Faradays	Ampere-hours	26.8	3.731×10^{-2}
Farads	Microfarads	10^6	10^{-6}
Farads	Micromicrofarads	10^{12}	10^{-12}
Farads	Millifarads	10^3	10^{-3}
Fathoms	Feet	6	0.16667
Feet	Centimeters	30.48	3.281×10^{-2}
Feet	Meters	0.3048	3.281
Feet	Mils	1.2×10^4	8.333×10^{-5}
Foot-pounds	Gram-centimeters	1.383×10^4	1.235×10^{-5}
Foot-pounds	Horsepower-hours	5.05×10^{-7}	1.98×10^6
Foot-pounds	Kilogram-meters	0.1383	7.233
Foot-pounds	Kilowatt-hours	3.766×10^{-7}	2.655×10^6

Table 8. (continued)

To Convert	Into	Multiply by	Conversely, Multiply by
Foot-pounds	Ounce-inches	192	5.208×10^{-3}
Gallons (liq. U.S.)	Cubic meters	3.785×10^{-3}	264.2
Gallons (liq. U.S.)	Gallons (liq. Br. Imp.)	0.8327	1.201
Gausses	Lines per sq. cm.	1.0	1.0
Gausses	Lines per sq. in.	6.452	0.155
Gausses	Webers per sq. in.	6.452×10^{-8}	1.55×10^{7}
Grams	Dynes	980.7	1.02×10^{-3}
Grams	Grains	15.43	6.481×10^{-2}
Grams	Ounces (avdp.)	3.527×10^{-2}	28.35
Grams	Poundals	7.093×10^{-2}	14.1
Grams per cm.	Pounds per in.	5.6×10^{-3}	178.6
Grams per cu. cm.	Pounds per cu. in.	3.613×10^{-2}	27.68
Henries	Microhenries	10^{6}	10^{-6}
Henries	Millihenries	10^{3}	10^{-3}
Horsepower	Btu per minute	42.418	2.357×10^{-2}
Horsepower	Foot-lbs. per minute	3.3×10^{4}	3.03×10^{-5}
Horsepower	Foot-lbs. per second	550	1.182×10^{-3}
Horsepower	Horsepower (metric)	1.014	0.9863
Horsepower	Kilowatts	0.746	1.341
Inches	Centimeters	2.54	0.3937
Inches	Feet	8.333×10^{-2}	12
Inches	Meters	2.54×10^{-2}	39.37
Inches	Miles	1.578×10^{-5}	6.336×10^{4}
Inches	Mils	10^{3}	10^{-3}
Inches	Yards	2.778×10^{-2}	36
Joules	Foot-pounds	0.7376	1.356
Joules	Ergs	10^{7}	10^{-7}
Joules	Watt-hours	2.778×10^{-4}	3600
Kilograms	Tonnes	10^{3}	10^{-3}
Kilograms	Tons (long)	9.842×10^{-4}	1016
Kilograms	Tons (short)	1.102×10^{-3}	907.2
Kilograms	Pounds (avdp.)	2.205	0.4536
Kilograms per sq. meter	Pounds per sq. feet	0.2048	4.882
Kilometers	Feet	3281	3.408×10^{-4}
Kilometers	Inches	3.937×10^{4}	2.54×10^{-5}
Kilometers	Light years	1.0567×10^{-13}	9.4637×10^{12}
Kilometers per hr.	Feet per minute	54.68	1.829×10^{-2}
Kilometers per hr.	Knots	0.5396	1.8532
Kilowatt-hours	Btu	3413	2.93×10^{-4}
Kilowatt-hours	Foot-pounds	2.655×10^{6}	3.766×10^{-7}
Kilowatt-hours	Joules	3.6×10^{6}	2.778×10^{-7}
Kilowatt-hours	Horsepower-hours	1.341	0.7457
Kilowatt-hours	Pounds water evaporated from and at 212°F.	3.53	0.284
Kilowatt-hours	Watt-hours	10^{3}	10^{-3}

Table 8. (continued)

To Convert	Into	Multiply by	Conversely, Multiply by
Knots	Feet per second	1.688	0.5925
Knots	Meters per minute	30.87	0.0324
Knots	Miles per hour	1.1508	0.869
Lamberts	Candles per sq. cm.	0.3183	3.142
Lamberts	Candles per sq. in.	2.054	0.4869
Leagues	Miles	3	0.33
Links	Chains	0.01	100
Links (surveyor's)	Inches	7.92	0.1263
Liters	Bushels (dry U.S.)	2.838×10^{-2}	35.24
Liters	Cubic centimeters	10^3	10^{-3}
Liters	Cubic meters	10^{-3}	10^3
Liters	Cubic inches	61.02	1.639×10^{-2}
Liters	Gallons (liq. U.S.)	0.2642	3.785
Liters	Pints (liq. U.S.)	2.113	0.4732
$\text{Log}_\epsilon N$	$\text{Log}_{10} N$	0.4343	2.303
Lumens per sq. ft.	Foot-candles	1	1
Lux	Foot-candles	0.0929	10.764
Maxwells	Kilolines	10^{-3}	10^3
Maxwells	Megalines	10^{-6}	10^6
Maxwells	Webers	10^{-8}	10^8
Meters	Centimeters	10^2	10^{-2}
Meters	Feet	3.28	30.48×10^{-2}
Meters	Inches	39.37	2.54×10^{-2}
Meters	Kilometers	10^{-3}	10^3
Meters	Miles	6.214×10^{-4}	1609.35
Meters	Yards	1.094	0.9144
Meters per minute	Feet per minute	3.281	0.3048
Meters per minute	Kilometers per hour	0.06	16.67
Mhos	Micromhos	10^6	10^{-6}
Mhos	Millimhos	10^3	10^{-3}
Microfarads	Micromicrofarads	10^6	10^{-6}
Miles (nautical)	Feet	6076.1	1.646×10^{-4}
Miles (nautical)	Meters	1852	5.4×10^{-4}
Miles (statute)	Feet	5280	1.894×10^{-4}
Miles (statute)	Kilometers	1.609	0.6214
Miles (statute)	Light years	1.691×10^{-13}	5.88×10^{12}
Miles (statute)	Miles (nautical)	0.869	1.1508
Miles (statute)	Yards	1760	5.6818×10^{-4}
Miles per hour	Feet per minute	88	1.136×10^{-2}
Miles per hour	Feet per second	1.467	0.6818
Miles per hour	Kilometers per hour	1.609	0.6214
Miles per hour	Knots	0.8684	1.152
Milliamperes	Microamperes	10^3	10^{-3}
Millihenries	Microhenries	10^3	10^{-3}
Millimeters	Centimeters	0.1	10
Millimeters	Inches	3.937×10^{-2}	25.4
Millimeters	Microns	10^3	10^{-3}

Table 8. (continued)

To Convert	Into	Multiply by	Conversely, Multiply by
Millivolts	Microvolts	10^3	10^{-3}
Mils	Minutes	3.438	0.2909
Minutes (angle)	Degrees	1.666×10^{-2}	60
Nepers	Decibels	8.686	0.1151
Newtons	Dynes	10^5	10^{-5}
Newtons	Pounds (avdp.)	0.2248	4.448
Ohms	Milliohms	10^3	10^{-3}
Ohms	Micro-ohms	10^6	10^{-6}
Ohms	Micromicro-ohms	10^{12}	10^{-12}
Ohms	Megohms	10^{-6}	10^6
Ohms	Ohms (International)	0.99948	1.00052
Ohms per foot	Ohms per meter	0.3048	3.281
Ounces (fluid)	Quarts	3.125×10^{-2}	32
Ounces (avdp.)	Pounds	6.25×10^{-2}	16
Picofarad	Micromicrofarad	1	1
Pints	Quarts (liq. U.S.)	0.50	2
Pounds (force)	Newtons	4.4482	0.2288
Pounds carbon oxidized	Btu	14,544	6.88×10^{-5}
Pounds carbon oxidized	Horsepower-hours	5.705	0.175
Pounds carbon oxidized	Kilowatt-hours	4.254	0.235
Pounds of water (dist.)	Cubic feet	1.603×10^{-2}	62.38
Pounds of water (dist.)	Gallons	0.1198	8.347
Pounds per sq. in.	Dynes per sq. cm.	6.8946×10^4	1.450×10^{-5}
Poundals	Dynes	1.383×10^4	7.233×10^{-5}
Poundals	Pounds (avdp.)	3.108×10^{-2}	32.17
Quadrants	Degrees	90	11.111×10^{-2}
Quadrants	Radians	1.5708	0.637
Radians	Mils	10^3	10^{-3}
Radians	Minutes	3.438×10^3	2.909×10^{-4}
Radians	Seconds	2.06265×10^5	4.848×10^{-6}
Rods	Feet	16.5	6.061×10^{-2}
Rods	Miles	3.125×10^{-3}	320
Rods	Yards	5.5	0.1818
Rpm	Degrees per second	6.0	0.1667
Rpm	Radians per second	0.1047	9.549
Rpm	Rps	1.667×10^{-2}	60
Square feet	Acres	2.296×10^{-5}	43,560
Square feet	Square centimeters	929.034	1.076×10^{-3}
Square feet	Square inches	144	6.944×10^{-3}
Square feet	Square meters	9.29×10^{-2}	10.764
Square feet	Square miles	3.587×10^{-8}	27.88×10^6
Square feet	Square yards	11.11×10^{-2}	9

Table 8. (continued)

To Convert	Into	Multiply by	Conversely, Multiply by
Square inches	Circular mils	1.273×10^{11}	7.854×10^{-7}
Square inches	Square centimeters	6.452	0.155
Square inches	Square mils	10^{6}	10^{-6}
Square inches	Square millimeters	645.2	1.55×10^{-3}
Square kilometers	Square miles	0.3861	2.59
Square meters	Square yards	1.196	0.8361
Square miles	Acres	640	1.562×10^{-3}
Square miles	Square yards	3.098×10^{6}	3.228×10^{-7}
Square millimeters	Circular mils	1973	5.067×10^{-4}
Square millimeters	Square centimeters	.01	100
Square mils	Circular mils	1.273	0.7854
Tons (long)	Pounds (avdp.)	2240	4.464×10^{-4}
Tons (short)	Pounds	2,000	5×10^{-4}
Tonnes	Pounds	2204.63	4.536×10^{-4}
Varas	Feet	2.7777	0.36
Volts	Kilovolts	10^{-3}	10^{3}
Volts	Microvolts	10^{6}	10^{-6}
Volts	Millivolts	10^{3}	10^{-3}
Watts	Btu per hour	3.413	0.293
Watts	Btu per minute	5.689×10^{-2}	17.58
Watts	Ergs per second	10^{7}	10^{-7}
Watts	Foot-lbs per minute	44.26	2.26×10^{-2}
Watts	Foot-lbs per second	0.7378	1.356
Watts	Horsepower	1.341×10^{-3}	746
Watts	Kilogram-calories per minute	1.433×10^{-2}	69.77
Watts	Kilowatts	10^{-3}	10^{3}
Watts	Microwatts	10^{6}	10^{-6}
Watts	Milliwatts	10^{3}	10^{-3}
Watt-seconds	Joules	1	1
Webers	Maxwells	10^{8}	10^{-8}
Webers per sq. meter	Gausses	10^{4}	10^{-4}
Yards	Feet	3	.3333
Yards	Varas	1.08	0.9259

Index

591

The Audel® Mail Order Bookstore

Here's an opportunity to order the valuable books you may have missed before and to build your own personal, comprehensive library of Audel books. You can choose from an extensive selection of technical guides and reference books. They will provide access to the same sources the experts use, put all the answers at your fingertips, and give you the know-how to complete even the most complicated building or repairing job, in the same professional way.

Each volume:

- **Fully illustrated**
- **Packed with up-to-date facts and figures**
- **Completely indexed for easy reference**

APPLIANCES

REFRIGERATION: HOME AND COMMERCIAL

Covers the whole realm of refrigeration equipment from fractional-horsepower water coolers, through domestic refrigerators to multi-ton commercial installations. 656 pages; 5½ x 8¼; hardbound. **Cat. No. 23286 Price: $9.50.**

AIR CONDITIONING: HOME AND COMMERCIAL

A concise collection of basic information, tables, and charts for those interested in understanding, troubleshooting, and repairing home air conditioners and commercial installations. 464 pages; 5½ x 8¼; hardbound. **Cat. No. 23288 Price: $7.50.**

HOME APPLIANCE SERVICING, 3rd Edition

A practical book for electric & gas servicemen, mechanics & dealers. Covers the principles, servicing, and repairing of home appliances. 592 pages; 5¼ x 8¼; hardbound. **Cat. No. 23214 Price: $12.95.**

REFRIGERATION AND AIR CONDITIONING LIBRARY—2 Vols.
Cat. No. 23305   Price: $15.95

OIL BURNERS, 3rd Edition

Provides complete information on all types of oil burners and associated equipment. Discusses burners—blowers—ignition transformers—electrodes—nozzles—fuel pumps—filters—Controls. Installation and maintenance are stressed. 320 pages; 5½ x 8¼; hardbound. **Cat. No. 23277 Price: $7.50.**

Use the order coupon on the back page of this book.

AUTOMOTIVE

AUTO BODY REPAIR FOR THE DO-IT-YOURSELFER

Shows how to use touch-up paint; repair chips, scratches, and dents; remove and prevent rust; care for glass, doors, locks, lids, and vinyl tops; and clean and repair upholstery. 96 pages; 8½ x 11; softcover. **Cat. No. 23238 Price: $5.95.**

AUTOMOBILE REPAIR GUIDE, 4th Edition

A practical reference for auto mechanics, servicemen, trainees, and owners Explains theory, construction, and servicing of modern domestic motorcars. 800 pages; 5½ x 8¼; hardbound. **Cat. No. 23291 Price: $12.95.**

CAN-DO TUNE-UP™ SERIES

Each book in this series comes with an audio tape cassette. Together they provide an organized set of instructions that will show you and talk you through the maintenance and tune-up procedures designed for your particular car. All books are softcover.

AMERICAN MOTORS CORPORATION CARS

(The 1964 thru 1974 cars covered include: Matador. Rambler. Gremlin. and AMC Jeep (Willys).). 112 pages; 5½ x 8½; softcover. **Cat. No. 23843 Price: $7.95.**
Cat. No. 23851 Without Cassette   **Price: $4.95**

CHRYSLER CORPORATION CARS

(The 1964 thru 1974 cars covered include: Chrysler, Dodge, and Plymouth.) 112 pages; 5½ x 8½; softcover. **Cat. No. 23825 Price $7.95.**
Cat. No. 23846 Without Cassette   **Price: $4.95**

FORD MOTOR COMPANY CARS

(The 1954 thru 1974 cars covered include: Ford, Lincoln, and Mercury.) 112 pages; 5½ x 8½; softcover. **Cat. No. 23827 Price: $7.95.**
Cat. No. 23848 Without Cassette   **Price: $4.95**

GENERAL MOTORS CORPORATION CARS

(The 1964 thru 1974 cars covered include: Buick, Cadillac, Chevrolet, Oldsmobile, and Pontiac.) 112 pages; 5½ x 8½; softcover. **Cat. No. 23824 Price: $7.95.**
Cat. No. 23845 Without Cassette   **Price: $4.95**

PINTO AND VEGA CARS,

1971 thru 1974. 112 pages· 5½ x 8½; softcover. **Cat. No. 23831 Price: $7.95.**
Cat. No. 23849 Without Cassette   **Price: $4.95**

TOYOTA AND DATSUN CARS,

1964 thru 1974. 112 pages; 5½ x 8½; softcover. **Cat. No. 23835 Price: $7.95.**
Cat. No. 23850 Without Cassette   **Price: $4.95**

VOLKSWAGEN CARS

(The 1964 thru 1974 cars covered include: Beetle. Super Beetle. and Karmann Ghia.) 96 pages; 5½ x 8½; softcover. **Cat. No. 23826 Price: $7.95.**
Cat. No. 23847 Without Cassette   **Price: $4.95**

AUTOMOTIVE AIR CONDITIONING

You can easily perform most all service procedures you've been paying for in the past. This book covers the systems built by the major manufacturers, even after-market installations. Contents: introduction—refrigerant—tools—air conditioning circuit—general service procedures—electrical systems—the cooling system—system diagnosis—electrical diagnosis—troubleshooting. 232 pages; 5½ x 8½; softcover. **Cat. No. 23318 Price: $5.95.**

Use the order coupon on the back page of this book.

DIESEL ENGINE MANUAL, 3rd Edition

A practical guide covering the theory, operation, and maintenance of modern diesel engines. Explains diesel principles—valves—timing—fuel pumps—pistons and rings—cylinders—lubrication—cooling system—fuel oil and more. 480 pages; 5½ x 8¼; hardbound. **Cat. No. 23199 Price: $8.95.**

GAS ENGINE MANUAL, 2nd Edition

A completely practical book covering the construction, operation, and repair of all types of modern gas engines. 400 pages; 5½ x 8¼; hardbound. **Cat. No. 23245 Price: $7.95.**

OUTBOARD MOTORS & BOATING, 3rd Edition

Provides the information you need to maintain, troubleshoot, repair, and adjust all types of outboard motors. Explains the basic principles of outboard motors and the functions of the various engine parts. 464 pages; 5½ x 8¼; softcover. **Cat. No. 23279 Price: $6.95.**

BUILDING AND MAINTENANCE

ANSWERS ON BLUEPRINT READING, 3rd Edition

Covers all types of blueprint reading for mechanics and builders. This book reveals the secret language of blueprints, step-by-step in easy stages. 312 pages; 5½ x 8¼; hardbound. **Cat. No. 23283 Price: $6.95.**

BUILDING A VACATION HOME

From selecting a building site to driving in the last nail, this book explains the entire process, with fully illustrated step-by-step details. Includes a complete set of drawings for a two-story vacation and/or retirement home. Softcover. 192 pages; 8½ x 11; softcover. **Cat. No. 23222 Price: $7.95.**

BUILDING MAINTENANCE, 2nd Edition

Covers all the practical aspects of building maintenance. Painting and decorating; plumbing and pipe fitting; carpentry; heating maintenance; custodial practices and more. (A book for building owners, managers, and maintenance personnel.) 384 pages; 5½ x 8¼; hardbound. **Cat. No. 23278 Price: $7.50.**

COMPLETE BUILDING CONSTRUCTION

At last—a *one-volume* instruction manual to show you how to construct a frame or brick building from the footings to the ridge. Build your own garage, tool shed, other outbuilding—even your own house or place of business. Building construction tells you how to lay out the building and excavation lines on the lot; how to make concrete forms and pour the footings and foundation; how to make concrete slabs, walks, and driveways; how to lay concrete block, brick and tile; how to build your own fireplace and chimney. It's one of the newest Audel books, clearly written by experts in each field and ready to help you every step of the way. 800 pages; 5½ x 8¼; hardbound. **Cat. No. 23323 Price: $19.95.**

GARDENING & LANDSCAPING

A comprehensive guide for homeowners and for industrial, municipal, and estate groundskeepers. Gives information on proper care of annual and perennial flowers; various house plants; greenhouse design and construction; insect and rodent controls; and more. 384 pages; 5½ x 8¼; hardbound. **Cat. No. 23229 Price: $7.95.**

CARPENTERS & BUILDERS LIBRARY, 4th Edition (4 Vols.)

A practical, illustrated trade assistant on modern construction for carpenters, builders, and all woodworkers. Explains in practical, concise language and illustrations all the principles, advances, and shortcuts based on modern practice. How to calculate various jobs. **Cat. No. 23244 Price: $24.50**

Vol. 1—Tools, steel square, saw filing, joinery cabinets. 384 pages; 5½ x 8¼; hardbound. **Cat. No. 23240 Price: $6.50.**

Vol. 2—Mathematics, plans, specifications, estimates 304 pages; 5½ x 8¼; hardbound. **Cat. No. 23241 Price: $6.50.**

Vol. 3—House and roof framing, laying out foundations. 304 pages; 5½ x 8¼; hardbound. **Cat. No. 23242 Price: $6.50.**

Vol. 4—Doors, windows, stairs, millwork, painting. 368 pages; 5½ x 8¼; hardbound. **Cat. No. 23243 Price: $6.50.**

Use the order coupon on the back page of this book.

CARPENTRY AND BUILDING

Answers to the problems encountered In today's building trades. The actual questions asked of an architect by carpenters and builders are answered in this book. 448 pages; 5½ x 8¼; hardbound. **Cat. No. 23142 Price: $7.95.**

WOOD STOVE HANDBOOK

The wood stove handbook shows how wood burned in a modern wood stove offers an immediate, practical, low-cost method of full-time or part-time home heating. The book points out that wood is plentiful, low in cost (sometimes free), and nonpolluting, especially when burned in one of the newer and more efficient stoves. In this book, you will learn about the nature of heat and its control, what happens inside and outside a stove, how to have a safe and efficient chimney, and how to install a modern wood burning stove. You will also learn about the different types of firewood and how to get it, cut it, split it, and store it. 128 pages; 8½ x 11; softcover. **Cat. No. 23319 Price: $6.95.**

HEATING, VENTILATING, AND AIR CONDITIONING LIBRARY (3 Vols.)

This three-volume set covers all types of furnaces, ductwork, air conditioners, heat pumps, radiant heaters, and water heaters, including swimming-pool heating systems. **Cat. No. 23227 Price: $25.50.**

Volume 1

Partial Contents: Heating Fundamentals . . . Insulation Principles . . . Heating Fuels . . . Electric Heating System . . . Furnace Fundamentals . . . Gas-Fired Furnaces . . . Oil-Fired Furnaces . . . Coal-Fired Furnaces . . . Electric Furnaces. **Cat. No. 23248 Price: $8.95**

Volume 2

Partial Contents: Oil Burners . . . Gas Burners . . . Thermostats and Humidistats . . . Gas and Oil Controls . . . Pipes, Pipe Fitting, and Piping Details . . . Valves and Valve Installations. 560 pages; 5½ x 8¼; hardbound.   **Cat. No. 23249 Price: $8.95.**

Volume 3

Partial Contents: Radiant Heating . . . Radiators, Convectors, and Unit Heaters . . . Stoves, Fireplaces, and Chimneys . . . Water Heaters and Other Appliances . . . Central Air Conditioning Systems . . . Humidifiers and Dehumidifiers.   544 pages; 5½ x 8¼; hardbound. **Cat. No. 23250 Price: $8.95.**

HOME MAINTENANCE AND REPAIR: Walls, Ceilings, and Floors

Easy-to-follow instructions for sprucing up and repairing the walls, ceiling, and floors of your home. Covers nail pops, plaster repair, painting, paneling, ceiling and bathroom tile, and sound control, 80 pages; 8½ x 11; softcover. **Cat. No. 23281 Price: $5.95.**

HOME PLUMBING HANDBOOK , 2nd Edition

A complete guide to home plumbing repair and installation. 200 pages; 8½ x 11; softcover. **Cat. No. 23321 Price: $6.95.**

HOME WORKSHOP & TOOL HANDY BOOK

Tells how to set up your own home workshop (basement, garage, or spare room) and explains the various hand and power tools (when, where, and how to use them). 464 pages; 5½ x 8¼; hardbound. **Cat. No. 23208 Price $6.50.**

MASONS AND BUILDERS LIBRARY—2 Vols.

A practical, illustrated trade assistant on modern construction for bricklayers, stone-masons, cement workers, plasterers, and tile setters. Explains all the principles, advances, and shortcuts based on modern practice—including how to figure and calculate various jobs. **Cat. No. 23185   Price: $13.95**

Vol. 1—Concrete, Block, Tile, Terrazzo. 368 pages; 5½ x 8¼; hardbound. **Cat. No. 23182 Price: $7.50.**

Vol. 2—Bricklaying, Plastering, Rock Masonry, Clay Tile. 384 pages; 5½ x 8¼; hardbound. **Cat. No. 23183 Price: $7.50.**

Use the order coupon on the back page of this book.

PLUMBERS AND PIPE FITTERS LIBRARY—3 Vols.

A practical, illustrated trade assistant and reference for master plumbers, journeymen and apprentice pipe fitters, gas fitters and helpers, builders, contractors, and engineers. Explains in simple language, illustrations, diagrams, charts, graphs, and pictures, the principles of modern plumbing and pipe-fitting practices. **Cat. No. 23255 Price $19.95**

Vol. 1—Materials, tools, roughing-in. 320 pages; 5½ x 8¼; hardbound. **Cat. No. 23256 Price: $6.95.**

Vol. 2—Welding, heating, air-conditioning. 384 pages; 5½ x 8¼; hardbound. **Cat. No. 23257 Price: $6.95.**

Vol. 3—Water supply, drainage, calculations. 272 pages; 5½ x 8¼; hardbound. **Cat. No. 23258 Price: $6.95.**

PLUMBERS HANDBOOK

A pocket manual providing reference material for plumbers and/or pipe fitters. General information sections contain data on cast-iron fittings, copper drainage fittings, plastic pipe, and repair of fixtures. 288 pages; 4 x 6; softcover. **Cat. No. 23339 Price: $5.95.**

QUESTIONS AND ANSWERS FOR PLUMBERS EXAMINATIONS,

2nd Edition

Answers plumbers' questions about types of fixtures to use, size of pipe to install, design of systems, size and location of septic tank systems, and procedures used in installing material. 256 pages; 5½ x 8¼; softcover. **Cat. No. 23285 Price: $5.50.**

TREE CARE MANUAL

The conscientious gardener's guide to healthy, beautiful trees. Covers planting, grafting, fertilizing, pruning, and spraying. Tells how to cope with insects, plant diseases, and environmental damage. 224 pages; 8½ x 11; softcover. **Cat. No. 23280 Price: $8.95.**

UPHOLSTERING

Upholstering is explained for the average householder and apprentice upholsterer. From repairing and reglueing of the bare frame, to the final sewing or tacking, for antiques and most modern pieces, this book covers it all. 400 pages; 5½ x 8¼; hardbound. **Cat. No. 23189 Price: $7.95.**

WOOD FURNITURE: Finishing, Refinishing, Repairing

Presents the fundamentals of furniture repair for both veneer and solid wood. Gives complete instructions on refinishing procedures, which includes stripping the old finish, sanding, selecting the finish and using wood fillers. 352 pages; 5½ x 8¼; hardbound. **Cat. No. 23216 Price: $8.50.**

ELECTRICITY/ELECTRONICS

ELECTRICAL LIBRARY

If you are a student of electricity or a practicing electrician, here is a very important and helpful library you should consider owning. You can learn the basics of electricity, study electric motors and wiring diagrams, learn how to interpret the NEC, and prepare for the electrician's examination by using these books. **Cat. No. 23324 Price: $43.95.**

Electric Motors, 3rd Edition. 528 pages; 5½ x 8¼; hardbound. **Cat. No. 23264 Price: $8.95.**

Guide to the 1978 National Electrical Code. 672 pages; 5½ x 8¼; hardbound. **Cat. No. 23308 Price: $9.95.**

House Wiring, 4th Edition. 256 pages; 5½ x 8¼; hardbound. **Cat. No. 23315 Price: $6.95.**

Practical Electricity, 3rd Edition. 496 pages; 5½ x 8¼; hardbound. **Cat. No. 23218 Price: $8.95**

Questions and Answers for Electricians Examinations, 6th Edition. 288 pages; 5½ x 8¼; hardbound. **Cat. No. 23307 Price: $6.95.**

Wiring Diagrams for Light and Power, 3rd Edition. 400 pages; 5½ x 8¼; hardbound. **Cat. No. 23232 Price: $6.95.**

ELECTRICAL COURSE FOR APPRENTICES AND JOURNEYMEN

A study course for apprentice or journeymen electricians. Covers electrical theory and its applications. 448 pages; 5½ x 8¼; hardbound. **Cat. No. 23209 Price: $7.95**

Use the order coupon on the back page of this book.

RADIOMANS GUIDE, 4th Edition

Contains the latest information on radio and electronics from the basics through transistors. 480 pages; 5½ x 8¼; hardbound. **Cat. No. 23259 Price: $7.50.**

TELEVISION SERVICE MANUAL, 4th Edition

Provides the practical information necessary for accurate diagnosis and repair of both black-and-white and color television receivers. 512 pages; 5½ x 8¼; hardbound. **Cat. No. 23247 Price: $8.95.**

ENGINEERS/MECHANICS/ MACHINISTS

MACHINISTS LIBRARY, 2nd Edition

Covers modern machine-shop practice. Tells how to set up and operate lathes, screw and milling machines, shapers, drill presses, and all other machine tools. A complete reference library. **Cat. No. 23300   Price: $23.00**

Vol. 1—Basic Machine Shop. 352 pages; 5½ x 8¼; hardbound. **Cat. No. 23301 Price: $7.95.**

Vol. 2—Machine Shop. 480 pages; 5½ x 8¼; hardbound. **Cat. No. 23302 Price: $7.95.**

Vol. 3—Toolmakers Handy Book. 400 pages; 5½ x 8¼; hardbound. **Cat. No. 23303 Price: $7.95.**

MECHANICAL TRADES POCKET MANUAL

Provides practical reference material for mechanical tradesmen. This handbook covers methods, tools, equipment, procedures, and much more. 256 pages; 4 x 6; softcover. **Cat. No. 23215 Price: $4.50.**

MILLWRIGHTS AND MECHANICS GUIDE, 2nd Edition

Practical information on plant installation, operation, and maintenance for millwrights, mechanics, maintenance men, erectors, riggers, foremen, inspectors, and superintendents. 960 pages; 5½ x 8¼; hardbound. **Cat. No. 23201 Price: $11.95.**

POWER PLANT ENGINEERS GUIDE, 2nd Edition

The complete steam or diesel power-plant engineer's library. 816 pages; 5½ x 8¼; hardbound. **Cat. No. 23220 Price: $12.95.**

QUESTIONS AND ANSWERS FOR ENGINEERS AND FIREMANS EXAMINATIONS, 3RD EDITION

Presents both legitimate and "catch" questions with answers that may appear on examinations for engineers and firemans licenses for stationary, marine, and combustion engines. 496 pages; 5½ x 8¼; hardbound. **Cat. No. 23327   Price $8.95.**

WELDERS GUIDE, 2nd Edition

This new edition is a practical and concise manual on the theory, practical operation, and maintenance of all welding machines. Fully covers both electric and oxy-gas welding. 928 pages; 5½ x 8¼; hardbound. **Cat. No. 23202 Price: $11.95.**

WELDER/FITTERS GUIDE

Provides basic training and instruction for those wishing to become welder/fitters. Step-by-step learning sequences are presented from learning about basic tools and aids used in weldment assembly, through simple work practices, to actual fabrication ‹ weldments. 160 pages· 8½ x 11; softcover; **Cat. No. 23325 Price: $7.95.**

se the order coupon on the back page of this book.

FLUID POWER

PNEUMATICS AND HYDRAULICS, 3rd Edition

Fully discusses installation, operation, and maintenance of both HYDRAULIC AND PNEUMATIC (air) devices. 496 pages; 5½ x 8¼; hardbound. **Cat. No. 23237 Price: $8.50.**

PUMPS, 3rd Edition

A detailed book on all types of pumps from the old-fashioned kitchen variety to the most modern types. Covers construction, application, installation, and troubleshooting. 480 pages; 5½ x 8¼; hardbound. **Cat. No. 23292 Price: $8.95.**

HYDRAULICS FOR OFF-THE-ROAD EQUIPMENT

Everything you need to know from basic hydraulics to troubleshooting hydraulic systems on off-the-road equipment. Heavy-equipment operators, farmers, fork-lift owners and operators, mechanics—all need this practical, fully illustrated manual. 272 pages; 5½ x 8¼; hardbound. **Cat. No. 23306 Price: $6.95.**

HOBBY

COMPLETE COURSE IN STAINED GLASS

Written by an outstanding artist in the field of stained glass, this book is dedicated to all who love the beauty of the art. Ten complete lessons describe the required materials, how to obtain them, and explicit directions for making several stained glass projects. 80 pages; 8½ x 11; softbound. **Cat. No. 23287   Price: $4.95**

BUILD YOUR OWN AUDEL
DO-IT-YOURSELF LIBRARY AT HOME!

Use the handy order coupon today to gain the valuable information you need in all the areas that once required a repairman. Save money and have fun while you learn to service your own air conditioner, automobile, and plumbing. Do your own professional carpentry, masonry, and wood furniture refinishing and repair. Build your own security systems. Find out how to repair your TV or Hi-Fi. Learn landscaping, upholstery, electronics and much, much more.